Choosing the most efficient statistical test is one of the basic problems of statistics. Asymptotic efficiency is an indispensable technique for comparing and ordering statistical tests in large samples. It is especially useful in nonparametric statistics where there exist numerous heuristic tests such as the Kolmogorov–Smirnov, Cramér–von Mises, and linear rank tests.

This monograph discusses the analysis and calculation of the asymptotic efficiencies of nonparametric tests. Powerful methods based on Sanov's theorem together with the techniques of limit theorems, variational calculus, and nonlinear analysis are developed to evaluate explicitly the large-deviation probabilities of test statistics. This makes it possible to find the Bahadur, Hodges–Lehmann, and Chernoff efficiencies for the majority of nonparametric tests for goodness-of-fit, homogeneity, symmetry, and independence hypotheses.

Of particular interest is the description of domains of the Bahadur local optimality and related characterization problems, based on recent research by the author. The general theory is applied to a classical problem of statistical radio physics: signal detection in noise of unknown level. Other results previously published only in Russian journals are also published here for the first time in English.

Researchers, professionals, and students in statistics will find this unified treatment of the subject invaluable.

ASYMPTOTIC EFFICIENCY OF NONPARAMETRIC TESTS

ASYMPTOTIC EFFICIENCY OF NONPARAMETRIC TESTS

YAKOV NIKITIN

St. Petersburg State University

CAMBRIDGE
UNIVERSITY PRESS

CAMBRIDGE UNIVERSITY PRESS
Cambridge, New York, Melbourne, Madrid, Cape Town, Singapore, São Paulo, Delhi

Cambridge University Press
The Edinburgh Building, Cambridge CB2 8RU, UK

Published in the United States of America by Cambridge University Press, New York

www.cambridge.org
Information on this title: www.cambridge.org/9780521115926

First published 1995
This digitally printed version 2009

A catalogue record for this publication is available from the British Library

Library of Congress Cataloguing in Publication data
Nikitin, Yakov.
Asymptotic efficiency of nonparametric tests / Yakov Nikitin.
p.   cm.
Includes bibliographical references and index.
ISBN 0-521-47029-3
1. Nonparametric statistics.   2. Asymptotic efficiencies
(Statistics)   I. Title.
QA278.8.N55   1995
519.5′4 – dc20   94-8820
CIP

ISBN 978-0-521-47029-2 hardback
ISBN 978-0-521-11592-6 paperback

Contents

v

Introduction

Choosing the most efficient statistical test of several ones that are at the disposal of the statistician is regarded as one of the basic problems of statistics. According to the classical Neyman–Pearson theory the uniformly most powerful tests are considered the best. However, it is well known that they exist merely for a narrow class of statistical models which do not fully cover the diversity of problems arising in theory and practice. One can still say that within the framework of parametric statistics this problem is not at all crucial. The point is that quite formal methods of constructing tests have been developed, for example, Bayes or likelihood ratio tests. They possess a number of remarkable properties and usually turn out to be asymptotically optimal in the sense of one or another definition of this concept.

The situation is quite different under the nonparametric approach. There exist numerous statistical tests proposed as a rule for heuristic reasons. The Kolmogorov–Smirnov and omega-square tests can serve as classical examples for goodness-of-fit testing. In other cases nonparametric procedures arise as simple substitutes of computationally complicated parametric procedures. The Wilcoxon rank test has been proposed in exactly this way. One more reason for using nonparametric tests is concerned with unreliable information on the distribution of observations in cases when it is reasonable to use, instead of the highly suitable parametric test, a nonparametric one, which is possibly less efficient but more robust with respect to changes of this distribution.

Because of these and some other reasons there has resulted an extraordinary diversity of nonparametric procedures. In the early sixties the well-known bibliography of nonparametric statistics by Savage (1962) contained already about 3,000 entries. The idea of this diversity was also presented in the three-volume Walsh handbook (1962, 1965, 1968) and the monograph by Hollander and Wolfe (1973) as well as in the reference *Handbook of Statistics* (Krishnaiah and Sen 1984).

Therefore the problem of comparing nonparametric tests on the basis of some quantitative characteristic that will make it possible to order these tests and recommend the proper test one should use in the given problem becomes extremely important. The asymptotic efficiency is just the most known and useful characteristic of such kind.

Kendall and Stuart (1967) pointed out that the notion of the asymptotic efficiency of tests is more complicated than the asymptotic efficiency of estimates. Various approaches to this notion were identified only in the late forties and early fifties, hence, 20 to 25 years later than for the estimation theory. We proceed now to their description.

Let $\{T_n\}$ and $\{V_n\}$ be two sequences of statistics based on n observations and assigned for testing the null or basic hypothesis H against the alternative A. We assume that the alternative is characterized by parameter θ and for $\theta = \theta_0$ turns into H. Denote by $N_T(\alpha, \beta, \theta)$ the sample size necessary for the sequence $\{T_n\}$ in order to attain the power β under the level α and the alternative value of parameter θ. The number $N_V(\alpha, \beta, \theta)$ is defined in the same way. The relative efficiency of the sequences $\{T_n\}$ with respect to the sequence $\{V_n\}$ is specified as the quantity

$$e_{T,V}(\alpha, \beta, \theta) = N_V(\alpha, \beta, \theta)/N_T(\alpha, \beta, \theta),$$

so it is the reciprocal ratio of sample sizes N_T and N_V.

The merits of the relative efficiency as a means for comparing the tests are beyond any doubt. Unfortunately it is extremely difficult to explicitly compute $N_T(\alpha, \beta, \theta)$ even for the simplest sequences of statistics $\{T_n\}$. At present it is recognized that it is possible to avoid this difficulty by calculating the limiting values $e_{T,V}(\alpha, \beta, \theta)$ as $\theta \to \theta_0$, as $\beta \to 1$, and as $\alpha \to 0$, keeping the two other parameters fixed. These limiting values are called respectively the Pitman, Hodges–Lehmann, and Bahadur asymptotic relative efficiency (ARE).

Only close alternatives, high powers, and small levels are of the most interest from the practical point of view. This assures one that the knowledge of these ARE types will facilitate comparing concurrent tests, thus producing well-founded application recommendations.

Coupled with the three "basic" approaches to the ARE calculation just described, intermediate approaches are also possible if the transition to the limit occurs simultaneously for two parameters in a controlled way.

Thus emerged the Chernoff ARE introduced by Chernoff (1952) and the intermediate, or Kallenberg, ARE introduced by Kallenberg (1983a).

The history of nonparametric statistics contains plenty of good examples when the calculation of the AREs of some sequences of tests has resulted in the reappraisal of their merits. For instance, in the forties and fifties a very popular homogeneity test of two samples was the run test of Wald and Wolfowitz (see, e.g., Wald and Wolfowitz (1940) or Noether (1950)). But Mood (1954) proved that its Pitman ARE with respect to parametric tests is equal to 0. Later Bahadur (1960b) stated that the same is also true in the case of the Bahadur efficiency. In subsequent years the run test was gradually discarded from the statistical literature and replaced by more efficient rank tests. The same has occurred in the case of goodness-of-fit tests based on spacings according to Chibisov (1961).

The Pitman efficiency introduced in the late forties by Pitman (1949) is the most well-known and carefully studied type of ARE (see also Pitman (1979)). The principal condition ensuring the possibility of its computation is the asymptotic normality of a given sequence of statistics under the null hypothesis and the alternative. In this case the Pitman ARE is a number not depending on α and β. It has been computed in numerous papers. The results obtained there are partly referred to in corresponding sections of the monographs by Kendall and Stuart (1967), Lehmann (1959, 1975), and Hollander and Wolfe (1973). At present one can say that the Pitman ARE is computed for the majority of pairs of nonparametric statistics having asymptotically normal distributions.

The situation becomes much more complicated when a given sequence of statistics has different limiting distributions under the null hypothesis and the alternative such that at least one of them is nonnormal. These are the cases of the Kolmogorov–Smirnov and ω^2 statistics, when the Pitman ARE can depend on α and β and it is difficult to determine its value. The most promising results have been obtained by Wieand (1976) who modified the definition of the ARE and found its limiting (as $\alpha \to 0$) value for most frequently used statistics. For that reason we do not pay much attention to the Pitman ARE in the present book.

The present monograph deals with the calculation and analysis in the Bahadur, Hodges–Lehmann, and Chernoff senses of the ARE of Kolmogorov–Smirnov and Cramér–von Mises nonparametric tests and their variants as well as linear rank tests for testing goodness-of-fit, homogeneity, symmetry, and independence. The conditions of the local asymptotic optimality of these tests are also considered here.

The calculation of the Bahadur, Hodges–Lehmann, and Chernoff efficiencies, unlike the Pitman efficiency, is based on rough asymptotics of large-deviation probabilities of a given sequence of statistics under the null hypothesis and the alternative. The approaches to calculating the ARE dealing with the analysis of large deviations have become widespread. The following citation from Dacunha-Castelle (1979) is significant: "The merit of large deviations consists in that they give the solution for the problems of classification of tests and the search of optimal solutions when ... classical methods do not give any answer. The application of such a technique seems methodologically quite justified." Convincing arguments in favor of the use of large deviations by asymptotic comparison of tests are given also by Bahadur (1960), Neyman (1980), Serfling (1980), Borovkov and Mogulskii (1992), and others.

The theory of large deviations has reached by now a state of considerable development that has been stimulated in many respects by the demands of statistics. Two lines of research in the study of large deviations of nonparametric statistics are evident among the multitude of publications in this area (see Chapter 1 for more a detailed exposition).

The first considers various generalizations and continuations of results by Cramér (1938) and Chernoff (1952) who analyzed sums of independent identically distributed random variables. It turned out that the achievements in this direction made it possible to find large-deviation asymptotics of the Kolmogorov–Smirnov statistics and their variants under the null hypothesis. See Abrahamson (1967), Bahadur (1971), Groeneboom and Shorack (1981), Nikitin (1987b), and Podkorytova (1990).

The second line is based on the fundamental paper by Sanov (1957) exploring large deviations of empirical distribution functions. The Sanov results were generalized and developed later by Hoadley (1967); Groeneboom, Oosterhoff, and Ruymgaart (1979); Fu (1985); and others. It is important to mention the paper of Borovkov (1967) where the author noted that the Sanov-type theorems "permit estimating the probabilities of large deviations, for instance, of ω^2 statistics and comparing the power of corresponding tests under various alternatives."

The rough asymptotics of large-deviation probabilities for the classical Cramér–von Mises–Smirnov goodness-of-fit statistic ω_n^2 under the null hypothesis have been determined by Mogulskii (1977) by applying the results of Borovkov (1967). This statistic is as follows:

$$\omega_n^2 = \int\limits_0^1 \big(F_n(t) - t \big)^2 \, dt, \tag{1}$$

where F_n is the empirical distribution function based on a sample from the uniform distribution on $[0, 1]$. It is well known that the Sanov theorem and its generalizations reduce the problem of large deviations to the minimization problem of the Kullback–Leibler information on the corresponding set of distribution functions. The latter belongs to the class of variational problems on conditional extremum. The extremal may be found from the Euler–Lagrange equation which is a nonlinear differential equation of the second order having the form

$$x'' - \lambda x - \lambda \varepsilon x x' = 0 \tag{2}$$

and should be considered jointly with the conditions:

$$x(0) = x(1) = 0, \qquad \int_0^1 x^2(t)\, dt = 1. \tag{3}$$

In equation (2) λ is an indeterminate Lagrange multiplier and ε is a small numerical parameter.

Mogulskii (1977) proved by using the specific form of equation (2) that the solution of problem (2)–(3) may be found in the class of functions

$$x(t) = \sum_{n=0}^{\infty} x_n(t)\, \varepsilon^{n/2}, \tag{4}$$

where

$$x_n(t) = \sum_{k=1}^{\infty} a_k^{(n)} \sin k\pi t,$$

and the series in (4) converges absolutely and uniformly for sufficiently small $\varepsilon > 0$. This fact enables one to prove that

$$\lim_{n \to \infty} n^{-1} \ln P\left(\omega_n^2 \geq \varepsilon\right) = -\tfrac{1}{2}\pi^2 \varepsilon + \sum_{k=3}^{\infty} c_k\, \varepsilon^{k/2}, \tag{5}$$

where the series on the right-hand side converges for sufficiently small positive ε.

However, the arguments of Mogulskii (1977) cannot be adapted even for the case of weighted statistics, namely,

$$\omega_{n,q}^2 = \int_0^1 \left(F_n(t) - t\right)^2 q(t)\, dt, \tag{6}$$

where q is a summable nonnegative weight function on $[0, 1]$. This is concerned, to a considerable extent, with the complicated integral statistics for testing homogeneity, symmetry, and independence when the Euler–Lagrange equations are much more involved.

Therefore we propose to use a more general method to analyze these equations. The essence of the matter is to consider equation (2) or its more complicated counterparts together with the boundary conditions as an implicit analytic operator acting in a suitably chosen Banach space. It is quite natural to expect the solutions of such equations, as (4), to be analytic in a small parameter ε or in a certain power of it. The important obstacle in proving this is the fact that the linear part (the Fréchet derivative for $\varepsilon = 0$) of all operators under consideration is noninvertible. Therefore, we use the theory of branching of the solutions of nonlinear equations (see Vainberg and Trenogin (1974)), due to Lyapunov and Schmidt, to rule out the existence of bifurcations. The joint analysis of the so-called branching equation with the normalization condition enables one to construct the solutions of the Euler–Lagrange equations in the form of power series in a small parameter. Moreover, their convergence in a neighborhood of zero follows automatically from the general theorems proved by Vainberg and Trenogin (1974).

After the substitution of these solutions into the minimized functional one obtains the expected asymptotics. Further we compute the local Bahadur efficiency of various integral statistics by using the indicated method as a basis.

The computation of the Hodges–Lehmann and Chernoff AREs meets with considerable difficulties because of the necessity to study large deviations of nonparametric statistics under the alternative when they are no longer "distribution-free." However, it has become possible to overcome these difficulties by modifying the method described earlier. This enables one to study the behavior of the Hodges–Lehmann and Chernoff indices of a large class of linear rank statistics and to prove their local coincidence with the Bahadur exact slopes. Taking into account that the local Bahadur efficiency is coincident with respect to these statistics with the Pitman and intermediate efficiencies (see Kremer (1979a,b) and Kallenberg (1983a)), one discovers an interesting phenomenon, namely, that the local asymptotic ordering of a large class of linear rank tests does not depend on the ARE type.

The situation turns out to be entirely different for the Kolmogorov–Smirnov and omega-square statistics. Combining general results on large deviations of empirical measures under the alternative with purely statistical reasons, we prove that the two-sided tests of Kolmogorov–Smirnov and ω^2 type attain the maximum possible Hodges–Lehmann ARE in all problems under consideration and thus are somewhat unexpectedly asymptotically optimal. Even the one-sided Smirnov test may possess this property under appropriate conditions. Local Chernoff indices of these statistics have complicated expressions and their values are not, as a rule, asymptotically optimal.

It is well known that the Bahadur exact slopes and Hodges–Lehmann indices of any sequence of statistics are bounded from above by a positive quantity whose value can be specified in terms of the Kullback–Leibler information. This was discovered by Stein for simple hypotheses; later, various generalizations were made by Rao (1962), Raghavachari (1970), Bahadur and Raghavachari (1972), Nikitin (1986c), and Kourouklis (1988). In other words a statistical test cannot be altogether "too good": Just like in regular experiments the estimates of a parameter cannot be "too precise" (the Cramér–Rao inequality). It has been proved that under some natural auxiliary conditions the likelihood ratio tests are asymptotically optimal in the Bahadur sense (see Bahadur (1965) and Rublik (1989)) as well as in the Hodges–Lehmann sense (see Brown (1971)) and Chernoff sense (see Kallenberg (1982)). It remains to be seen if this is possible for nonparametric tests.

We investigate this problem with regard to the Kolmogorov–Smirnov-type tests and ω^2-type tests as well as for linear rank tests. We find that these tests may possess the property of Bahadur *local* optimality. Relying on expressions for local exact slopes one succeeds in obtaining the description of the structure of those families of distributions for which a given nonparametric test is locally asymptotically optimal in the Bahadur sense (the domain of the local asymptotic optimality will be given in our terminology). A key part in this description is played by the "leading" sets of functions which are specific for each test and indicate, roughly speaking, those "directions" in the set of alternatives where a given test is most sensitive.

If some auxiliary information is available on the structure of the initial family of distributions (i.e., location, scale, Lehmann families, etc), the conditions of local asymptotic optimality lead to original characterization theorems. For instance, for the location families the Kolmogorov goodness-of-fit test is locally asymptotically optimal in the Bahadur sense only for the Laplace distribution. Likewise, the ω^2 test is Bahadur locally optimal only for the "hyperbolic cosine" distribution and the Wilcoxon two-sample rank test possesses this property only for the logistic one. These results are partially preserved for other types of asymptotic efficiency.

The nonparametric tests under consideration are widely used in solving numerous problems of natural sciences and engineering. Their asymptotic efficiency facilitates the choice of the most efficient test and therefore meets the requirements of statistical practice. The asymptotic comparison of tests in such a high-priority problem of statistical radio physics as the problem of detection of signals in Gaussian noise of unknown level in order to apply them in concrete conditions is considered at the end of Chapter 3.

The monograph consists of the Introduction and six chapters divided into sections. We adopt a triple numbering convention, with formulas, theorems, and lemmas being independent in each section. Chapter 1 has an auxiliary character. There the mathematical results used subsequently for the calculation and analysis of the asymptotic efficiency are put together. The next four chapters deal with large deviations and asymptotic efficiency of nonparametric tests. We consider goodness-of-fit, homogeneity, symmetry, and independence tests in Chapters 2, 3, 4, and 5, respectively. The final chapter contains the description of the local asymptotic optimality domains of nonparametric statistics and corresponding characterization results.

The author expresses his profound gratitude to Professor I. A. Ibragimov whose constant support and interest in the problems of asymptotic efficiency have greatly contributed to the appearance of this book. He is also sincerely thankful to the late Professor N. N. Chentsov for his well-disposed attention to the problems discussed here and for the idea of preparing this monograph, which he expressed together with Professor I. A. Ibragimov.

1

Asymptotic Efficiency of Statistical Tests and Mathematical Means for Its Computation

1.1 General Approach to Computation of Asymptotic Efficiency

Let $(\mathfrak{X}, \mathfrak{A})$ be a sample space corresponding to the observation X. It is assumed that the distribution P_θ of this observation is determined by parameter θ taking on values in a parametric set Θ. Let $s = \{X_1, X_2, \ldots\}$ be a sequence of independent identically distributed random variables with values in $\mathfrak{X}$ and having the distribution P_θ on $\mathfrak{A}$. For any positive integer n put $\mathbf{X}^{(n)} := (X_1, X_2, \ldots, X_n)$ and denote by $(\mathfrak{X}^{(n)}, \mathfrak{A}^{(n)})$ the corresponding sample space and by $P_\theta^{(n)}$ the distribution of $\mathbf{X}^{(n)}$ on $\mathfrak{A}^{(n)}$. In the sequel $P_\theta^{(n)}$, $1 \leq n \leq \infty$, will be usually abbreviated to P_θ.

Consider the problem of testing the hypothesis

$$H\colon \theta \in \Theta_0 \subset \Theta$$

against the alternative

$$A\colon \theta \in \Theta_1 = \Theta \setminus \Theta_0$$

on the basis of observations $X_1, X_2, \ldots, X_n$. For this purpose we use a sequence of statistics $\{T_n\}$, $T_n(s) := T_n(X_1, X_2, \ldots, X_n)$, assuming (without essential loss of generality) large values of T_n to be significant. Thus the critical or rejection region of H is given by

$$\{\, s\colon T_n(s) \geq c \,\},$$

where c is some real number.

The *power function* of this test is the quantity $P_\theta(T_n \geq c)$ considered as a function of θ and its *size* is equal to

$$\sup\{\, P_\theta\,(T_n \geq c)\colon \theta \in \Theta_0 \,\}.$$

1

Now define for any $\beta \in (0,\ 1)$ and $\theta \in \Theta_1$ a real sequence $c_n := c_n(\beta, \theta)$ with the aid of double inequality

$$P_\theta\left(T_n > c_n\right) \leq \beta \leq P_\theta\left(T_n \geq c_n\right). \qquad (1.1.1)$$

Then

$$\alpha_n(\beta, \theta) := \sup\left\{\ P_{\theta'}(T_n \geq c_n)\colon\ \theta' \in \Theta_0\ \right\}$$

is the minimal size of the test based on $\{T_n\}$ for which the power at the point θ is not less than β. Let us define for any given level of significance α, $0 < \alpha < \beta$, the positive integer

$$N_T(\alpha, \beta, \theta) := \min\left\{n\colon\ \alpha_m(\beta, \theta) \leq \alpha\quad \text{for all } m \geq n\right\}.$$

It is clear that $N_T(\alpha, \beta, \theta)$ is the minimal sample size necessary for the test at a level α, based on $\{T_n\}$, to have the power not less than β at the point θ.

Suppose that for testing H against A we have two sequences of test statistics $\{T_n\}$ and $\{V_n\}$. Define by $e_{V,T}(\alpha, \beta, \theta)$ the *relative efficiency* of the sequence $\{V_n\}$ with respect to $\{T_n\}$ in the following way:

$$e_{V,T}(\alpha, \beta, \theta) := N_T(\alpha, \beta, \theta)\big/ N_V(\alpha, \beta, \theta). \qquad (1.1.2)$$

A value $e_{V,T}(\alpha, \beta, \theta)$ larger than 1 indicates that for given α, β, and θ one should prefer the sequence $\{V_n\}$ to $\{T_n\}$ because the first sequence requires less observations for reaching the power β for the level α and the alternative value θ.

As has been already noted in the Introduction, relative efficiency (1.1.2) has indisputable merits. Unfortunately it also has two substantial drawbacks. One consists in that the value of $e_{V,T}(\alpha, \beta, \theta)$ depends on three arguments (and two sequences of statistics), the other is connected with the fact that it is extremely difficult or simply impossible to calculate this value. It is possible at present to overcome these difficulties by calculating the limiting values of $e_{V,T}(\alpha, \beta, \theta)$ as $\alpha \to 0$, as $\beta \to 1$, and as $\theta \to \theta_0 \in \partial\Theta_0$ (in a certain topology on Θ) keeping fixed the values of two remaining parameters. As a result one obtains three fundamental types of the asymptotic relative efficiency (ARE).

If for $\beta \in (0,\ 1)$ and $\theta \in \Theta_1$ there exists the limit

$$e_{V,T}^{B}(\beta, \theta) := \lim_{\alpha \downarrow 0} e_{V,T}(\alpha, \beta, \theta), \qquad (1.1.3)$$

it is called the *Bahadur ARE of the sequence $\{V_n\}$ with respect to $\{T_n\}$.*

If for $\alpha \in (0,\ 1)$ and $\theta \in \Theta_1$ there exists the limit

$$e_{V,T}^{HL}(\alpha, \theta) := \lim_{\beta \uparrow 1} e_{V,T}(\alpha, \beta, \theta)\,, \qquad (1.1.4)$$

it is called the *Hodges–Lehmann ARE of the sequence $\{V_n\}$ with respect to $\{T_n\}$*.

If for $0 < \alpha < \beta < 1$ and $\theta \to \theta_0 \in \partial\Theta_0$ (in a certain topology on Θ) there exists the limit

$$e_{V,T}^{P}(\alpha, \beta, \theta_0) := \lim_{\theta \to \theta_0} e_{V,T}(\alpha, \beta, \theta)\,, \qquad (1.1.5)$$

it is called the *Pitman ARE of the sequence $\{V_n\}$ with respect to $\{T_n\}$*.

It is also difficult to calculate these three types of the ARE, but it is still much easier than relative efficiency (1.1.2). Moreover, the Bahadur ARE usually does not depend on β, the Hodges–Lehmann ARE does not depend on α, and the Pitman ARE in most cases depends neither on α nor on β and turns out to be a constant. We emphasize once again that from the practical point of view the very cases of small levels, high powers, and close alternatives are the most important. That is why one may suppose that the knowledge of three types of the ARE such as (1.1.3)–(1.1.5) will in a sense help to put in order principal tests used in a concrete problem and will permit the well-founded recommendations for their applications in practice.

Note that there exist the intermediate approaches to measuring the ARE not coinciding with the approaches of Bahadur, Hodges and Lehmann, and Pitman. The typical examples are the Chernoff ARE (see Chernoff (1952), Kallenberg (1982), and Ronzhin (1985)) when for a fixed θ the other parameters α and β tend to 0, and the case of the intermediate or Kallenberg ARE (see Kallenberg (1983a)), when β is fixed, but θ and α tend to θ_0 and 0 at a controlled rate.

The different definitions of the ARE belong also to Rubin and Sethuraman (1965b) and to Borovkov and Mogulskii (1992). The first exploits the notion of the Bayes risk whereas the second deals with a certain modification of the number $N_T(\alpha, \beta, \theta)$, being more symmetric with respect to the errors of the first and second kinds. Their values for more or less complicated nonparametric statistics are unknown.

It would be most interesting to learn if the values of AREs calculated under different approaches are close to the values of the relative efficiency for finite samples and reasonable values of α, β, and θ arising in practical problems. The first such comparison has been realized by Groeneboom and Oosterhoff (1981) with the aid of statistical modeling. The results of Groeneboom and Oosterhoff (1981) are connected with some simplest examples only and for the present do not give any reasons for definite conclusions.

It may happen that the value of the ARE is equal to 1 and this circumstance prevents the asymptotic comparison of tests. In that case one may recommend following Hodges and Lehmann (1970) and using more sensitive means for comparing different tests, namely, the *deficiency*

$$\operatorname{def}(V, T; \alpha, \beta, \theta) := N_V(\alpha, \beta, \theta) - N_T(\alpha, \beta, \theta), \quad \theta \in \Theta_1.$$

Asymptotic approximations to the deficiency as $\alpha \to 0$ and $\theta \to \theta_0 \in \partial\Theta_0$ have been considered, mainly for parametric statistics, by Albers (1974), Chandra and Ghosh (1978), Groeneboom and Oosterhoff (1981), and Kallenberg (1981, 1982) as well as by Borovkov and Mogulskii (1992). Not much is known in the nonparametric case and we will not use the notion of the deficiency in the sequel.

1.2 Bahadur Asymptotic Relative Efficiency

The Bahadur approach to measuring the ARE is opposite to the classical approach of Neyman and Pearson and prescribes one to fix the power of tests and to compare the rate of decreasing of their sizes for the increasing number of observations. This point of view, expressed for the first time by Cochran (1952), has been deeply and systematically developed by Bahadur (1960a,b, 1967, 1971). Other expositions of the Bahadur theory with reviews of publications in this area may be found in Savage (1969), Groeneboom and Oosterhoff (1977), and Serfling (1980).

Denote for any θ, t and any sequence of statistics $\{T_n\}$

$$F_n(t; \theta) := P_\theta\left(s: T_n(s) < t\right), \qquad G_n(t) := \inf\left\{F_n(t; \theta): \theta \in \Theta_0\right\}.$$

The quantity

$$L_n(s) := 1 - G_n(T_n(s))$$

is called the *attained level* or the *P-value*. This is a random variable representing the degree to which the test statistic T_n rejects H.

For $\theta \in \Theta_0$ the P-value is distributed approximately uniformly on $[0, 1]$. Anyway the following inequality is valid:

$$P_\theta\left(L_n \leq u\right) \leq u \qquad \text{for any } u \in [0, 1]. \tag{1.2.1}$$

In the case of continuous distribution function $F_n(t; \theta_0)$ this inequality follows directly from the definition of L_n; the general case is based on a suitable approximation (see Bahadur (1971), Theorem 7.4). Therefore working with finite samples one compares the P-value with the preassigned level α and rejects the basic hypothesis if $L_n < \alpha$. Substantial discussion and the interpretation of P-values were given by Gibbons and Pratt (1975), Serfling (1980), Lambert and Hall (1982), and Chandra (1989).

The asymptotic behavior of L_n under the alternative ($\theta \in \Theta_1$) is of considerable interest for comparing sequences of test statistics. It is usually the case that for $\theta \in \Theta_1$ the following convergence in P_θ-probability takes place:

$$\lim_{n \to \infty} n^{-1} \ln L_n = -\tfrac{1}{2} c_T(\theta), \qquad (1.2.2)$$

where $c_T(\theta)$ is a nonrandom positive function of parameter θ on Θ_1, that is called the *Bahadur exact slope of the sequence* $\{T_n\}$. The factor $\tfrac{1}{2}$ is present in (1.2.2) due to historical reasons. Some authors, for example, Groeneboom and Oosterhoff (1977) as well as Chandra and Ghosh (1978), have called $c_T(\theta)$ the *weak slope*, in contrast to the *strong slope* when the convergence in (1.2.2) takes place with P_θ-probability 1. In the sequel we will use the term slope mainly for weak slopes as it is sufficient to consider the convergence in probability in most statistical problems.

One may rewrite (1.2.2) in terms of sample sizes $N_T(\alpha, \beta, \theta)$.

Theorem 1.2.1 (Bahadur 1967; Groeneboom and Oosterhoff 1977) *If (1.2.2) is valid for a sequence of statistics $\{T_n\}$ with $c_T(\theta) > 0$, then*

$$N_T(\alpha, \beta, \theta) \sim \frac{2 \ln 1/\alpha}{c_T(\theta)} \qquad as \ \alpha \to 0. \qquad (1.2.3)$$

Proof Let us prove at first that for any $\beta \in (0, 1)$ we have

$$\lim_{n \to \infty} n^{-1} \ln \alpha_n(\beta, \theta) = -\tfrac{1}{2} c_T(\theta). \qquad (1.2.4)$$

Suppose that (1.2.4) fails for some β. Let us define for this β the sequence $\{c_n\}$ by formula (1.1.1). There exist an increasing subsequence $\{n_k\}$ and $\varepsilon > 0$ such that one of the following two inequalities

$$-n_k^{-1} \ln \alpha_{n_k}(\beta, \theta) < \tfrac{1}{2} c_T(\theta) - \varepsilon, \qquad (a)$$

$$-n_k^{-1} \ln \alpha_{n_k}(\beta, \theta) > \tfrac{1}{2} c_T(\theta) + \varepsilon \qquad (b)$$

is valid for all k.

In the case (a) we obtain, as a result of (1.2.2), that

$$
\begin{aligned}
P_\theta\big(-n_k^{-1}\ln L_{n_k} > \tfrac{1}{2}c_T(\theta) - \varepsilon\big) &\leq P_\theta\big(-n_k^{-1}\ln L_{n_k} > -n_k^{-1}\ln \alpha_{n_k}(\beta,\theta)\big) \\
&= P_\theta\big(L_{n_k} < \alpha_{n_k}(\beta,\theta)\big) \\
&= P_\theta\big(1 - G_{n_k}(T_{n_k}) < 1 - G_{n_k}(c_{n_k})\big) \\
&\leq P_\theta\,(T_{n_k} > c_{n_k}) \leq \beta \qquad \text{for all } k.
\end{aligned}
$$

The left-hand side of the considered inequality tends to 1 as $k \to \infty$, which contradicts the assumption $0 < \beta < 1$. The other case (b) may be examined analogously.

Now let us derive from (1.2.4) the conclusion of the theorem. Put for brevity $N_\alpha := N_T(\alpha,\beta,\theta)$. It follows from (1.2.4) that

$$
\alpha_n(\beta,\theta) > \exp\{-n\,c_T(\theta)\}
$$

for sufficiently large n. Therefore for any such n the inequality

$$
\alpha < \exp\{-n\,c_T(\theta)\}
$$

entails $N_\alpha > n$, ensuring that $N_\alpha \to \infty$ as $\alpha \to 0$.

The definition of N_α implies that

$$
\alpha_{N_\alpha}(\beta,\theta) \leq \alpha \leq \alpha_{N_\alpha-1}(\beta,\theta)
$$

or, equivalently,

$$
-N_\alpha^{-1}\ln \alpha_{N_\alpha-1}(\beta,\theta) \leq -N_\alpha^{-1}\ln \alpha \leq -N_\alpha^{-1}\ln \alpha_{N_\alpha}(\beta,\theta).
$$

The transition to the limit as $\alpha \to 0$ together with (1.2.4) completes the proof of (1.2.3). $\qquad\square$

Thus, if two sequences of statistics $\{V_n\}$ and $\{T_n\}$ are such that (1.2.2) holds, their Bahadur ARE $e_{V,T}^B(\beta,\theta)$ can be calculated, according to (1.1.2), by means of the formula

$$
e_{V,T}^B(\beta,\theta) = c_V(\theta) / c_T(\theta). \tag{1.2.5}
$$

If $e_{V,T}^B > 1$ for some θ then we should prefer the sequence $\{V_n\}$ to $\{T_n\}$.

The following theorem contains the most simple method for the calculation of exact slopes.

Theorem 1.2.2 (Bahadur 1967, 1971) *Let for a sequence $\{T_n\}$ the following two conditions be fulfilled:*

$$
T_n \xrightarrow{P_\theta} b(\theta), \qquad \theta \in \Theta_1, \tag{1.2.6}
$$

where $-\infty < b(\theta) < \infty$;

$$\lim_{n\to\infty} n^{-1}\ln\left[1 - G_n(t)\right] = -f(t) \qquad (1.2.7)$$

for each t from an open interval I on which f is continuous and $\{b(\theta),$
$\theta \in \Theta_1\} \subset I$. *Then (1.2.2) is valid and, moreover, for any* $\theta \in \Theta_1$

$$c_T(\theta) = 2\, f\big(b(\theta)\big)\,. \qquad (1.2.8)$$

Formula (1.2.8) plays an exceptionally important role in the Bahadur
theory. It shows that for the calculation of exact slopes it is necessary
to solve the problem of determining *large-deviation asymptotics* of a sequence $\{T_n\}$ under the null hypothesis. This problem is always nontrivial
as Bahadur (1971) himself notes. By contrast the verification of (1.2.6)
does not usually present any difficulties.

In the cases when a sequence $\{T_n\}$ does not satisfy the conditions
of Theorem 1.2.2, usually one succeeds in selecting strictly monotone
functions ψ_n such that a new sequence $\{T_n^*\}, T_n^* := \psi_n(T_n)$, already
satisfies these conditions. Since the P-value L_n stays invariable under
this transform, the exact slope of $\{T_n\}$ coincides with the exact slope of
$\{T_n^*\}$, which might be calculated by (1.2.8).

Proof of Theorem 1.2.2 Fix an arbitrary $\theta \in \Theta_1$ and $\varepsilon > 0$ such that
$(b - \varepsilon, b + \varepsilon) \subset I$. Put

$$\Omega := \big\{\, s \in \mathfrak{X}^{(\infty)} \colon\; b - \varepsilon < T_n(s) < b + \varepsilon \,\big\}\,.$$

It follows from (1.2.6) that for sufficiently large n the estimate $P_\theta\,(\Omega) >$
$1 - \delta$ is valid for any $\delta > 0$. As F_n is monotone, the following inequalities

$$1 - F_n(b + \varepsilon) \le L_n(s) \le 1 - F_n(b - \varepsilon)$$

are valid for the same $s \in \Omega$. Taking logarithms and passing to the limit
as $n \to \infty$, we obtain under condition (1.2.7) that

$$-f(b + \varepsilon) \le \varliminf_{n\to\infty} n^{-1}\ln L_n \le \varlimsup_{n\to\infty} n^{-1}\ln L_n \le -f(b - \varepsilon)$$

for each $s \in \Omega$. By virtue of the continuity of f and of ε being arbitrary
it follows now that in P_θ-probability

$$\lim_{n\to\infty} n^{-1}\ln L_n = -f(b)\,. \qquad \square$$

The right-hand side of (1.2.2) may be, generally speaking, a random variable. Note that such situations were discussed by Bahadur and

Raghavachari (1972). It is natural in that case to call the exact slope *stochastic* (see Berk and Brown (1978) and Kallenberg (1981)) as distinct from nonstochastic exact slopes defined by (1.2.4). Fortunately it follows from Theorem 1.2.1 that, if the limit in (1.2.2) is nonrandom, both notions coincide. In the sequel we shall meet only nonrandom limits in (1.2.2), the values of which may be calculated via Theorem 1.2.2.

Another fundamental result in the Bahadur theory is the existence of an upper bound for exact slopes that is sometimes compared in the literature with the Cramér–Rao inequality in the estimation theory.

Define for any two elements P_θ and $P_{\theta'}$ of the basic family of distributions the *Kullback–Leibler information number* (or simply *information*) by means of the formula

$$K(P_\theta, P_{\theta'}) := \begin{cases} \displaystyle\int_{\mathfrak{X}} \ln \frac{d\,P_\theta}{d\,P_{\theta'}}\, d\,P_\theta & \text{if } P_\theta \ll P_{\theta'}\,, \\ +\infty & \text{otherwise.} \end{cases} \tag{1.2.9}$$

We shall henceforth often write $K(\theta, \theta')$ instead of $K(P_\theta, P_{\theta'})$. The properties of the Kullback–Leibler information have been examined by Kullback (1959), Bahadur (1971), and Borovkov (1984) among others. It is well known that $K(\theta, \theta') \geq 0$ and $K(\theta, \theta') = 0$ only if $P_\theta = P_{\theta'}$.

Put for any $\theta \in \Theta_1$

$$K(\theta, \Theta_0) := \inf \left\{ K(\theta, \theta_0)\colon\ \theta_0 \in \Theta_0 \right\}. \tag{1.2.10}$$

Theorem 1.2.3 (Raghavachari 1970; Bahadur 1971) *For any $\theta \in \Theta_1$ with P_θ-probability 1 we have*

$$\varliminf_{n\to\infty} n^{-1} \ln L_n(s) \geq -K(\theta, \Theta_0). \tag{1.2.11}$$

Proof Only the case when $K(\theta, \Theta_0) < \infty$ is of interest. Fix a $\theta \in \Theta_1$ for which it is valid. For any $\varepsilon > 0$ there exists $\theta_0 \in \Theta_0$ such that

$$0 \leq K(\theta, \theta_0) < K(\theta, \Theta_0) + \varepsilon < +\infty. \tag{1.2.12}$$

For any fixed θ and θ_0 denote for brevity $K(\theta, \theta_0)$ by K. If $K < \infty$ we have $P_\theta \ll P_{\theta_0}$ and consequently

$$d\,P_\theta = r(x)\, d\,P_{\theta_0} \qquad \text{on } (\mathfrak{X}, \mathfrak{A})\,,$$

$$d\,P_\theta^{(n)} = r_n(s)\, d\,P_{\theta_0}^{(n)} \qquad \text{on } \left(\mathfrak{X}^{(n)}, \mathfrak{A}^{(n)}\right),$$

where $r_n(s) := \Pi_{i=1}^n r(X_i)$. Note that by the strong law of large numbers one can state that with P_θ-probability 1

$$\lim_{n\to\infty} n^{-1}\ln r_n(s) = K.\qquad(1.2.13)$$

For any positive integer n let us introduce the events

$$A_n := \left\{L_n < \exp\left[-n\left(K+2\varepsilon\right)\right]\right\},\qquad B_n := \left\{r_n < \exp\left[n\left(K+\varepsilon\right)\right]\right\}.$$

Then

$$P_\theta(A_n B_n) = \int_{A_n B_n} dP_\theta^{(n)} = \int_{A_n B_n} r_n\, dP_{\theta_0}^{(n)}$$

$$\leq \exp\left\{n\left(K+\varepsilon\right)\right\} \int_{A_n} dP_{\theta_0}^{(n)}$$

$$= \exp\left\{n\left(K+\varepsilon\right)\right\}\cdot P_{\theta_0}(A_n) \leq \exp\left\{-n\varepsilon\right\},\quad(1.2.14)$$

where the last inequality follows from (1.2.1).

It follows from (1.2.14) that $\sum_n P_\theta(A_n B_n) < \infty$. By the Borel–Cantelli lemma only a finite number of events $A_n B_n$ occurs with P_θ-probability 1. Taking into account (1.2.13) we obtain that the inequality

$$L_n(s) \geq \exp\left\{-n\left(K+2\varepsilon\right)\right\}$$

holds almost surely for sufficiently large n. Under condition (1.2.12) we establish the conclusion of Theorem 1.2.3 due to the arbitrary choice of ε. Some generalizations are contained in Bahadur and Raghavachari (1972) and Bahadur, Chandra, and Lambert (1982). $\qquad\square$

Theorem 1.2.3 implies that the exact slope $c_T(\theta)$ of any sequence of statistics $\{T_n\}$ satisfies the inequality

$$c_T(\theta) \leq 2\,K(\theta,\Theta_0).\qquad(1.2.15)$$

If equality takes place in (1.2.15) for all $\theta \in \Theta_1$, the sequence $\{T_n\}$ is said to be *asymptotically optimal (AO) in the Bahadur sense*. The class of such statistics is apparently rather narrow, though it contains under certain conditions the likelihood ratio statistics (see Bahadur (1965, 1967) and Rublik (1989)). But if for each $\theta_0 \in \partial\Theta_0$ the weaker condition holds, namely

$$c_T(\theta) \sim 2\,K(\theta,\Theta_0),\qquad \theta\to\theta_0,\qquad(1.2.16)$$

then the sequence $\{T_n\}$ is said to be *locally asymptotically optimal (LAO) in the Bahadur sense*.

In the initial papers on the Bahadur ARE it was impossible to find the function f in (1.2.7) because of insufficient development of large-deviation theory. In this connection it was proposed by Bahadur (1960b) that the exact distribution of $\{T_n\}$ in the definition of the P-value be replaced by its limiting distribution. Suppose that for all $\theta_0 \in \Theta_0$ and $t \in \mathbf{R}^1$ there exists continuous distribution function F such that

$$F_n(t, \theta_0) \longrightarrow F(t) \qquad \text{as } n \to \infty.$$

Then the substitute of $L_n(s)$ is

$$L_n^*(s) := 1 - F\big(T_n(s)\big),$$

and what's more in typical cases there exists the limit in P_θ-probability

$$\lim_{n\to\infty} \left(-n^{-1}\ln L_n^*\right) = \tfrac{1}{2}\,c_T^*(\theta) > 0. \qquad (1.2.17)$$

If (1.2.17) is actually valid, the function $c_T^*(\theta)$ is called the *approximate* (as opposed to *exact*) *slope of the sequence* $\{T_n\}$. The ratio of approximate slopes of two sequences of statistics $\{V_n\}$ and $\{T_n\}$ is called their *approximate Bahadur* ARE and is denoted by $e_{V,T}^{*B}(\theta)$.

The method of calculating approximate slopes analogous with Theorem 1.2.2 still applies (see Bahadur (1960b)): *Suppose a sequence $\{T_n\}$ satisfies (1.2.6) for some function b, and for some constant a, $0 < a < \infty$, the limiting distribution function F satisfies the condition that*

$$\ln\left[1 - F(t)\right] \sim -\tfrac{1}{2}\,a\,t^2, \qquad t \to \infty. \qquad (1.2.18)$$

Then (1.2.17) holds and, besides,

$$c_T^*(\theta) = a\,b^2(\theta).$$

Approximate slopes are not very reliable as means of comparing tests because monotone transforms of test statistics may lead to entirely different approximate slopes (see, e.g., Groeneboom and Oosterhoff (1977)). But nevertheless they are still used in the statistical literature, mainly for the following reasons:

The approximate Bahadur ARE may be more easily calculated than any other known type of AREs;

the approximate and exact slopes are often locally (as $\theta \to \theta_0$) equivalent, so the approximate ARE gives a notion of the local exact ARE;

the approximate slopes give a simple method for the calculation of the Pitman ARE (for more detail see Section 1.4).

A definite merit of the Bahadur ARE in the opinion of a number of authors (see Savage (1969), Groeneboom and Oosterhoff (1977), and Singh (1984)) lies in the fact that it permits one to distinguish tests in the cases when other types of AREs are useless. We quote some typical examples.

Let X_1, X_2, ..., X_n be a normally distributed sample with parameters $(\theta, 1)$. The null hypothesis H: $\theta = 0$ is tested against the alternative A: $\theta > 0$. With that end in mind two sequences of statistics are proposed: the sample means $\{\bar{X}_n\}$ and the Student ratios $\{t_n\}$. As the Student test does not use the information that the true variance is equal to 1, it should lose to the test based on $\{\bar{X}_n\}$. The Bahadur exact slopes prove this correct (see Bahadur (1971)) because for any $\theta > 0$

$$c_t(\theta) \equiv \ln\left(1 + \theta^2\right) < c_{\bar{X}}(\theta) \equiv \theta^2.$$

Meanwhile these tests are indistinguishable from the point of view of the Hodges–Lehmann and Pitman AREs as

$$e_{t,\bar{X}}^{\mathrm{HL}}(\alpha, \theta) = e_{t,\bar{X}}^{P}(\alpha, \beta) \equiv 1.$$

The other example is based on the paper by Mason (1984) where it has been shown that the statistics based on sequential ranks have lower Bahadur efficiency than usual rank statistics whereas they are equivalent from the standpoint of the Pitman efficiency.

Now we return to the definition of exact slopes. Formula (1.2.2) may be interpreted as the law of large numbers for logarithms of P-values. Lambert and Hall (1982) initiated a new stage in studying the asymptotics of P-values by proving that for large n and under appropriate conditions the logarithms of P-values have approximately the normal distribution. The following theorem is the precise formulation of their result.

Theorem 1.2.4 (Lambert and Hall 1982) *Suppose for $\theta \in \Theta_1$ there exist constants $b(\theta)$ and $\sigma(\theta)$, $-\infty < b(\theta) < \infty$, $0 < \sigma(\theta) < \infty$, such that for any $z \in \mathbf{R}^1$*

(I) $P_\theta\left(\sqrt{n}\,(T_n - b(\theta)) < z\right) \to \Phi\left(z/\sigma(\theta)\right)$, $n \to \infty$, *where Φ is the standard normal distribution function.*

   In addition, assume that I is an open interval containing $b(\theta)$ and there exists on I a real function g possessing the property that if $b \in I$ and $\{b_n\}$ is a sequence of numbers, $b_n = b + O\left(n^{-1/2}\right)$, then under the hypothesis H we have

(II) $\ln\left[1 - G_n(b_n)\right] = -n\,g(b_n) + o(\sqrt{n})$, $\qquad n \to \infty$.

Moreover, assume that

(III) the function g is continuously differentiable on I.

Then for all $\theta \in \Theta_1$ and $z \in \mathbf{R}^1$

$$P_\theta\left(\frac{\ln L_n + \frac{1}{2} n\, c(\theta)}{\sqrt{n}} < z \right) \longrightarrow \Phi\left(\frac{z}{\tau(\theta)} \right) \qquad \text{as } n \to \infty,$$

where

$$c(\theta) := 2\, g\big(b(\theta) \big), \qquad\qquad \tau(\theta) := \sigma(\theta)\, g'\big(b(\theta) \big).$$

Here $c(\theta)$ is the exact slope of the sequence $\{T_n\}$ and $\tau^2(\theta)$ is the asymptotic variance of $\ln L_n$. The arguments of Lambert and Hall (1982) show that the pair $\big(c(\theta),\, \tau^2(\theta) \big)$ may serve as a more sensitive measure of test comparison than the single exact slope $c(\theta)$. This corresponds to the well-known situation in the estimation theory when the quality and the choice of the unbiased estimator depend on its variance. If two different tests have the same or very close exact slopes, but the asymptotic variances of logarithms of their P-values are distinct, then just the comparison of the latter enables one to make the well-founded choice of a suitable test. Moreover it is possible to obtain the following refinement of Theorem 1.2.1. Put for brevity

$$c := c(\theta), \qquad\qquad A := -2\ln\alpha, \qquad\qquad \tau^2 := \tau^2(\theta).$$

Lambert and Hall (1982) proved that

$$N_T(\alpha, \beta, \theta) = \frac{A}{c}\left\{ 1 + \frac{\tau\, \Phi^{-1}(\beta)}{(A\,c)^{1/2}} + o\big(A^{-1/2} \big) \right\}, \qquad A \to \infty.$$

So the asymptotic variance τ^2 precisely determines the correction term with respect to the main term A/c from Theorem 1.2.1. Generalizations of these results may be found in Bahadur, Chandra, and Lambert (1982); Kallenberg (1983b); Berk (1984); Bahadur and Gupta (1985); and Chandra (1989).

The verification of the conditions of Theorem 1.2.4, especially of condition II, is a rather complicated task. It has been done by Lambert and Hall (1982) for some important parametric statistics. The Ph.D. Thesis by Leont'yev (1990) was devoted to the proof of assertions close to Theorem 1.2.4 for nonparametric and asymptotically nonparametric goodness-of-fit statistics; specifically the Kolmogorov statistic was considered by Leont'yev (1987) and the ω^2 statistic also by Leont'yev (1988).

In concluding this section we note that an attempt at constructing the *sequential* variant of the Bahadur theory has been made by Berk and Brown (1978) and Kourouklis (1984). It turns out, however, that the Bahadur efficiency is not well adapted to the discrimination of sequential tests.

1.3 Hodges–Lehmann Asymptotic Relative Efficiency

This type of ARE proposed by Hodges and Lehmann (1956) is at most in conformity with the classical approach of Neyman and Pearson, but at the same time has been studied inadequately. This is apparent because of the purely technical difficulties connected with the calculation of large-deviation asymptotics of a given sequence of statistics not under the null hypothesis (as in the case of the Bahadur ARE) but under the alternative. These difficulties are pointed out by Bahadur (1967) and Singh (1984) and also by Rao (1962) through the discussion of different measures of efficiency. After the paper of Hodges and Lehmann (1956) the study of the Hodges–Lehmann ARE remained at a standstill until the appearance of the paper of Brown (1971) on the conditions of asymptotic optimality of likelihood ratio tests. Then there later appeared papers of Raghavachari (1983); Brown, Ruymgaart, and Truax (1984); Baringhaus (1987); and Kourouklis (1988) that were also devoted essentially to parametric tests. As a matter of fact the Hodges–Lehmann efficiency has not been used for comparing nonparametric tests.

Let, in the framework of Section 1.1, $\beta_n(\alpha, \theta)$ be the power of the significance test based on the sequence $\{T_n\}$ having size not exceeding $\alpha \in (0, 1)$ and considered under the alternative $\theta \in \Theta_1$. If there exists a function $d_T(\theta)$, $0 < d_T(\theta) < \infty$, such that

$$\lim_{n \to \infty} n^{-1} \ln \left(1 - \beta_n(\alpha, \theta) \right) = -\tfrac{1}{2} d_T(\theta), \qquad (1.3.1)$$

then $d_T(\theta)$ is called the *Hodges–Lehmann index* (or simply *index*) of the sequence $\{T_n\}$. For two such sequences the Hodges–Lehmann ARE (1.1.4) is equal to the ratio of corresponding indices.

Indeed, it follows from the definition of $N := N_T(\alpha, \beta, \theta)$ and condition (1.3.1) that $N \to \infty$ as $\beta \to 1$ and

$$\beta_N(\alpha, \theta) \geq \beta \geq \beta_{N-1}(\alpha, \theta).$$

This implies

$$N^{-1} \ln \left(1 - \beta \right) \sim -\tfrac{1}{2} d_T(\theta) \qquad \text{as } \beta \to 1,$$

or, equivalently,

$$N_T(\alpha, \beta, \theta) \sim -2 \ln(1 - \beta) \, / \, d_T(\theta) \qquad \text{as } \beta \to 1. \tag{1.3.2}$$

It remains to use (1.3.2) and the definition of the Hodges–Lehmann ARE.

Thus the indices serve as a measure of the Hodges–Lehmann ARE. Formula (1.3.1) shows that their computation is based on the large-deviation asymptotics of $\{T_n\}$ under the alternative.

There exists an upper bound for the Hodges–Lehmann indices analogous to the upper bound for exact slopes stated by inequality (1.2.15). Introduce for any $\theta \in \Theta_1$

$$K(\Theta_0, \theta) := \inf \left\{ K(\theta_0, \theta) \colon \theta_0 \in \Theta_0 \right\}. \tag{1.3.3}$$

The quantity $K(\Theta_0, \theta)$ is "dual" to the quantity specified in (1.2.10), but does not coincide with it since the Kullback–Leibler information is nonsymmetric.

Theorem 1.3.1 *For any sequence of statistics $\{T_n\}$ and any $\theta \in \Theta_1$ we have*

$$\lim_{n \to \infty} \left(1 - \beta_n(\alpha, \theta) \right) \geq -K(\Theta_0, \theta). \tag{1.3.4}$$

The proof of Theorem 1.3.1 is very close to the proof of Theorem 1.2.3 and may be found in Nikitin (1986c, 1987a) or Kourouklis (1988). Generalization for several samples will be given subsequently in Theorem 2.7.1.

It follows from (1.3.4) that the index of $\{T_n\}$ should satisfy the inequality

$$d_T(\theta) \leq 2 \, K(\Theta_0, \theta). \tag{1.3.5}$$

As in the Bahadur theory the sequence of statistics $\{T_n\}$ is said to be *asymptotically optimal in the Hodges–Lehmann sense* if equality takes place in (1.3.5). If

$$d_T(\theta) \sim 2 \, K(\Theta_0, \theta) \qquad \text{as } \theta \to \theta_0 \in \partial \Theta_0, \tag{1.3.6}$$

this sequence is said to be *locally asymptotically optimal (LAO) in the Hodges–Lehmann sense*.

Hettmansperger (1973) has made the attempt to define the notion of the approximate Hodges–Lehmann ARE by analogy with the approximate Bahadur ARE. This concept has, however, only limited value because in a number of important cases the exact and approximate

Hodges–Lehmann AREs do not coincide even locally. Some considerations in this connection may be found in Mikulski (1976). The Hodges–Lehmann ARE of sequential tests was touched upon by Berk (1976).

1.4 Pitman Asymptotic Relative Efficiency

The classical Pitman efficiency is very often used for the purpose of the asymptotic comparison of various tests. It was introduced at the end of the forties by Pitman (1949) in his unpublished lectures. Therefore the calculations of this type of the ARE have been for a long time based on the Pitman results stated in the publications by Noether (1950, 1955). The main Pitman result looks as follows.

Theorem 1.4.1 *Let* $\Theta = \mathbf{R}^1$ *and* $\Theta_0 = (-\infty, \theta_0]$. *Suppose a sequence of statistics* $\{T_n\}$ *possesses the following three properties.*

(1) There exist functions μ and σ such that

$$\lim_{n \to \infty} P_{\theta_n}\left(\frac{T_n - \mu(\theta_n)}{\sigma(\theta_n)} < z \right) = \Phi(z) \qquad (1.4.1)$$

for all $z \in \mathbf{R}^1$ *and* $\theta_n = \theta_0 + k\,n^{-1/2}$, $k \geq 0$.

(2) There exists the right-hand derivative $\mu'(\theta_0) > 0$.

(3) $\displaystyle \lim_{n \to \infty} \frac{\mu'(\theta_n)}{\mu'(\theta_0)} = 1, \qquad \lim_{n \to \infty} \frac{\sigma(\theta_n)}{\sigma(\theta_0)} = 1.$

Let $\{\widetilde{T}_n\}$ *be another sequence of statistics satisfying the same conditions with functions $\tilde{\mu}$ and $\tilde{\sigma}$. Then the Pitman ARE exists for all* $0 < \alpha < \beta < 1$ *and may be calculated by the formula*

$$e^P_{T,\tilde{T}}(\alpha,\beta,\theta_0) \equiv e^P_{T,\tilde{T}}(\theta_0) = \left\{ \frac{\mu'(\theta_0)}{\sigma(\theta_0)} \Big/ \frac{\tilde{\mu}'(\theta_0)}{\tilde{\sigma}(\theta_0)} \right\}^2. \qquad (1.4.2)$$

More general results may be found in Kendall and Stuart (1967) and Serfling (1980).

We quote now as an example one of the first Pitman results that stimulated the development of nonparametric statistics. Consider the classical problem of testing the hypothesis about the mean of the Gaussian law under the location alternative. Let $e^P_{W,t}$ be the Pitman ARE of the one-sample Wilcoxon rank test with respect to the Student test. It turns out that

$$e^P_{W,t} = 3/\pi \approx 0.955,$$

and this shows that the ARE of the Wilcoxon test in comparison with the Student test that is optimal for this problem is unexpectedly high. Then Hodges and Lehmann (1956) proved that

$$0.864 \leq e^{P}_{W,t} \leq +\infty$$

if one rejects the assumption of normality and, moreover, the lower bound is attained at the density

$$f(x) = \begin{cases} 3(5-x^2)/(20\sqrt{5}) & \text{if } |x| \leq \sqrt{5}, \\ 0 & \text{otherwise.} \end{cases}$$

The question of the ARE bounds for various pairs of nonparametric statistics is of interest and is closely connected with the tail ordering of distributions. Having no opportunity to touch upon this topic, we address readers to the references of Hodges and Lehmann (1961), Sinha and Wieand (1977), and Weissfeld and Wieand (1984) as well as Loh (1984) and Caperaà (1988), which contain promising advancements.

Returning to Theorem 1.4.1 we note that under its conditions the Pitman ARE is a constant independent of α and β. It is ensured by the asymptotic normality of statistics T_n and $\tilde{T}_n$ both under the null hypothesis and the alternative, that is, by condition (1.4.1). If this condition fails, considerable difficulties arise when calculating the Pitman ARE as the latter may not at all exist or may depend on α and β. The results by Rothe (1981) give an example of such a situation for the limiting χ^2 distribution. In the case of the Kolmogorov–Smirnov and ω^2 statistics or their variants, which, as is well known, have nonnormal limiting distributions under the null hypothesis, the values of the Pitman ARE are either unknown or very crude bounds are obtained for them. See Yu (1971), Mikulski (1976), Archambault and Mikulski (1979), and Neuhaus (1982).

Important progress in this field has been made by Wieand (1976), who has established the correspondence between the limit of the Pitman efficiency as $\alpha \to 0$ and the limit as $\theta \to \theta_0$ of the approximate Bahadur efficiency which is easy to calculate. Results of such nature for asymptotically normal statistics have been obtained by Bahadur (1960b). The merit of the paper by Wieand (1976) consists in the possibility of considering much more general situations.

To formulate the main result of Wieand (1976) we need to strengthen condition (1.2.6) in Theorem 1.2.2. Let us say that the sequence of statistics $\{T_n\}$ is a *Wieand sequence* if there exists θ^* such that for any $\varepsilon > 0$

and $\delta \in (0, 1)$ there exists a constant $\mathcal{C}$ with the following property: For any $\theta \in (\theta_0, \theta^*)$ and $n > \mathcal{C} \, b^{-2}(\theta)$

$$P_\theta\left(\left|T_n - b(\theta)\right| > \varepsilon \, b(\theta)\right) < \delta. \tag{1.4.3}$$

Therefore for any Wieand sequence the convergence T_n to $b(\theta)$ in probability takes place at a certain rate.

We quote an auxiliary statement by Wieand (1976) that often simplifies the verification of the previous condition.

Theorem 1.4.2 *Let $\{U_{n,\theta}\}$ be a family of sequences of statistics satisfying for all $\theta \in \Theta$ and $z \in \mathbf{R}^1$ the relation*

$$\lim_{n\to\infty} P_\theta\left(U_{n,\theta} < z\right) = \mathcal{Q}(z) \tag{1.4.4}$$

where $\mathcal{Q}$ is a continuous distribution function and the rate of convergence in (1.4.4) is independent of θ. Let $d(\theta) \in (0, 1)$ be an arbitrary function on (θ_0, θ^*). Then for any $\varepsilon > 0$ and $\delta \in (0, 1)$ there exists a number C' such that for all $\theta_0 < \theta < \theta^*$ and $n > C'd^{-2}(\theta)$ the following estimate holds:

$$P_\theta\left(\left|U_{n,\theta}\right|/\sqrt{n} < \varepsilon \, d(\theta)\right) > 1 - \delta.$$

Recall that the Pitman ARE $e_{V,T}^P(\alpha, \beta)$ of the sequence $\{V_n\}$ with respect to the sequence $\{T_n\}$ has been defined as the limit of the relative efficiency $e_{V,T}(\alpha, \beta, \theta)$ as $\theta \to \theta_0$ (see (1.1.5)). As this limit may not necessarily exist, let us introduce the *upper* and *lower Pitman AREs* as

$$e_{V,T}^{+P}(\alpha, \beta) := \sup_{(\mathbf{\Pi})} \varlimsup_{j\to\infty} e_{V,T}(\alpha, \beta, \theta_j),$$

$$e_{V,T}^{-P}(\alpha, \beta) := \inf_{(\mathbf{\Pi})} \varliminf_{j\to\infty} e_{V,T}(\alpha, \beta, \theta_j),$$

where the symbol $(\mathbf{\Pi})$ denotes the set of all sequences $\{\theta_j\}$, $\theta_j \in \Theta_1$, $\theta_j \to \theta_0$. The main result by Wieand (1976) is as follows.

Theorem 1.4.3 *Let $\{V_n\}$ and $\{T_n\}$ be two Wieand sequences of statistics having under H the limiting distribution functions satisfying (1.2.18) and being strictly increasing for all large values of their arguments. If there exists the finite limit $\lim_{\theta\to\theta_0} e_{V,T}^{*B}(\theta)$, then*

$$\lim_{\theta\to\theta_0} e_{V,T}^{*B}(\theta) = \lim_{\alpha\to 0} e_{V,T}^{+P}(\alpha, \beta) = \lim_{\alpha\to 0} e_{V,T}^{-P}(\alpha, \beta). \tag{1.4.5}$$

The common value of finite limits in (1.4.5) is called the *limiting* (as $\alpha \to 0$) *Pitman* ARE. Under the conditions of Theorem 1.4.3 it is independent of β.

The absolute majority of statistics considered in this book satisfy the conditions of the theorems of the present section and this enables one to calculate their Pitman or limiting Pitman AREs.

Examples of such calculations may be found in Kendall and Stuart (1967), Lehmann (1975), Wieand (1976), Pitman (1979), and Pratt and Gibbons (1981). A survey of results concerning the Pitman ARE for parametric statistics, taking into account the effects of "second order," has been presented by Chibisov (1983).

In conclusion we touch upon the problem of an upper bound for the Pitman ARE. As stated earlier, under certain conditions including the asymptotic normality under the null hypothesis and the alternative, the Pitman ARE depends on the so-called *efficacy* $\left(\mu'(\theta_0)/\sigma(\theta_0)\right)^2$. Under some additional conditions Rao (1963) proved that

$$\left(\mu'(\theta_0)/\sigma(\theta_0)\right)^2 \leq I(\theta_0), \tag{1.4.6}$$

where $I(\theta_0)$ is the Fisher information at the point θ_0, corresponding to the distribution of initial observations. If equality in (1.4.6) holds, the sequence of statistics is said to be *asymptotically optimal in the Pitman sense*. Any variants of (1.4.6) for statistics having nonnormal limit distribution are unknown.

1.5 Different Approaches to the Definition of the ARE

In this section we shall describe briefly the different approaches to the definition and the calculation of AREs. Some authors believe that the Bahadur and Hodges–Lehmann approaches contain a certain "lack of balance," as when determining the corresponding efficiencies one of the error probabilities is kept fixed and the other tends to zero. The Chernoff ARE introduced by Chernoff (1952) is free from this drawback as here both error probabilities are tending to zero. The theory of the Chernoff efficiency was later developed by Kallenberg (1982).

Let, as in Section 1.3, $\beta_n(\alpha, \theta)$ be the power at the point $\theta \in \Theta_1$ of the test based on a sequence of statistics $\{T_n\}$ with large critical values and size not exceeding $\alpha \in [0, 1]$. Put

$$\rho_n(\alpha, \theta) := \max\left\{\alpha, 1 - \beta_n(\alpha, \theta)\right\}$$

and further

$$\rho_n(\theta) := \inf \left\{ \rho_n(\alpha, \theta) \colon 0 \le \alpha \le 1 \right\}.$$

If there exists a function $\rho_T(\theta)$, $0 < \rho_T(\theta) < \infty$, such that for each $\theta \in \Theta_1$

$$\lim_{n \to \infty} n^{-1} \ln \rho_n(\theta) = -\rho_T(\theta), \qquad (1.5.1)$$

it is called the *Chernoff index* of a given sequence $\{T_n\}$. It is clear that $\rho_T(\theta)$ describes the rate of exponential decrease for the maximum of the probabilities of the error of first and second kind.

For two sequences of statistics $\{T_n\}$ and $\{V_n\}$ with corresponding Chernoff indices $\rho_T(\theta)$ and $\rho_V(\theta)$ the Chernoff ARE is defined by

$$e_{T,V}^C(\theta) := \rho_T(\theta) / \rho_V(\theta). \qquad (1.5.2)$$

The following theorem proved by Kallenberg (1982) shows that the computation of the Chernoff indices needs rough large-deviation asymptotics for test statistics both under the null hypothesis and the alternative.

Theorem 1.5.1 *Let for some $c^* \in \mathbf{R}^1$ and $\theta \in \Theta_1$*

$$\lim_{n \to \infty} n^{-1} \ln \sup \left\{ P_{\theta_0}(T_n > c^*) \colon \theta_0 \in \Theta_0 \right\} = \lim_{n \to \infty} n^{-1} \ln P_\theta \left(T_n \le c^* \right).$$
$$(1.5.3)$$
Then $\rho_T(\theta) = -u(c^*)$, where $u(c^*)$ is the common value of both limits in (1.5.3).

There exists an upper bound for the Chernoff indices, too. We describe it now in the simplest case. Assume we are testing the simple hypothesis $H \colon \theta = \theta_0$ against the alternative $A \colon \theta = \theta_1$ and the distributions P_{θ_0} and P_{θ_1} have their common support in $\mathbf{R}^1$. Denote by $f(x; \theta_0)$ and $f(x; \theta_1)$ their densities with respect to the measure

$$\mu = \tfrac{1}{2} \left(P_{\theta_0} + P_{\theta_1} \right)$$

and consider the function

$$\psi(t) := \ln \int_{\mathbf{R}^1} \exp \left[t \ln \left\{ f(x; \theta_1) / f(x; \theta_0) \right\} \right] d P_{\theta_0}(x).$$

It is obvious that ψ is strictly convex on $[0, 1]$ and $\psi(0) = \psi(1) = 0$. Hence, there exists a unique point $t^* \in [0, 1]$ such that

$$\psi(t^*) = \min \left\{ \psi(t) \colon t \in [0, 1] \right\}.$$

It turns out (see Kallenberg (1982)) that the Chernoff index $\rho_T(\theta_1)$ of any sequence of statistics $\{T_n\}$ used for this problem satisfies the inequality

$$\rho_T(\theta_1) \le -\psi(t^*).\qquad(1.5.4)$$

More general variants of this inequality may also be found in Kallenberg (1982). In the aforementioned paper it has been proved that the likelihood ratio test is Chernoff asymptotically optimal if the distribution of observations belongs to an exponential family.

Ronzhin (1985) introduced a certain generalization of the Chernoff efficiency where the error probabilities α and β are decreasing in such manner that

$$\ln \alpha / \ln \beta \longrightarrow \mu \in [0, +\infty].$$

Kallenberg (1983a) presented another approach to measuring the ARE, taking some intermediate position between the Bahadur and Pitman approaches. Under this approach the level α_n tends to zero, but not too fast. The alternative value of parameter θ_n also tends to θ_0 at a controlled rate. Besides, θ_n and α_n are coordinated in such a way that the power of the test at the point θ_n is separated from 0 and 1. Kallenberg (1983a) has shown that for calculating this "intermediate," or Kallenberg ARE, one needs information about large deviations of moderate or Cramér type. Moderate deviations have been used also by Rubin and Sethuraman (1965b) for the definition of the *Bayes risk efficiency*. (See also Serfling (1980).)

Recently Borovkov and Mogulskii (1992) proposed another definition of efficiency where the probabilities of errors are used in a more symmetric way. Denote for any $\alpha \in \mathbf{R}^1$ and any sequence of suitably normalized test statistics $\{T_n\}$

$$\varepsilon_{n,\alpha}(T) := \sup\left\{ P_\theta\big(T_n > \alpha\sqrt{n}\big)\colon \theta \in \Theta_0 \right\},$$

$$\delta_{n,\alpha}(T) := \sup\left\{ P_\theta\big(T_n \le \alpha\sqrt{n}\big)\colon \theta \in \Theta_1 \right\},$$

and let us introduce for any given $\varepsilon > 0$ and $\delta \in (0, 1)$ a new characteristic as

$$n(\varepsilon,\delta) := \min\left\{ n \;\middle|\; \exists \alpha_0 \text{ such that } \varepsilon_{n,\alpha_0}(T) \le \varepsilon,\ \delta_{n,\alpha_0}(T) \le \delta \right\}.$$

$$(1.5.5)$$

It is clear that $n(\varepsilon, \delta)$ is the minimal sample size, ensuring for $\{T_n\}$ the given probabilities of errors ε and δ. An application of (1.5.5) makes it possible to define the relative efficiency and furthermore the ARE with the aid of the appropriate limit.

The last three types of the ARE are of indisputable interest, however their study and calculation are beyond the framework of this book.

1.6 Some Results on Probabilities of Large Deviations

As we have seen in the previous sections the calculation of the Bahadur, Hodges–Lehmann, and Chernoff AREs is closely related to the evaluation of rough asymptotics for the large-deviation probabilities both under the null hypothesis and the alternative. Besides we are interested in the first place in the deviations of order $O(\sqrt{n})$. In the classification by Wentzell (1990, Sec. 4.2) these are "very large" deviations. Sometimes they are called "Chernoff-type" to distinguish them from "Cramér-type" deviations of order $o(\sqrt{n})$ and moderate deviations of order $O(\sqrt{\ln n})$ for which the analysis requires a somewhat different technique. The literature connected with large deviations is extremely vast and numbers in the hundreds. We point out among them the fundamental papers by Chernoff (1952) and Sanov (1957) as well as the papers by Bahadur and Ranga Rao (1960); Hoeffding (1965); Borovkov (1967); Petrov (1975); Donsker and Varadhan (1975a, 1975b, 1976); Bahadur and Zabell (1979); Borovkov and Mogulskii (1978, 1980); Groeneboom, Oosterhoff, and Ruymgaart (1979); Azencott (1980); Bolthausen (1984); Ellis (1984); Mogulskii (1984); Stroock (1984); Varadhan (1984); Deuschel and Stroock (1989); Wentzell (1990); and Saulis and Statulevicius (1991).

We quote in this section some of the most frequently used results in this field necessary for the next chapters. They may be symbolically divided into the following two groups: one connected with the Chernoff theorem and the other connected with the Sanov theorem.

Let Y be a real random variable with d.f. F. Denote by ψ the *moment generating function* of Y:

$$\psi(t) := E \exp\{t\, Y\} = \int_{-\infty}^{\infty} e^{ty}\, dF(y), \qquad -\infty < t < \infty, \qquad (1.6.1)$$

and put

$$\rho := \inf\{\psi(t) \colon\, t \geq 0\}. \qquad (1.6.2)$$

It is clear that $0 \leq \rho \leq 1$.

The following elementary result ascribed to S. N. Bernstein appears very useful in the future.

Theorem 1.6.1

$$P(Y \geq 0) \leq \rho.$$

For any sequence $\{Y_j\}$ of independent identically distributed (i.i.d.) random variables having the same distribution as Y this theorem yields

$$P(Y_1 + \cdots + Y_n \geq 0) \leq \rho^n, \qquad n \geq 1. \tag{1.6.3}$$

But in this respect profound results have been obtained with the following theorem.

Theorem 1.6.2 *Let $\{u_n\}$ be a real sequence such that $u_n \to u$, $-\infty < u < +\infty$, and*

$$P(Y > u) > 0. \tag{1.6.4}$$

Then

$$\lim_{n \to \infty} n^{-1} \ln P(Y_1 + Y_2 + \cdots + Y_n \geq n u_n) = -f(u), \tag{1.6.5}$$

$$\lim_{n \to \infty} n^{-1} \ln P(Y_1 + Y_2 + \cdots + Y_n > n u_n) = -f(u) \tag{1.6.6}$$

where $f(u)$ is determined by the equality

$$\exp[-f(u)] = \inf \{e^{-tu} \psi(t) : t \geq 0\}. \tag{1.6.7}$$

If one assumes additionally that $\psi(t) < \infty$ in a neighborhood of zero, $EY_i = 0$ and $\operatorname{Var} Y_i = \sigma^2 > 0$, then

$$f(u) = \frac{u^2}{2\sigma^2} \left(1 + o(1)\right) \qquad as \ u \to 0.$$

Note that the function f is continuous in a neighborhood of zero.

Theorem 1.6.2 belongs to Chernoff (1952) who used as a base the earlier results by Cramér (1938). Variants of the proof may be found, for example, in Bahadur (1971) and Steinebach (1980).

There exist a lot of generalizations of the Chernoff theorem (Borovkov and Mogulskii (1978, 1980), Bahadur and Zabell (1979), Bretagnolle (1979), and others). We confine ourselves to the following result by Sethuraman (1964) (see also Rao (1972)) dealing with random variables taking on values in a separable Banach space $\mathcal{Y}$ with the corresponding norm $\| \cdot \|$. Denote by $\mathcal{Y}_1^*$ the space of all continuous linear functionals y^* on $\mathcal{Y}$ with the unit norm.

Theorem 1.6.3 *Let $\{Y_n(\omega)\}$ be a sequence of independent identically distributed random variables defined on a probability space $(\Omega, \mathfrak{F}, P)$ with values in $\mathcal{Y}$. Suppose that*

$$\int_\Omega y^*\big(Y_1(\omega)\big)\, dP(\omega) = 0, \qquad\qquad \forall y^* \in \mathcal{Y}_1^*, \qquad (1.6.8)$$

$$\int_\Omega \exp\big\{\, z\, \|Y_1(\omega)\|\,\big\}\, dP(\omega) < \infty, \qquad \forall z \in \mathbf{R}^1. \qquad (1.6.9)$$

Then for any $\varepsilon > 0$ one has

$$\lim_{n\to\infty} n^{-1} \ln P\big\{\|Y_1(\omega) + \cdots + Y_n(\omega)\| \ge n\,\varepsilon\big\} = \ln \rho(\mathcal{Y}_1^*, \varepsilon),$$

where

$$\rho(\mathcal{Y}_1^*, \varepsilon) := \sup\big\{\rho(y^*, \varepsilon)\colon\, y^* \in \mathcal{Y}_1^*\big\},$$

and

$$\rho(y^*, \varepsilon) := \max\Big[\rho_1(y^*, \varepsilon) := \min\big\{e^{-t\varepsilon}\, E\big(\exp\big[t\,y^*(Y_1(\omega))\big]\big)\colon\, t \ge 0\big\},$$

$$\rho_2(y^*, \varepsilon) := \min\big\{e^{t\varepsilon}\, E\big(\exp\big[t\,y^*(Y_1(\omega))\big]\big)\colon\, t \le 0\big\}\Big].$$

Moreover,

$$\rho(\mathcal{Y}_1^*, \varepsilon) = -\frac{\varepsilon^2}{2\,\sigma^2}\,\big(1 + o(1)\big) \qquad as\ \varepsilon \to 0,$$

where

$$\sigma^2 = \sup\big\{\,\mathrm{Var}\, y^*(Y_1(\omega))\colon\, y^* \in \mathcal{Y}_1^*\big\}.$$

Note that the function $\rho(\mathcal{Y}_1^*, \varepsilon)$ is continuous in ε in a neighborhood of zero.

Other generalizations of the Chernoff theorem are connected with the transition from sums of random variables to U-*statistics*. Let $Y_1, Y_2, \ldots,$ Y_n be i.i.d. random variables and $\Phi\colon \mathbf{R}^m \to \mathbf{R}^1$ be a symmetric function of m variables, $m < n$. We recall that a U-statistic is a random variable given by

$$U_n := \binom{n}{m}^{-1} \sum_{1 \le i_1 < \cdots < i_m \le n} \Phi\big(Y_{i_1}, \ldots, Y_{i_m}\big).$$

The number m and Φ are defined as the *degree* of the U-statistic and its *kernel*, respectively. The problem of the Chernoff-type large deviations of U-statistics has been partially solved in the following theorem by Dasgupta (1984). Denote first

$$\varphi(t) := E\big\{\Phi(Y_1,\ldots,Y_m)\,|\,Y_1 = t\big\}$$

and suppose that

$$\delta^2 := E\varphi^2(Y_1) > 0. \tag{1.6.10}$$

Usually U-statistics satisfying (1.6.10) are said to be *nondegenerate*.

Theorem 1.6.4 *Let (1.6.10) and the conditions*

$$E\Phi(Y_1,\ldots,Y_m) = 0, \quad E\exp\{t\,\Phi^2(Y_1,\ldots,Y_m)\} < \infty \quad \text{for any } t \in \mathbf{R}^1,$$

and

$$\lim_{n\to\infty} n^{-1}\ln P\left(\sum_{i=1}^n \varphi(Y_i) > n\varepsilon_0\right) < \infty \qquad \text{for some } \varepsilon_0 > 0$$

be fulfilled. Then for a sequence $\{\gamma_n\}$ such that $\gamma_n \to 0$ as $n \to \infty$ we have

$$P\left(\frac{U_n}{2\delta} \geq \varepsilon + \gamma_n\right) = P\left(\sum_{i=1}^n \varphi(Y_i) > n\varepsilon\right)(1 + o(1)).$$

Combining Theorems 1.6.4 and 1.6.2 we come to the following corollary.

Corollary *Under conditions of Theorems 1.6.4 and 1.6.2 with respect to the random variables $\varphi(Y_i)$, $i = 1,\ldots,n$, we have*

$$\lim_{n\to\infty} n^{-1}\ln P\big(U_n \geq a + \gamma_n\big) = -\frac{a^2}{8\,\delta^2}\,(1 + o(1)) \qquad \text{as } a \to 0.$$
$$\tag{1.6.11}$$

Some variants of such results were described in Korolyuk and Borovskikh (1993).

Numerous nonparametric statistics cannot be represented in the form of U-statistics but in the form of *von Mises functionals* intimately connected with them. Recall that these r.v.s look like

$$V_n := n^{-m}\sum_{i_1=1}^n \cdots \sum_{i_m=1}^n \Phi_{mk}\big(Y_{i_1},\,\ldots,\,Y_{i_m}\big),$$

where the kernel Φ is defined as before. Consider for $k = 1, \ldots, m$ U-statistics

$$U_{nk} := \binom{n}{k}^{-1} \sum_{1 \le i_1 < \cdots < i_k \le n} \Phi_{mk}\big(Y_{i_1}, \ldots, Y_{i_k}\big), \qquad k = 1, 2, \ldots, m,$$

with the symmetric kernels

$$\Phi_{mk}(x_1, \ldots, x_k) := \sum_{\substack{\nu_1 + \cdots + \nu_k = m \\ \nu_j \ge 1}} \frac{m!}{\nu_1! \cdots \nu_k!} \; \Phi\big(\mathbf{x}_1^{\nu_1}, \ldots, \mathbf{x}_k^{\nu_k}\big),$$

where $\mathbf{x}_j^{\nu_j} = (x_j, \ldots, x_j)$ is a vector with ν_j identical coordinates.

The following Bönner and Kirschner (1977) representation is valid:

$$V_n = n^{-m} \sum_{k=1}^{m} \binom{n}{k} U_{nk}. \tag{1.6.12}$$

This formula enables one in many cases to reduce large deviations for von Mises functionals to large deviations of U-statistics. For example, if the kernels of U-statistics U_{nk} are bounded then the main asymptotic contribution to (1.6.12) belongs to the summand $n^{-m}\binom{n}{m}U_{nm}$ whereas the others may be neglected. To study the large-deviation asymptotics of the random variables $n^{-m}\binom{n}{m}U_{nm}$ one may apply Theorem 1.6.4.

The Sanov theorem presents another fundamental result in the theory of large deviations. Again let $\{Y_j\}$ be a sequence of i.i.d. random variables with the common d.f. F_0 and

$$F_n(t) := n^{-1} \sum_{i=1}^{n} \mathbf{1}_{\{Y_i < t\}} \tag{1.6.13}$$

be the *empirical d.f.* based on the sample $Y_1, Y_2, \ldots, Y_n$. The main result by Sanov (1957) is as follows: *For a sufficiently regular set Ω in the space of d.f.s on the real line (F_0-distinguishable in his terminology)*

$$\lim_{n \to \infty} n^{-1} \ln P\left(F_n \in \Omega\right) = -\inf\left\{ \int_{-\infty}^{\infty} \ln \frac{dF}{dF_0} \, dF : F \in \Omega \right\}. \tag{1.6.14}$$

The integral on the right-hand side coincides with the Kullback–Leibler information introduced in (1.2.9).

As it is hard to verify the condition of F_0-distinguishability, numerous papers have been connected with sufficient conditions for the validity of (1.6.14). We present now some results in this field.

Let us denote, in agreement with (1.2.10), the right-hand side of (1.6.14) by $K(\Omega, F_0)$. Let $\mathcal{E}$ be the set of d.f.s g on $\mathbf{R}^1$ being absolutely continuous with respect to F_0 and possessing the property that *there exists a finite division into intervals* $\Delta_1, \ldots, \Delta_n$ *of the real line on which the derivative* dg/dF_0 *is monotone and bounded.* We denote by ρ the uniform metric in the space Λ_1 of d.f.s on $\mathbf{R}^1$ and by Cl_ρ the closure in this metric.

Theorem 1.6.5 (Borovkov 1967) *Suppose F_0 is a continuous distribution function and Ω is a ρ-open set in Λ_1 such that*

$$K\left(\Omega \cap \mathcal{E}, F_0\right) = K\left(\mathrm{Cl}_\rho\left(\Omega\right) \cap \mathcal{E}, F_0\right).$$

Then equality (1.6.14) is valid.

When studying large deviations of concrete statistics they may often be represented as functionals $T(F_n)$ of empirical d.f.s. In this connection let us introduce for any $r \in \mathbf{R}^1$ the sets

$$\Omega_r^T := \left\{F \in \Lambda_1 \colon T(F) \geq r\right\}.$$

Theorem 1.6.6 (Hoadley 1967) *Let F_0 be a continuous distribution function, $T(F)$ be a uniformly continuous in the ρ-topology functional, and $\{u_n\}$ be a real sequence such that $u_n \to 0$ as $n \to \infty$. If the function $t \mapsto K(\Omega_t^T, F_0)$ is continuous at the point $t = a$ then*

$$\lim_{n \to \infty} n^{-1} \ln P\left(T(F_n) \geq a + u_n\right) = -K(\Omega_a^T, F_0). \qquad (1.6.15)$$

Hoadley (1967) also presented the generalization of this theorem for several independent samples.

Later Groeneboom et al. (1979) gave generalizations of Theorems 1.6.6 and 1.6.5 simultaneously in three directions: They considered the space of measures on a Hausdorff space $\mathbf{S}$ instead of the space of d.f.s on the real line, the uniform continuity of the functional $T(F)$ was replaced by continuity in some suitable topology, and finally, they showed that the distribution of the initial sample could be not only continuous, but also atomic.

Let $\mathbf{S}$ be a Hausdorff space, $\mathfrak{B}$ be the Borel σ-algebra of sets in $\mathbf{S}$, and Λ be the space of all probability measures on $\mathfrak{B}$. Define for any

$P, Q \in \Lambda$, in agreement with (1.2.9),

$$
K(Q, P) := \begin{cases} \displaystyle\int_{\mathbf{S}} \ln \frac{dQ}{dP} \, dQ, & \text{if } Q \ll P, \\[2mm] +\infty, & \text{otherwise,} \end{cases}
\tag{1.6.16}
$$

and put for any subset Ω of Λ

$$
K(\Omega, P) := \inf \left\{ K(Q, P): \ Q \in \Omega \right\},
\tag{1.6.17}
$$

keeping in mind that $K(\emptyset, P) = +\infty$.

Let $\{Y_j\}$ be a sequence of i.i.d. random variables with values in $\mathbf{S}$ and having the distribution $P \in \Lambda$. Denote by $\mathcal{P}_n$ the empirical distribution based on $Y_1, Y_2, \ldots, Y_n$, that is, for any $B \in \mathfrak{B}$ put

$$
\mathcal{P}_n(B) := n^{-1} \sum_{i=1}^{n} \mathbf{1}_{\{Y_i \in B\}} \,.
$$

Consider, finally, in the space Λ the topology τ of convergence on all Borel sets, that is, the "coarsest" topology for which the application $Q \mapsto Q(B)$, $Q \in \Lambda$, is continuous for all $B \in \mathfrak{B}$.

The main result by Groeneboom at al. (1979) is as follows.

Theorem 1.6.7 *Let $P \in \Lambda$ and $T \colon \Lambda \mapsto \bar{\mathbf{R}}^1$ be a τ-continuous functional on each*

$$
Q \in \Gamma := \left\{ R \in \Lambda: \ K(R, P) < \infty \right\}.
$$

If the function $t \mapsto K(\Omega_t^T, P)$, where $\Omega_t^T := \left\{ Q \in \Lambda : T(Q) \geq t \right\}$ and $t \in \mathbf{R}^1$, is right-continuous at the point $t = r$, then

$$
\lim_{n \to \infty} n^{-1} \ln P \left(T(\mathcal{P}_n) \geq r + u_n \right) = -K(\Omega_r^T, P)
\tag{1.6.18}
$$

for any real sequence $\{u_n\}$ such that $u_n \to 0$ as $n \to \infty$.

To compute the Hodges–Lehmann ARE one needs the asymptotics for the probability that $\mathcal{P}_n$ hits the sets of the form

$$
\Delta_z^T := \left\{ Q \in \Lambda: \ T(Q) \leq z \right\}.
$$

But for many important examples the function $z \mapsto K(\Delta_z^T, P)$ is right-continuous, due to Lemma 3.3 of Groeneboom et al. (1979), and is not left-continuous. This property does not allow one to obtain any analog of (1.6.18) for Δ_z^T. In this connection we formulate here an asymptotic inequality that may be derived easily from Lemmas 2.4 and 3.1 of Groeneboom et al. (1979).

Theorem 1.6.8 *Let* $P \in \Lambda$, $\Omega \subset \Lambda$ *and*

$$K(\Omega, P) = K\big(\mathrm{Cl}_\tau(\Omega), P\big). \qquad (1.6.19)$$

Then

$$\varlimsup_{n \to \infty} n^{-1} \ln P\big(\mathcal{P}_n \in \Omega\big) \leq -K(\Omega, P). \qquad (1.6.20)$$

It is well known (see Groeneboom et al. (1979)) that the τ-topology is stronger than the ρ-topology. Consequently if a functional T is ρ-continuous then the corresponding set $\mathbf{\Delta}_z^T$ is ρ-closed and a fortiori τ-closed. This enables one to apply Theorem 1.6.8 as most functionals encountered in nonparametric statistics are usually ρ-continuous.

Other conditions supporting the validity of Sanov's theorem may be found in Stone (1974), Sievers (1976), and Fu (1985).

Ermakov (1990) formulated a "uniform" variant of Theorem 1.6.7. Put for any sets Ω and $\mathbf{\Pi} \subseteq \Lambda$

$$K(\Omega, \mathbf{\Pi}) := \inf \big\{ K(Q, P) \colon Q \in \Omega,\ P \in \mathbf{\Pi} \big\}.$$

Theorem 1.6.9 *Assume* $\mathbf{\Pi}$ *is compact in the* τ-*topology. If*

$$K\big(\mathrm{Int}_\tau(\Omega), P\big) = K\big(\mathrm{Cl}_\tau(\Omega), P\big) \qquad \text{for all } P \in \mathbf{\Pi},$$

then

$$\lim_{n \to \infty} \sup_{P \in \mathbf{\Pi}} \left| n^{-1} \ln P\big(\mathcal{P}_n \in \Omega\big) + K(\Omega, P) \right| = 0.$$

If, however,

$$K\big(\mathrm{Int}_\tau(\Omega), \mathbf{\Pi}\big) = K\big(\mathrm{Cl}_\tau(\Omega), \mathbf{\Pi}\big)$$

then

$$\lim_{n \to \infty} \sup_{P \in \mathbf{\Pi}} n^{-1} \ln P\big(\mathcal{P}_n \in \Omega\big) = -K(\Omega, P).$$

This result was applied by Ermakov (1990) to the analysis of asymptotic minimaxity of some nonparametric tests.

The theorems on large deviations for the Markov processes with close formulations are presented in the book of Wentzell (see Wentzell (1990), Sec. 4.3). However they do not imply the Sanov theorem and its generalizations.

We emphasize also that there exists a very close connection between the results of the Chernoff and Sanov types allowing one to derive one from the other. (See Bahadur and Zabell (1979), Bretagnolle (1979),

Groeneboom et al. (1979), Azencott (1980), Fu (1985), and Deuschel and Stroock (1989).)

Inglot and Ledwina (1990) proposed another method to obtain the information on asymptotics of large-deviation probabilities based on the idea of the strong approximation.

Let $G_n(t)$ be the empirical d.f. based on a sample of size n from the uniform distribution on $[0, 1]$ and

$$\alpha_n(t) = \sqrt{n}\left(G_n(t) - t\right), \qquad t \in [0, 1],$$

be the usual empirical process. The well-known result by Komlós, Major, and Tusnády (1975, 1976) is as follows:

There exist a probability space and the sequences of processes $\{\alpha_n^*\} \overset{D}{=} \{\alpha_n\}$ and Brownian bridges $\{B_n\}$ defined on this probability space such that the following inequality

$$P\left(\sup_{0 \le t \le 1} \left|\alpha_n^*(t) - B_n(t)\right| > n^{-1/2}\left(C \ln n + x\right)\right) \le L\, e^{-lx} \quad (1.6.21)$$

holds for all n and x with some positive constants C, L, and l.

Consider now a sequence of statistics $\{T_n\}$ representable in the form $T_n = T(\alpha_n^*)$ where $T\colon D[0,1] \mapsto \mathbf{R}^1$ is some functional satisfying the following conditions:

there exists a constant c such that

$$\left|T(\alpha_n^*) - T(B_n)\right| \le c \sup_{0 \le t \le 1} \left|\alpha_n^*(t) - B_n(t)\right| \qquad a.s., \qquad (1.6.22)$$

$$\ln P\left(T(B_1) \ge y\right) = -\tfrac{1}{2}\,a\,y^2\left(1 + o(1)\right), \qquad y \to \infty, \qquad (1.6.23)$$

for some constant $a > 0$.

Theorem 1.6.10 (Inglot and Ledwina 1990) *Suppose the conditions (1.6.22) and (1.6.23) are fulfilled. Then for any $d > 1$, $K = ac/(2l)$, and $x \in \left(0, (\sqrt{d} - 1)/(dK)\right)$ the following inequalities are valid:*

$$-\tfrac{1}{2}\,a\,x^2(1 + d\,K\,x)^2 \le \underline{\lim}\, n^{-1} \ln P\left(T(\alpha_n) \ge x\sqrt{n}\right)$$

$$\le \overline{\lim}\, n^{-1} \ln P\left(T(\alpha_n) \ge x\sqrt{n}\right)$$

$$\le -\tfrac{1}{2}\,a\,x^2(1 - K\,x)^2\,. \qquad (1.6.24)$$

Proof Denote

$$H_n := \sup_{0 \le t \le 1} \left| \alpha_n^*(t) - B_n(t) \right|.$$

Using (1.6.21)–(1.6.23) we obtain

$$P\big(T(\alpha_n) \ge x\sqrt{n}\big) \le P\big(T(B_n) \ge x(1 - Kx)\sqrt{n}\big)$$
$$+ P\big(H_n \ge c^{-1}Kx^2\sqrt{n}\big)$$
$$\le \exp\left\{ -\tfrac{1}{2}a\,x^2(1 - Kx)^2 n\,(1 + o(1)) \right\}$$
$$+ L\,\exp\left\{ -l\left(c^{-1}Kx^2n - C\ln n\right) \right\}.$$

The lower estimate may be found in the same way:

$$P\big(T(\alpha_n) \ge x\sqrt{n}\big) \ge P\big(T(B_n) \ge x(1 + dKx)\sqrt{n}\big)$$
$$- P\big(H_n \ge c^{-1}dKx^2\sqrt{n}\big)$$
$$\ge \exp\left\{ -\tfrac{1}{2}a x^2(1 + dKx)^2 n(1 + o(1)) \right\}$$
$$- L\,\exp\left\{ -l(c^{-1}dKx^2n - C\ln n) \right\}.$$

Taking logarithms, using the definition of the constant K, and passing to the limit we get (1.6.24). □

If for one reason or another it is known that the usual limit

$$I(x) := \lim_{n \to \infty} n^{-1}\ln P\big(T(\alpha_n) \ge x\sqrt{n}\big)$$

exists then (1.6.24) yields

$$\lim_{x \to 0} \lim_{n \to \infty} (n\,x^2)^{-1}\ln P\big(T(\alpha_n) \ge x\sqrt{n}\big) = -\tfrac{1}{2}a. \qquad (1.6.25)$$

Formula (1.6.25) gives useful information about the leading term of large-deviation asymptotics. It is important that the problem of the calculation of the constant a is connected with the well-studied problem of finding the tail asymptotics for the distributions of Brownian bridge functionals (see Fernique (1971), Marcus and Shepp (1972), Borell (1975), Kallianpur and Oodaira (1978), and Aki and Kashiwagi (1989)).

It should be emphasized, however, that the proof of the existence of the aforementioned limit $I(x)$ and its continuity in x that is important for the calculation of the Bahadur efficiency (cf. Theorem 1.2.2) is a serious problem itself. Therefore the possibilities of the method under discussion are seriously limited when studying the Chernoff-type deviations.

By contrast this method has proved utile for the study of Cramér-type and moderate deviations (Inglot, Kallenberg, and Ledwina 1992; Inglot and Ledwina 1993) where the problem of existence of this limit is unessential.

In conclusion we present some results on large deviations under the null hypothesis of linear rank statistics. Let $R_1, R_2, \ldots, R_n$ be the *ranks* of n random variables $Z_1, Z_2, \ldots, Z_n$, that is, R_i be the number of Z_i in the variational series based on $Z_1, Z_2, \ldots, Z_n$. Consider as the null hypothesis that $(R_1, R_2, \ldots, R_n)$ is equally likely to be any of $n!$ permutations of $(1, 2, \ldots, n)$. This supposition is in agreement with special hypotheses considered in the sequel. A *linear rank statistic* is one of the form

$$T_n = \sum_{i=1}^{n} J_n\left(\frac{R_i}{n+1} , \frac{i}{n+1} \right), \tag{1.6.26}$$

where $J_n(u, v)$ is a function on the unit square. Let

$$\mathcal{H} := \left\{ h \colon h \geq 0, \ \int_0^1 h(u, v)\, du = \int_0^1 h(u, v)\, dv = 1 \right\}$$

be the set of all bivariate densities on the unit square with uniform marginals. It will be assumed that the sequence of functions $\{J_n\}$ satisfies Woodworth's (1970) property A, that is,

(1) for each n the function J_n is constant on the rectangles

$$\left\{ i - 1 \leq nu < i, \ j - 1 \leq nv < j \right\}, \qquad 1 \leq i, j \leq n;$$

(2) there exists a function J on the unit square such that

$$\sup_{h \in \mathcal{H}} \left\{ \left| \int_0^1 \int_0^1 (J_n - J) h\, du\, dv \right| \colon h \in \mathcal{H} \right\} \longrightarrow 0 \qquad \text{as } n \to \infty.$$

Woodworth (1970) noted that property A is fulfilled for almost all known linear rank statistics. He also gave simple sufficient conditions for its validity.

Without loss of generality we may assume

$$\int_0^1 \int_0^1 J(u, v)\, du\, dv = 0. \tag{1.6.27}$$

The first papers on large deviations of linear rank statistics belong to Klotz (1965), Stone (1967, 1968), and Hoadley (1967), who considered only the most known representatives of this class: the Wilcoxon and normal scores statistics. The first general result was obtained by Woodworth (1970).

Theorem 1.6.11 *Let* $\{T_n\}$ *be a sequence of linear rank statistics (1.6.26) satisfying property* A *and* $\{\varepsilon_n\}$ *be a sequence of real numbers such that* $\varepsilon_n \to \varepsilon$ *as* $n \to \infty$. *Then for* $0 < \varepsilon < \varepsilon(J)$

$$\lim_{n\to\infty} n^{-1} \ln P(T_n \geq n\,\varepsilon_n) = -I(\varepsilon, J),\qquad(1.6.28)$$

where

$$\varepsilon(J) := \sup\left\{\int_0^1\!\!\int_0^1 J\,h\,du\,dv\colon\; h \in \mathcal{H}\right\},$$

$$I(\varepsilon, J) := \inf\left\{\int_0^1\!\!\int_0^1 h\,\ln h\,du\,dv\colon\; \int_0^1\!\!\int_0^1 J\,h\,du\,dv \geq \varepsilon,\; h \in \mathcal{H}\right\},$$

$$(1.6.29)$$

and the function $\varepsilon \mapsto I(\varepsilon, J)$ *is continuous for sufficiently small* $\varepsilon > 0$.

Further efforts have been directed to the computation of $I(\varepsilon, J)$ for small ε and under appropriate conditions imposed on a *score function* J. The papers of Woodworth (1970), Hwang (1976), Kremer (1979a,b, 1981, 1982), and Ledwina (1987) have been connected with this problem. The most general result belongs to Kallenberg and Ledwina (1987).

Theorem 1.6.12 *Let*

$$\tilde{J}(u, v) := J(u, v) - \int_0^1 J(u, y)\,dy - \int_0^1 J(x, v)\,dx + \int_0^1\!\!\int_0^1 J(u, v)\,du\,dv,$$

$$b^2 := \int_0^1\!\!\int_0^1 \tilde{J}^{\,2}(u, v)\,du\,dv.$$

Suppose that $b > 0$ *and for some* $r > 0$

$$\int_0^1\!\!\int_0^1 \exp\{r\,\tilde{J}(u, v)\}\,du\,dv < \infty.\qquad(1.6.30)$$

Then

$$I(\varepsilon, J) = \frac{\varepsilon^2}{2\,b^2} + o(\varepsilon^2)\qquad as\;\varepsilon \to 0.$$

We note that condition (1.6.30) is much less restrictive than the conditions in Woodworth (1970) and Kremer (1979a, 1981) and permits us to use unbounded score functions.

1.7 On Large Deviations of Empirical Measures in the Case of Several Independent Samples

Many problems of statistics are in a natural way connected with the simultaneous use of several independent samples. Statistical tests proposed for the solution of these problems (two-sample tests of homogeneity or, say, multisample tests of symmetry) are usually based on functionals of corresponding empirical measures. This section contains some results on large deviations of such functionals from Groeneboom et al. (1979) and Nikitin (1987a). We preserve the notations introduced in Section 1.6.

Denote by Λ^c the c-fold Cartesian product of space Λ (c is an arbitrary positive integer). The space Λ^c is endowed with the product topology generated by the τ-topology. Put for brevity

$$\mathbf{P} := (P_1, P_2, \ldots, P_c) \in \Lambda^c, \qquad \rho := (\rho_1, \rho_2, \ldots, \rho_c),$$

where $0 < \rho_i < 1$ and $\sum_{i=1}^c \rho_i = 1$. Denote also for any $\mathbf{P} = (P_1, P_2, \ldots, P_c)$ and $\mathbf{Q} = (Q_1, Q_2, \ldots, Q_c)$ from Λ^c and any $\Omega \subset \Lambda^c$

$$J_\rho(\mathbf{Q}, \mathbf{P}) := \sum_{i=1}^c \rho_i K(Q_i, P_i), \qquad (1.7.1)$$

$$J_\rho(\mathbf{\Omega}, \mathbf{P}) := \inf\left\{ J_\rho(\mathbf{Q}, \mathbf{P}) \colon \mathbf{Q} \in \mathbf{\Omega} \right\}. \qquad (1.7.2)$$

Consider now $c \geq 1$ independent samples $Y_{i,1}, \ldots, Y_{i,n_i}$, $1 \leq i \leq c$, taking values in $\mathbf{S}$ and having distributions $P_1, \ldots, P_c$. Assume that sample sizes n_i tend to infinity in such a way that

$$\lim_{N \to \infty} \frac{n_i}{N} = \rho_i > 0, \qquad (1.7.3)$$

where $N = n_1 + \cdots + n_c$. Let $P_{1,n_1}, \ldots, P_{c,n_c}$ be the empirical measures based on these samples and put for brevity

$$\mathcal{P}_N := (P_{1,n_1}, \ldots, P_{c,n_c}).$$

For any functional $T \colon \Lambda^c \mapsto \bar{\mathbf{R}}^1$ and any $z \in \mathbf{R}^1$ consider the set

$$\Omega_z^T := \left\{ \mathbf{Q} \in \Lambda^c \colon T(\mathbf{Q}) \geq z \right\}. \qquad (1.7.4)$$

Theorem 1.7.1 (Groeneboom et al. 1979) *Let* $\mathbf{P} \in \mathbf{\Lambda}^c$ *and* T: $\mathbf{\Lambda}^c \mapsto \bar{\mathbf{R}}^1$ *be a* τ-*continuous functional on each* $\mathbf{Q} \in \Gamma := \{\mathbf{R} \in \mathbf{\Lambda}^c$: $J_\rho(\mathbf{R}, \mathbf{P}) < \infty\}$. *If the function* $z \mapsto J(\mathbf{\Omega}_z^T, \mathbf{P})$ *is right-continuous at the point* $z = a$ *and* $\{u_N\}$ *is a real sequence, such that* $u_N \to 0$ *as* $N \to \infty$, *then*

$$\lim_{N \to \infty} N^{-1} \ln P\big(T(\mathcal{P}_N) \geq a + u_N\big) = -J_\rho(\mathbf{\Omega}_a^T, \mathbf{P}).$$

As in the case of one sample, the τ-continuity follows from the ρ-continuity, which may be verified in a much easier way, and holds for the majority of functionals encountered in nonparametric statistics.

To calculate the Hodges–Lehmann ARE of a number of nonparametric statistics we are interested in special sets of the form

$$\mathbf{\Delta}_z^T := \Big\{ \mathbf{Q} \in \mathbf{\Lambda}^c \colon T(\mathbf{Q}) \leq z \Big\}, \qquad (1.7.5)$$

where T is a functional on $\mathbf{\Lambda}^c$ and $z \in \mathbf{R}^1$. The attempt to obtain for such sets the precise analog of Theorem 1.7.1 fails as the function $z \mapsto J_\rho(\mathbf{\Delta}_z^T, \mathbf{P})$ is not obliged to be left-continuous (though the right-continuity holds and may be proved in the same way as in Lemma 3.3 from Groeneboom et al. (1979)).

In this connection we formulate and outline the proof of the c-sample variant of Theorem 1.6.8.

Theorem 1.7.2 (Nikitin 1987a) *Let* $\mathbf{P} \in \mathbf{\Lambda}^c$, $\mathbf{\Omega} \subset \mathbf{\Lambda}^c$, *and*

$$J_\rho(\mathbf{\Omega}, \mathbf{P}) = J_\rho(\mathrm{Cl}_\tau(\mathbf{\Omega}), \mathbf{P}). \qquad (1.7.6)$$

Then the estimate

$$\varlimsup_{N \to \infty} N^{-1} \ln P\big\{ \mathcal{P}_N \in \mathbf{\Omega} \big\} \leq -J_\rho(\mathbf{\Omega}, \mathbf{P}) \qquad (1.7.7)$$

is valid.

The proof of Theorem 1.7.2 basically repeats the arguments of Groeneboom et al. (1979) and we give only the main stages of the proof.

For any $\mathbf{P}, \mathbf{Q} \in \mathbf{\Lambda}$ and any finite *partition* π of the space $\mathbf{S}$ consisting of the Borel sets $B_1, B_2, \ldots, B_m$ put

$$K_\pi(\mathbf{Q}, \mathbf{P}) := \sum_{j=1}^m Q(B_j) \ln \frac{Q(B_j)}{P(B_j)}.$$

For c partitions $\pi_1, \pi_2, \ldots, \pi_c$ of $\mathbf{S}$ define by $\boldsymbol{\pi} := \pi_1 \times \cdots \times \pi_c$ a decomposition of $\mathbf{S}^c$ consisting of direct products of elements of corresponding partitions. Denote for any $\mathbf{P} = (P_1, P_2, \ldots, P_c)$ and $\mathbf{Q} = (Q_1, Q_2, \ldots, Q_c)$ from $\mathbf{\Lambda}^c$

$$J_{\rho,\,\pi}(\mathbf{Q}, \mathbf{P}) := \sum_{i=1}^{c} K_{\pi_i}(Q_i, P_i).$$

Lemma 1.7.1 *The function* $\mathbf{Q} \mapsto J_\rho(\mathbf{Q}, \mathbf{P})$ *is* τ*-lower semicontinuous on* $\mathbf{\Lambda}^c$.

This lemma is a generalization of Lemma 2.2 from Groeneboom et al. (1979) and has been proved by Nikitin (1987a).

Lemma 1.7.2 *Let* $\mathbf{P} \in \mathbf{\Lambda}^c$ *and* $\Gamma := \{\mathbf{Q} \in \mathbf{\Lambda}^c : J_\rho(\mathbf{Q}, \mathbf{P}) \leq \alpha\}$ *for some finite constant* α. *The set* Γ *is compact and sequentially compact in the product topology* τ *on* $\mathbf{\Lambda}^c$.

This lemma generalizes again the corresponding statement by Groeneboom et al. (1979), namely, Lemma 2.3; the proof has been given by Nikitin (1987a).

Lemma 1.7.3 *Let* π *be an arbitrary partition of* $\mathbf{S}^c$. *Under the conditions of Theorem 1.7.2 we have*

$$J_\rho(\mathbf{\Omega}, \mathbf{P}) = \sup_{\text{all } \pi} \; J_{\rho,\pi}(\mathbf{\Omega}, \mathbf{P}). \qquad (1.7.8)$$

The proof relies on Lemma 1.7.2 and has no essential differences from the proof of Lemma 2.4 by Groeneboom et al. (1979).

Lemma 1.7.4 *Let conclusion (1.7.8) of Lemma 1.7.3 be fulfilled, then*

$$\varlimsup_{N \to \infty} \; N^{-1} \ln P\{\mathcal{P}_N \in \mathbf{\Omega}\} \leq -J_\rho(\mathbf{\Omega}, P). \qquad (1.7.9)$$

This lemma generalizes Lemma 3.1 from Groeneboom et al. (1979) and uses in the proof a similar technique based on the asymptotical analysis of polynomial distributions.

Combining Lemmas 1.7.3 and 1.7.4 we establish the conclusion of Theorem 1.7.2.

Let us quote some more auxiliary statements from this class of problems necessary for the sequel.

Lemma 1.7.5 *Let Ω be a nonempty τ-closed set of probability measures in Λ^c and $\mathbf{P} \in \Lambda^c$. Then there exists $Q \in \Omega$ such that*

$$J_\rho\left(\mathbf{Q}, \mathbf{P}\right) = J_\rho\left(\Omega, \mathbf{P}\right).$$

The proof is similar to the proof of Lemma 3.2 from Groeneboom et al. (1979) and is based on Lemmas 1.7.1 and 1.7.2.

Lemma 1.7.6 *Let $\mathbf{P} \in \Lambda^c$ and T be a τ-continuous functional on Λ^c. Consider for any real z the sets Ω_z^T and Δ_z^T introduced by (1.7.4) and (1.7.5). Then the function $z \mapsto J_\rho\left(\Omega_z^T, \mathbf{P}\right)$ is left-continuous and the function $z \mapsto J_\rho\left(\Delta_z^T, \mathbf{P}\right)$ is right-continuous.*

The proof is again analogous to the proof of Lemma 3.3 from Groeneboom et al. (1979), but one should use Lemmas 1.7.1–1.7.5 instead of those results that these lemmas generalize.

1.8 Some Results from the Theory of Extremal Problems and the Theory of Implicit Operators

As we have seen in Sections 1.5–1.7, studying the large-deviation probabilities entails extremal problems connected with the minimization of the Kullback–Leibler information on sets of special type that depend on the structure of the sequences of statistics under consideration. We present in this section some results that are useful for the solution of these problems in subsequent chapters.

Let $\mathcal{X}$ and $\mathcal{Y}$ be normed spaces, f_i, $i = 0, 1, \ldots, m$, be smooth functions on $\mathcal{X}$, and F be a smooth mapping of $\mathcal{X}$ into $\mathcal{Y}$. Consider the following extremal problem:

$$\text{extremize } f_0(x), \text{ subject to}$$

$$F(x) = 0, \tag{1.8.1}$$

$$f_i(x) \gtrless 0, \qquad i = 1, \ldots, m,$$

where the symbol $f_i(x) \gtrless 0$ means that the ith constraint is $f_i(x) = 0$, $f_i(x) \geq 0$, or $f_i(x) \leq 0$.

Problems of type (1.8.1) are usually called *smooth extremal problems with constraints of the equality and inequality type* (Ioffe and Tikhomirov 1979; Alekseev, Tikhomirov, and Fomin 1979). For such problems the Lagrange multiplier rule is true. Let us call the *Lagrange function* of problem (1.8.1) the function

$$\mathcal{L}(x, y^*, \boldsymbol{\lambda}, \lambda_0) := \sum_{i=0}^{m} \lambda_i \, f_i(x) + \langle y^*, F(x) \rangle,$$

where $\boldsymbol{\lambda} = (\lambda_1, \ldots, \lambda_m) \in \mathbf{R}^m$, $\lambda_0 \in \mathbf{R}^1$, $y^* \in \mathcal{Y}^*$, and the symbol $\langle y^*, F(x) \rangle$ signifies the value of the functional y^* acting on the element $F(x) \in \mathcal{Y}$ and $\mathcal{Y}^*$ is the dual space with respect to $\mathcal{Y}$.

Theorem 1.8.1 (Alexeev et al. 1979, Sec. 3.2) *Let $\mathcal{X}$ and $\mathcal{Y}$ be Banach spaces, $\mathcal{U}$ be an open set in $\mathcal{X}$, the functions $f_i: \mathcal{U} \mapsto \mathbf{R}^1$, $i = 0, 1, \ldots, m$, and the mapping $F: \mathcal{U} \mapsto \mathcal{Y}$ be strictly differentiable at the point $\hat{x}$. If $\hat{x}$ is a local extremum in problem (1.8.1) and if the image $\text{Im}\, F'(\hat{x})$ is a closed subspace in $\mathcal{Y}$, then there exist Lagrange multipliers $\hat{y}^*$, $\hat{\boldsymbol{\lambda}}$, $\hat{\lambda}_0$, not equal to 0 simultaneously, for which hold:*

(a) the condition of stationarity of the Lagrange function in x:

$$\mathcal{L}_x(\hat{x}, \hat{y}^*, \hat{\boldsymbol{\lambda}}, \hat{\lambda}_0) = 0; \qquad (1.8.2)$$

(b) the condition of coordination of signs: $\hat{\lambda}_0 \geq 0$ if $\hat{x}$ is a minimum, $\hat{\lambda}_0 \leq 0$ if $\hat{x}$ is a maximum, and

$$\hat{\lambda}_i \gtrless 0, \qquad i = 1, 2, \ldots, m; \qquad (1.8.3)$$

(c) the complementary condition:

$$\hat{\lambda}_i \, f(\hat{x}) = 0, \qquad i = 1, 2, \ldots, m. \qquad (1.8.4)$$

Conditions (1.8.3) mean that if in (1.8.1) $f_i(x) \geq 0$ then $\hat{\lambda}_i \leq 0$, if $f_i(x) \leq 0$ then $\hat{\lambda}_i \geq 0$, and finally if $f_i(x) = 0$ then $\hat{\lambda}_i$ may have arbitrary sign.

We recall that the mapping $F: \mathcal{X} \mapsto \mathcal{Y}$ is said to be *strictly differentiable* at the point $\hat{x}$ if there exists a linear operator $\boldsymbol{\Upsilon}: \mathcal{X} \mapsto \mathcal{Y}$ such that for each $\varepsilon > 0$ there exists $\delta > 0$ such that for any x_1 and x_2 satisfying the conditions

$$\| x_1 - \hat{x} \|_{\mathcal{X}} < \delta, \qquad \| x_2 - \hat{x} \|_{\mathcal{X}} < \delta,$$

the following inequality holds:

$$\| F(x_1) - F(x_2) - \boldsymbol{\Upsilon}(x_1 - x_2) \|_{\mathcal{Y}} < \varepsilon \| x_1 - x_2 \|_{\mathcal{X}}. \qquad (1.8.5)$$

This condition may be verified using the result given by Alexeev et al. (1979, Sec. 2.2.3).

Theorem 1.8.2 *Let $\mathcal{X}$ and $\mathcal{Y}$ be normed spaces, $\mathcal{U}$ be a neighborhood of $\hat{x}$ in $\mathcal{X}$ and the mapping $F \colon \mathcal{U} \longmapsto \mathcal{Y}$ be Gâteaux differentiable in each $x \in \mathcal{U}$. If the mapping $x \longmapsto F'(x)$ is continuous (in the uniform-operator topology of the space $\mathcal{L}(\mathcal{X}, \mathcal{Y})$) at the point $\hat{x}$, then the mapping F is strictly differentiable in $\hat{x}$ (and consequently Fréchet differentiable at the same point). (Here $\mathcal{L}(\mathcal{X}, \mathcal{Y})$ is the usual space of linear continuous mappings of $\mathcal{X}$ into $\mathcal{Y}$.)*

Let also quote three results concerning the existence of implicit functions and operators and being of essential importance in the sequel.

Theorem 1.8.3 (Vainberg and Trenogin 1974) *Let $F_i(y_1, \ldots, y_p; x_1, \ldots, x_s)$, $i = 1, 2, \ldots, p$, be real continuous functions of real arguments vanishing at the point $M = (y_1^0, \ldots, y_p^0; x_1^0, \ldots, x_s^0)$ that in some ball with center M admit convergent series expansions in the powers of $y_i - y_i^0$ and $x_k - x_k^0$, $i = 1, 2, \ldots, p$, $k = 1, 2, \ldots, s$. Suppose that these series do not contain free members. Then, if the Jacobian*

$$\mathcal{J} = \frac{\partial(F_1, \ldots, F_p)}{\partial(y_1, \ldots, y_p)}$$

is not equal to 0 at the point M, the system of equations

$$F_i(y_1, \ldots, y_p; x_1, \ldots, x_s) = 0, \qquad i = 1, 2, \ldots, p,$$

has the unique solution $y_i = \varphi_i(x_1, \ldots, x_s)$, satisfying the condition

$$\varphi_i(x_1^0, \ldots, x_s^0) = y_i^0,$$

and, moreover, in some ball with center at the point $(x_1^0, \ldots, x_s^0)$ the functions φ_i may be expanded in convergent series in powers of $x_k - x_k^0$, $k = 1, 2, \ldots, s$.

Now denote by $\mathbf{E}_1$, $\mathbf{E}_2$, $\mathbf{E}_3$ three Banach spaces and by $\mathcal{D}_r(x_0, \mathbf{E})$ the ball of radius r in the space $\mathbf{E}$ with center at the point x_0. Consider the problem of finding the solutions $x = x(y)$ (y plays the role of a parameter) of the operator equation

$$F(x, y) = 0, \qquad\qquad (1.8.6)$$

satisfying the condition

$$x(y) = x_0, \qquad\qquad (1.8.7)$$

provided that

$$F(x_0, y_0) = 0. \qquad\qquad (1.8.8)$$

Theorem 1.8.4 (Vainberg and Trenogin 1974, Sec. 22.3) *Let $F(x, y)$ be an analytic operator in $\mathcal{D}_r(x_0, \mathbf{E}_1) \times \mathcal{D}_\rho(y_0, \mathbf{E}_2)$ taking values in $\mathbf{E}_3$ such that the operator*

$$B = F'_x(x_0, y_0) \tag{1.8.9}$$

(the Fréchet derivative) has a bounded inverse operator. Then there exist positive numbers r_1 and ρ_1 such that in the ball $\mathcal{D}_{r_1}(x_0, \mathbf{E}_1)$ there exists a unique solution

$$x = f(y)$$

of equation (1.8.6). This solution is defined in $\mathcal{D}_{\rho_1}(y_0, \mathbf{E}_2)$, is analytic in this ball, and satisfies condition (1.8.7).

Theorem 1.8.5 (Alexeev et al. 1979) *Let $\mathcal{W}$ be a neighborhood in $\mathbf{E}_1 \times \mathbf{E}_2$ and $F \colon \mathcal{W} \mapsto \mathbf{E}_3$ be a mapping of the class $C^1(\mathcal{W})$. If condition (1.8.8) is fulfilled and the operator B given by (1.8.9) is invertible, then there exist $\varepsilon > 0$ and $\delta > 0$ and a mapping $\varphi \colon \mathcal{D}_\delta(y_0, \mathbf{E}_2) \mapsto \mathbf{E}_1$ of the class $C^1(\mathcal{D}_\delta(y_0, \mathbf{E}_2))$ such that $\varphi(y_0) = x_0$ and that the condition $\| y - y_0 \|_{\mathbf{E}_2} < \delta$ implies $\| \varphi(y) - x_0 \|_{\mathbf{E}_1} < \varepsilon$ and $F(\varphi(y), y) = 0$. This mapping φ is locally unique.*

We will constantly use these results in subsequent chapters.

2

Asymptotic Efficiency of Nonparametric Goodness-of-Fit Tests

2.1 Problem Statement and Tests under Consideration

Let X_1, X_2, ..., X_n be a sample with continuous distribution function F and F_n be the empirical distribution function based on this sample. The problem consists in testing the *goodness-of-fit hypothesis* H_0: $F = F_0$ where F_0 is a known continuous distribution function against the general alternative A_0: $F \neq F_0$. In the sequel the structure of the alternative hypothesis generally will be specified.

Various statistics used to verify H_0 against A_0 are based on the difference between F_n and F_0. Statistics of such kind are the main subject of this chapter.

The most famous and well-studied statistic of this type is obviously the Kolmogorov statistic (see Kolmogorov (1933))

$$D_n = \sup_{-\infty < t < \infty} |F_n(t) - F_0(t)| \qquad (2.1.1)$$

and its variations, namely, the one-sided statistics of Smirnov (see Smirnov (1944))

$$D_n^+ = \sup_{-\infty < t < \infty} [F_n(t) - F_0(t)] , \qquad (2.1.2)$$

$$D_n^- = \sup_{-\infty < t < \infty} [F_0(t) - F_n(t)] , \qquad (2.1.3)$$

as well as the Kuiper statistic (see Kuiper (1960))

$$V_n = \sup_{-\infty < t < \infty} [F_n(t) - F_0(t)] - \inf_{-\infty < t < \infty} [F_n(t) - F_0(t)] . \qquad (2.1.4)$$

It is evident that

$$D_n = \max\left(D_n^+, D_n^-\right) \qquad \text{and} \qquad V_n = D_n^+ + D_n^- .$$

Watson (1976) and Darling (1983a, 1983b) have introduced the centered versions of the Kolmogorov–Smirnov statistics:

$$G_n = \sup_{-\infty < t < \infty} \left| F_n(t) - F_0(t) - \int_{-\infty}^{\infty} \left(F_n(s) - F_0(s) \right) dF_0(s) \right|,$$

(2.1.5)

$$G_n^+ = \sup_{-\infty < t < \infty} \left[F_n(t) - F_0(t) - \int_{-\infty}^{\infty} \left(F_n(s) - F_0(s) \right) dF_0(s) \right],$$

(2.1.6)

$$G_n^- = \sup_{-\infty < t < \infty} \left[\int_{-\infty}^{\infty} \left(F_n(s) - F_0(s) \right) dF_0(s) + F_0(t) - F_n(t) \right].$$

(2.1.7)

In the sequel G_n, $G_n^{\pm}$ will be called the Watson–Darling statistics. As with the Kuiper statistic V_n, they are rather convenient for testing the hypotheses on the circle, being invariant to the location of origin, but may be used also for testing on the real line.

Another group of statistics is based on the integral distance between F_n and F_0. The best known among them is the celebrated statistic of Cramér–von Mises–Smirnov (see, e.g., Darling (1957) and Martynov (1978)):

$$\omega_{n,1}^2 = \int_{-\infty}^{\infty} \left(F_n(t) - F_0(t) \right)^2 dF_0(t).$$

(2.1.8)

Anderson and Darling (1952) proposed to improve properties of this statistic by introducing a nonnegative weight function q. Another generalization is connected with the consideration of arbitrary positive integer powers of $F_n - F_0$. As a result we obtain the statistics

$$\omega_{n,q}^k = \int_{-\infty}^{\infty} \left(F_n(t) - F_0(t) \right)^k q\left(F_0(t) \right) dF_0(t),$$

(2.1.9)

where k is a positive integer. Obviously, $\omega_{n,q}^k$ are not consistent against all alternatives for odd k; however, for one-sided alternatives these statistics may turn out to be serious competitors of classic statistics and deserve careful examination.

The best known weighted integral statistic is the Anderson–Darling statistic (see Anderson and Darling (1952, 1954)):

$$A_n^2 = \int_{-\infty}^{\infty} \frac{\left(F_n(t) - F_0(t)\right)^2}{F_0(t)\left(1 - F_0(t)\right)} \, dF_0(t) . \qquad (2.1.10)$$

The Watson statistic (Watson 1961) is sometimes used as a centered variant of the statistic $\omega_{n,1}^2$:

$$U_n^2 = \int_{-\infty}^{\infty} \left[F_n(t) - F_0(t) - \int_{-\infty}^{\infty} \left(F_n(s) - F_0(s) \right) dF_0(s) \right]^2 dF_0(t) .$$
$$(2.1.11)$$

Clearly $U_n^2 = \omega_{n,1}^2 - (\omega_{n,1}^1)^2$.

Still another type of statistic, based on the martingale part of the empirical process, has been studied by Khmaladze (1981) and Aki (1986) (see also Shorack and Wellner (1986), and Koziol (1989)):

$$K_n = \sup_{-\infty < t < \infty} \left| F_n(t) - \int_{-\infty}^{t} \frac{1 - F_n(s)}{1 - F_0(s)} \, dF_0(s) \right| \qquad (2.1.12)$$

(the one-sided statistics K_n^+ and K_n^- are introduced in a similar way);

$$L_n^2 = \int_{-\infty}^{\infty} \left| F_n(t) - \int_{-\infty}^{t} \frac{1 - F_n(s)}{1 - F_0(s)} \, dF_0(s) \right|^2 dF_0(t) . \qquad (2.1.13)$$

It follows from Theorem 1.2.2 that to calculate exact slopes of the statistics under consideration and to find their Bahadur ARE, it is necessary to know their rough large-deviation asymptotics under the null hypothesis. We shall study the asymptotics of such kind in the following sections.

2.2 Large Deviations of Kolmogorov–Smirnov-Type Statistics

In this section we consider the statistics D_n, $D_n^{\pm}$, V_n, G_n, $G_n^{\pm}$, and K_n. Without loss of generality we may assume that if H_0 is true, then $F_0(t) = t$ for $0 \le t \le 1$.

To formulate a theorem on large deviations of the statistics D_n, $D_n^{\pm}$, and V_n, we consider for $0 < a < 1$ the functions

$$
f(a,t) := \begin{cases} (a+t)\ln\dfrac{a+t}{t} + (1-a-t)\ln\dfrac{1-a-t}{1-t}, & 0 \le t \le 1-a, \\[2mm] +\infty, & 1-a < t \le 1, \end{cases}
$$

$$(2.2.1)$$

and

$$
g_1(a) := \inf_{0 \le t \le 1} f(a,t). \tag{2.2.2}
$$

Lemma 2.2.1 *The function g is continuous; moreover*

$$
g_1(a) = 2\,a^2\big(1+o(1)\big) \qquad \text{as } a \to 0.
$$

Proof It follows from (2.2.1) that the infimum of f is attained at point $t = t(a)$, a root of the equation

$$
\ln \frac{(1-t)\,(a+t)}{t\,(1-a-t)} = \frac{a}{t\,(1-t)}.
$$

By the implicit function theorem $t(a)$ depends continuously on a and, in addition

$$
t(a) = \tfrac{1}{2} + o(1) \qquad \text{as } a \to 0. \tag{2.2.3}
$$

Putting (2.2.3) in (2.2.1) we obtain the conclusion of the lemma. $\qquad\square$

Theorem 2.2.1 (Abrahamson 1967; Bahadur 1971) *Let T_n be any of the statistics D_n, $D_n^{\pm}$, or V_n. If the hypothesis H_0 is true then*

$$
\lim_{n \to \infty} n^{-1}\ln P\big(T_n \ge a\big) = -g_1(a). \tag{2.2.4}
$$

Proof Denote for brevity

$$
P_n^{+} := P\big(D_n^{+} \ge a\big), \qquad P_n^{-} := P\big(D_n^{-} \ge a\big), \qquad P_n := P\big(D_n \ge a\big).
$$

In view of the equality $P_n^{+} = P_n^{-}$ (Smirnov 1944), it is clear that

$$
P_n^{+} \le P_n \le P_n^{+} + P_n^{-} \le 2\,P_n^{+}. \tag{2.2.5}
$$

Therefore, if (2.2.4) is proved for D_n^{+}, it will be enough to use (2.2.5) in order to obtain the conclusion of Theorem 2.2.1 for D_n^{-} and D_n. First we estimate P_n^{+} from below.

For any $t \in [0, 1]$

$$P_n^+ \geq P\big(F_n(t) - t - a \geq 0\big). \qquad (2.2.6)$$

The difference $F_n(t) - t - a$ may be viewed as a sum of independent identically distributed random variables. Owing to this fact we conclude by Theorem 1.6.2 that

$$\lim_{n \to \infty} n^{-1} \ln P\big(F_n(t) - t - a \geq 0\big) = -f(a, t).$$

Since t is arbitrary, we have

$$\lim_{n \to \infty} n^{-1} \ln P_n^+ \geq -g_1(a). \qquad (2.2.7)$$

To obtain the same upper estimate we choose a positive integer m so large as to satisfy the inequalities $0 < a - \frac{1}{m} < 1$. Then for all $1 \leq i \leq m$

$$\max\Big\{F_n(t) - t : \frac{i-1}{m} \leq t \leq \frac{i}{m}\Big\} \leq F_n\Big(\frac{i}{m}\Big) - \frac{i-1}{m},$$

hence

$$D_n^+ \leq \max\Big\{F_n\Big(\frac{i}{m}\Big) - \frac{i-1}{m} : 1 \leq i \leq m\Big\}.$$

Now Theorem 1.6.1 yields

$$P_n^+ \leq \sum_{i=1}^{m} P\Big(F_n\Big(\frac{i}{m}\Big) - \frac{i}{m} - \Big(a - \frac{1}{m}\Big) \geq 0\Big) \leq m \exp\Big[-n\,g_1\Big(a - \frac{1}{m}\Big)\Big].$$

Taking logarithms and passing to the limit, we obtain

$$\overline{\lim_{n \to \infty}} \; n^{-1} \ln P_n^+ \leq -g_1\Big(a - \frac{1}{m}\Big).$$

Using the continuity of g_1 and passing now to the limit as $m \to \infty$, we find

$$\overline{\lim_{n \to \infty}} \; n^{-1} \ln P_n^+ \leq -g_1(a). \qquad (2.2.8)$$

Comparing (2.2.7) and (2.2.8) we deduce the statement of the theorem for the Kolmogorov–Smirnov statistics.

Siegmund (1982) has even found the exact asymptotics for large deviations of the one-sided statistics D_n^+ and D_n^-. Let θ_1 and θ_2 be real numbers such that $\theta_2 < 0 < \theta_1$, $\theta_1^{-1} + |\theta_2|^{-1} = a^{-1}$, and

$$\theta_1 - \theta_2 = \ln\big[(1 - \theta_2)/(1 - \theta_1)\big].$$

Then the following assertion holds:

$$P_n^+ \sim \frac{\exp\left\{-n\left[(\theta_1 - \theta_2)\,a + \theta_2 + \ln(1 - \theta_2)\right]\right\}}{\left\{a\,|\theta_2|^{-1}(1 - \theta_2)\left[1 + (\,|\theta_2|\,\theta_1^{-1})^3\,(1 - \theta_1)\,(1 - \theta_2)^{-1}\right]\right\}^{1/2}}$$

$$\text{as } n \to \infty.$$

It can be shown that this result yields (2.2.4) as far as D_n^+ and D_n^- are concerned. The statistic V_n requires some additional arguments. Denote

$$Q_n = P\big(V_n \geq a\big).$$

As $Q_n \geq P_n^+$, it suffices to prove

$$\varlimsup_{n \to \infty} n^{-1} \ln Q_n \leq -g_1(a). \tag{2.2.9}$$

Observe that

$$V_n = \sup\big\{F_n(t) - t + u - F_n(u) : 0 \leq t \leq 1, 0 \leq u \leq 1\big\}.$$

Let a positive integer m be so large that $0 < a - \frac{2}{m} < 1$ and i, j be positive integers, $1 \leq i, j \leq m$. Then

$$F_n(t) - t + u - F_n(u) \leq F_n\Big(\frac{i}{m}\Big) - \frac{i-1}{m} + \frac{j}{m} - F_n\Big(\frac{j-1}{m}\Big) := \Delta_n(i,j)$$

for $\frac{i-1}{m} \leq t \leq \frac{i}{m}$ and $\frac{j-1}{m} \leq u \leq \frac{j}{m}$. Applying Theorem 1.6.2 to $\Delta_n(i,j)$ we obtain

$$P\big(\Delta_n(i,j) \geq a\big) \leq \begin{cases} \exp\left[-n\,f\big(a - \frac{2}{m}, \frac{i-j+1}{m}\big)\right], & \text{if } i \geq j - 1, \\[2mm] \exp\left[-n\,f\big(a - \frac{2}{m}, 1 - \frac{j-1-i}{m}\big)\right], & \text{if } i \leq j - 1. \end{cases}$$

Hence it follows, for all i, j, that

$$P\big(\Delta_n(i,j) \geq a\big) \leq \exp\left[-n\,g_1(a - \tfrac{2}{m})\right],$$

and so

$$Q_n \leq \sum_{i,j} P\big(\Delta_n(i,j) \geq a\big) \leq m^2 \exp\left[-n\,g_1(a - \tfrac{2}{m})\right].$$

Taking logarithms of both sides of this inequality, multiplying by n^{-1}, and passing to the limit first as $n \to \infty$ and then as $m \to \infty$, we obtain estimate (2.2.9) with the aid of Lemma 2.2.1. $\qquad\square$

Now we proceed to consider the Watson–Darling statistics G_n and $G_n^\pm$. One can restrict study to rough asymptotics of large deviations for

G_n^+. For the remaining two statistics the corresponding asymptotics will be the same. We introduce the function

$$g_2(a) := \inf_{s \geq 0} \left\{ -a\,s - \tfrac{1}{2}\,s + \ln\left[(e^s - 1)/s \right] \right\}.$$

It is easy to show that g_2 is continuous and

$$g_2(a) = 6\,a^2 \big(1 + o(1) \big) \qquad \text{as } a \to 0. \qquad (2.2.10)$$

Theorem 2.2.2 (Nikitin 1987b) *If the hypothesis H_0 is true and $a \in (0, 1)$, then*

$$\lim_{n \to \infty} n^{-1} \ln P\big(G_n^+ \geq a \big) = -g_2(a). \qquad (2.2.11)$$

Proof We write G_n^+ in the form

$$G_n^+ = \sup_{0 \leq t \leq 1} \left[F_n(t) - t + \bar{x} - \tfrac{1}{2} \right].$$

For any $t \in [0, 1]$ we have

$$P\big(G_n^+ \geq a \big) \geq P\left(n^{-1} \sum_{j=1}^{n} Z_j \geq a \right),$$

where $Z_j := \mathbf{1}_{\{X_j < t\}} - t + X_j - \tfrac{1}{2}$ and the random variables $X_j, j = 1, \ldots, n$, are uniformly distributed on $[0, 1]$. Elementary calculations show that the moment generating function of Z_j does not depend on t and is equal to

$$E \exp\left(s\,Z_j \right) = e^{-s/2} \left(e^s - 1 \right)/s, \qquad s \geq 0.$$

Applying the Chernoff theorem (Theorem 1.6.2), we obtain

$$\lim_{n \to \infty} n^{-1} \ln P\big(G_n^+ \geq a \big) \geq - \inf_{s \geq 0} \ln \left[e^{-as} E \exp\left(s\,Z_j \right) \right] = -g_2(a).$$

An upper estimate of the same kind could be obtained in the same way as in Theorem 2.2.1. $\qquad \square$

Large deviations of the Khmaladze–Aki statistics are described in the following theorem.

Theorem 2.2.3 (Podkorytova 1990) *Let T_n be any of statistics K_n, K_n^+, or K_n^-. If the hypothesis H_0 is true and $a \in (0, 1)$, then*

$$\lim_{n \to \infty} n^{-1} \ln P\big(T_n \geq a \big) = g_3(a), \qquad (2.2.12)$$

where

$$g_3(a) := a + \ln(1 - a) = -\tfrac{1}{2}\,a^2\big(1 + o(1)\big) \qquad as\ a \to 0. \qquad (2.2.13)$$

Proof It suffices to consider only K_n^+. For any $t \in [0, 1]$ one has

$$P\big(K_n^+ \geq a\big) \ \geq \ P\bigg(n^{-1}\sum_{j=1}^{n} W_j \geq a\bigg),$$

where

$$W_j = \begin{cases} 1 + \ln(1 - X_j) & \text{if } X_j < t, \\[2mm] \ln(1 - t) & \text{if } X_j \geq t. \end{cases}$$

The moment generating function for W_1 is

$$\varphi_t(s) = E\exp(s\,W_1) = \frac{e^s - (1 - t)^{s+1}e^s + (1 - t)^{s+1}(s + 1)}{s + 1}, \quad s \geq 0.$$

By Theorem 1.6.2

$$\lim_{n\to\infty} n^{-1}\ln P\bigg(\sum_{j=1}^{n} W_j \geq na\bigg) \ = \ -\psi(a, t),$$

where $\exp\big[-\psi(a, t)\big] = \inf\big\{e^{-sa}\,\varphi_t(s)\colon\ s \geq 0\big\}$. Consequently

$$\varliminf_{n\to\infty} n^{-1}\ln P\big(K_n^+ \geq a\big) \geq \sup_{0\leq t\leq 1}\big\{-\psi(a, t)\big\}$$

$$= \sup_{0\leq t\leq 1}\inf_{s\geq 0}\big\{-s\,a + \ln\varphi_t(s)\big\}$$

$$= \sup_{0\leq t\leq 1}\inf_{s\geq 0}\big\{s(1 - a) - \ln(1 + s)$$

$$+ \ln\big(1 - (1 - t)^{s+1}\big(1 - (s + 1)e^{-s}\big)\big)\big\}.$$

We denote by $h(s, t, a)$ the function in the braces in the last formula. It increases in t for fixed s and a. Therefore the function $\inf[h(s, t, a)\colon s \geq 0]$ is nondecreasing in t also, so

$$\sup_{0\leq t\leq 1}\big\{-\psi(a, t)\big\} = \inf_{s\geq 0}h(s, 1, a) = \inf_{s\geq 0}\big\{s(1 - a) - \ln(1 + s)\big\}$$

$$= a + \ln(1 - a) \equiv g_3(a).$$

Hence the following lower estimate holds:

$$\varliminf_{n\to\infty} n^{-1}\ln P\big(K_n^+ \geq a\big) \geq g_3(a). \qquad (2.2.14)$$

To obtain an upper estimate we take a large positive integer k such that $a > \varepsilon_k = (\ln \ln k)^{-1}$. Then

$$
P(K_n^+ \geq a) \leq P\left(\max_{1 \leq i \leq k} \left[F_n(\tfrac{i-1}{k}) - \int_0^{(i-1)/k} \frac{1 - F_n(s)}{1 - s}\, ds \right] \geq a - \varepsilon_k \right)
$$
$$
+ P\left(\max_{1 \leq i \leq k} \left[F_n(\tfrac{i}{k}) - F_n(\tfrac{i-1}{k}) \right] \geq \varepsilon_k \right) = A_1 + A_2 \,.
$$

Now we estimate A_1 and A_2 with the aid of (1.6.3):

$$
A_1 \leq \sum_{i=1}^{k} P\left(F_n\left(\tfrac{i-1}{k}\right) - \int_0^{(i-1)/k} \frac{1 - F_n(s)}{1 - s}\, ds \geq a - \varepsilon_k \right)
$$
$$
\leq k \max_{1 \leq i \leq k} \exp\left[-n\,\psi(a - \varepsilon_k, \tfrac{i-1}{k}) \right]
$$
$$
\leq k \exp\left[n \sup_{0 \leq t \leq 1} \left\{ -\psi(a - \varepsilon_k, t) \right\} \right]
$$
$$
= k \exp\left[n\big(a - \varepsilon_k + \ln(1 - a - \varepsilon_k)\big) \right],
$$

$$
A_2 \leq \sum_{i=1}^{k} P\left(F_n(\tfrac{i}{k}) - F_n(\tfrac{i-1}{k}) \geq \varepsilon_k \right)
$$
$$
\leq 2k \exp\left[-n\big((1 - \varepsilon_k)\ln(1 - \varepsilon_k) \right.
$$
$$
\left. + \varepsilon_k \ln \varepsilon_k + \varepsilon_k \ln(k - 1) \big) \right]. \tag{2.2.15}
$$

The estimate (2.2.15) is contained also in Borovkov (1986, Ch. 5, Th. 10). Therefore for all sufficiently large k we have

$$
\varlimsup_{n \to \infty}\, n^{-1} \ln P(K_n^+ \geq a)
$$
$$
\leq \max\{ a - \varepsilon_k + \ln(1 - a - \varepsilon_k),\, \ln 2 - \varepsilon_k \ln(k - 1) \}.
$$

Passing to the limit as $k \to \infty$ we obtain

$$
\varlimsup_{n \to \infty}\, n^{-1} \ln P(K_n^+ \geq a) \leq g_3(a)\,. \tag{2.2.16}
$$

The conclusion of Theorem 2.2.3 follows now from (2.2.14) and (2.2.16).
□

Finally we consider the weighted variants of the Kolmogorov–Smirnov statistics. Suppose that q is a finite positive weight function, continuous on $(0, 1)$ and symmetric with respect to the point $t = \tfrac{1}{2}$ and such that

$\lim_{t \to 0} q(t)$ exists and belongs to the interval $[0, +\infty]$. Denote by $\mathfrak{Q}$ the class of such weights. Consider the statistics

$$D_{n,q} = \sup_{-\infty < t < \infty} |F_n(t) - F_0(t)|\, q(F_0(t)). \qquad (2.2.17)$$

The statistics $D_{n,q}^+$ and $D_{n,q}^-$ are defined similarly.
  Put

$$g_q(a) := \inf_{0 < t < 1} f\big(a/q(t), t\big),$$

where $f(a,t)$ was introduced in (2.2.1).

Theorem 2.2.4 (Groeneboom and Shorack 1981) *Let $q \in \mathfrak{Q}$ and $\{D_{n,q}^*\}$ be any of sequences $\{D_{n,q}\}, \{D_{n,q}^+\}$, or $\{D_{n,q}^-\}$. If the hypothesis H_0 is true then*

$$\lim_{n \to \infty} n^{-1} \ln P\big(D_{n,q}^* \geq a\big) = -g_q(a). \qquad (2.2.18)$$

Moreover, if

$$\lim_{t \to 0} \frac{\ln t}{q(t)} = 0 \ ,$$

then

$$g_q(a) = 0 \qquad \text{for all } a \geq 0.$$

The last condition shows that the weight function

$$q(t) = -\ln\big[t\,(1-t)\big] \qquad (2.2.19)$$

plays a key role in Theorem 2.2.4.
  It has been proved by Groeneboom and Shorack (1981) that in this case

$$g_q(a) = \tfrac{1}{8} e^2\, a^2 \big(1 + o(1)\big) \qquad \text{as } a \to 0. \qquad (2.2.20)$$

The paper of Groeneboom and Shorack (1981) contains also a generalization of Theorem 2.2.4 for nonsymmetric weights q. Such weights have been considered recently by Sinclair, Spurr, and Ahmad (1990) who have proposed the modified Anderson–Darling statistics with $q(t) = t^{-1}$ and $q(t) = (1-t)^{-1}$ and have found their limiting distributions.

2.3 Large Deviations of Integral Statistics. Euler–Lagrange Equations

We consider in this section the integral statistics $\omega^k_{n,q}$ introduced earlier in (2.1.9). The first result concerning large-deviation probabilities of such statistics for $k = 2$ and $q \equiv 1$ has been proved by Mogulskii (1977): *Under the hypothesis H_0*

$$\lim_{n \to \infty} n^{-1} \ln P\left(\omega^2_{n,1} \geq a\right) = -\frac{\pi^2}{2} a + \sum_{j=3}^{\infty} c_j \, a^{j/2}, \qquad (2.3.1)$$

where the series with numerical coefficients c_j converges for sufficiently small $a > 0$.

In the case of moderate deviations the assertion analogous to (2.3.1) has been obtained formerly by Rubin and Sethuraman (1965a). Mogulskii (1977) used Theorem 1.6.5 and applied variational calculus to obtain extremals of the corresponding extremal problem. The method of solution of the Euler–Lagrange equation has been based on a remarkably simple form of the linear part of this equation. For $\omega^2_{n,1}$ it takes the following form:

$$y'' - \lambda y = 0, \qquad\qquad y(0) = y(1) = 0. \qquad (2.3.2)$$

The solutions of (2.3.2) are the eigenfunctions $y_m = C \sin m\pi t$ corresponding to the eigenvalues $\lambda_m = -m^2\pi^2$, $m = 1, 2, \ldots$.

The arguments used by Mogulskii (1977) do not admit generalizations of this method for more complicated situations in the analysis of $\omega^2_{n,q}$ and especially in the case of $\omega^k_{n,q}$, $k > 2$. Even for $q \not\equiv 1$ one obtains instead of (2.3.2) the boundary-value problem for the Sturm–Liouville equation:

$$y'' - \lambda q y = 0, \qquad\qquad y(0) = y(1) = 0, \qquad (2.3.3)$$

solutions of which cannot be written explicitly. Therefore the method of Mogulskii (1977) cannot be used here. Much more complicated effects arise for $k > 2$ when the considered nonlinear boundary-value problems have the form:

$$y'' - \lambda q y^{k-1} = 0, \qquad y(0) = y(1) = 0, \qquad \int_0^1 y^k q \, dt = 1. \qquad (2.3.4)$$

In this connection we propose to apply here and in the sequel a much more general method of analysis of the Euler–Lagrange equations based on ideas of nonlinear functional analysis. The intention is to use the

Lyapunov–Schmidt theory of branching solutions of nonlinear equations in Banach spaces (see Vainberg and Trenogin (1974)). By applying this method we prove now the following basic result.

Theorem 2.3.1 (Nikitin 1979, 1980) *Suppose that the hypothesis H_0 is true. Let k be a positive integer and q be a positive weight function, summable on $(0, 1)$. Then*

$$\lim_{n \to \infty} n^{-1} \ln P\big(\omega_{n,q}^k \geq a\big) = \sum_{j=2}^{\infty} c_j \, a^{j/k}, \qquad (2.3.5)$$

where the series with numerical coefficients c_j converges for sufficiently small $a > 0$; moreover $c_2 = \frac{1}{2} \lambda_0(q; k)$, where $\lambda_0(q; k)$ is the "principal" eigenvalue of problem (2.3.4).

Remark In the special case $q \equiv 1$, $k = 2$ we obtain $\lambda_0(1; 2) = -\pi^2$, which implies (2.3.1). Note that $\lambda_0(q; k) < 0$ always.

Proof As in the previous section we can assume that the initial sample is taken from the uniform distribution on $[0, 1]$. The statistic

$$\omega_{n,q}^k = \int_0^1 \big(F_n(t) - t\big)^k q(t)\, dt$$

may be considered as a functional $\chi(F_n)$ of the empirical d.f. F_n. For any d.f.s F_1 and F_2 on $[0, 1]$ the elementary inequality

$$\big|\chi(F_1) - \chi(F_2)\big| \leq k \left\{ \sup_{0 \leq t \leq 1} |F_1(t) - F_2(t)| \right\} \int_0^1 q(t)\, dt$$

implies that functional χ is uniformly continuous in the ρ-topology and consequently in the τ-topology in the space Λ_1 of d.f.s on $[0, 1]$ (Groeneboom et al. 1979).

Introduce now for any $a > 0$ the set Ω_a of d.f.s F on $[0, 1]$ satisfying the inequality

$$\int_0^1 \big(F(t) - t\big)^k q(t)\, dt \geq a. \qquad (2.3.6)$$

For any absolutely continuous d.f. F on $[0, 1]$ denote by $K(F)$ the corresponding Kullback–Leibler information

$$K(F) := \int\limits_0^1 F'(t) \ln F'(t) \, dt \, . \qquad (2.3.7)$$

As in (1.2.10) put

$$K(\Omega_a) := \inf \left\{ K(F) : \ F \in \Omega_a \right\} .$$

Let us suppose that for sufficiently small $a > 0$

$$K(\Omega_a) = -\sum_{j=2}^{\infty} c_j \, a^{j/k}, \qquad (2.3.8)$$

where in the right-hand side is the same series as in (2.3.5). Then $K(\Omega_a)$ is a continuous function in a for all small $a > 0$. Since $w_{n,q}^k$ is a τ-continuous functional of F_n, we can apply here Theorem 1.6.7. Consequently, for small $a > 0$

$$\lim_{n \to \infty} n^{-1} \ln P\big(w_{n,q}^k \geq a\big) = -K(\Omega_a) \, . \qquad (2.3.9)$$

Note that the conclusion of Theorem 2.3.1 follows finally from (2.3.9) and (2.3.8).

Now we proceed to the proof of formula (2.3.8). Note that the set Ω_a is nonempty and τ-closed. The functional $F \to K(F)$ is τ-lower semicontinuous by Lemma 2.2 from Groeneboom et al. (1979). Consequently, according to Lemma 3.2 from the same paper this functional attains its lower bound on Ω_a. Therefore the solution of the extremal problem connected with the calculation of $K(\Omega_a)$ exists.

It should be natural to apply the apparatus of classical variational calculus for the calculation of $K(\Omega_a)$. But the latter is intended for discovering the extremals in the class of continuously differentiable functions for smooth functionals under minimization. Meanwhile when calculating $K(\Omega_a)$ we are forced to look for an extremum in the much broader class of absolutely continuous functions where even the variation of the corresponding functional may not exist. To overcome these difficulties we first restrict the class of admissible elements to an appropriate subset of the Banach space $\overset{\circ}{\mathbf{W}}_{\infty, 1}$ where the variations do exist. Then one can use the general form of the Lagrange principle for conditional extrema (Theorem 1.8.1).

It is clear that for the calculation of $K(\Omega_a)$ one can restrict consideration to the class $\mathcal{F}$ of absolutely continuous d.f.s F on $[0, 1]$ with corresponding densities f. Introduce the following subsets of the class $\mathcal{F}$. For any $\delta > 0$ put

$$V_\delta := \{ F \in \mathcal{F} : f \geq \delta \quad \text{a.e. on } [0, 1] \}, \qquad (2.3.10)$$

$$V'_\delta := \{ F \in \mathcal{F} : f > \delta \quad \text{a.e. on } [0, 1] \}. \qquad (2.3.11)$$

Our arguments are now for the proof that we shall commit only a small error by calculating $K(\Omega_a)$ if we restrict Ω_a to elements of $V_\delta \cap \Omega_a$.

Lemma 2.3.1 *There exists a nonnegative function α with $\alpha(\delta) \to 0$ as $\delta \to 0$ in such a way that for sufficiently small $\delta > 0$*

$$K(\Omega_a) \geq K\big(\Omega_{a-\alpha(\delta)} \textstyle\bigcap V_{\delta/2}\big) - \delta. \qquad (2.3.12)$$

Proof Suppose that the value of $K(\Omega_a)$ is attained for d.f. F with the corresponding density f. Consider the density

$$f_\delta := \frac{f + \delta}{1 + \delta} \geq \frac{\delta}{2}.$$

Then the corresponding d.f. F_δ belongs to $V_{\delta/2}$. The continuity of the functional χ implies that

$$\chi(F_\delta) \longrightarrow \chi(F) \geq a \qquad \text{as } \delta \to 0.$$

It follows that $F_\delta \in \Omega_{a-\alpha(\delta)} \cap V_{\delta/2}$ where $\alpha(\delta) \to 0$ as $\delta \to 0$. Now we prove the inequality

$$\int_0^1 f_\delta \ln f_\delta - \int_0^1 f \ln f \leq \delta \qquad (2.3.13)$$

(here and after we omit for brevity in integrals differentials and variables of integration). Using the formula of finite increments, we get

$$\int_0^1 f_\delta \ln f_\delta - \int_0^1 f \ln f = -\ln(1+\delta) - \frac{\delta}{1+\delta} \int_0^1 f \ln f$$

$$+ \frac{1}{1+\delta} \int_0^1 \big[(f+\delta)\ln(f+\delta) - f \ln f\big]$$

$$\leq \frac{\delta}{1+\delta} \int_0^1 \big(\ln(f+\delta^*) + 1\big), \qquad 0 < \delta^* < \delta.$$

The estimate (2.3.13) follows now from the elementary inequality $\ln x \le x - 1$ for $x > 0$.

Hence

$$K(\mathbf{\Omega}_a) = \int_0^1 f \ln f \ge \int_0^1 f_\delta \ln f_\delta - \delta,$$

and we establish the conclusion of Lemma 2.3.1.   □

Moreover, Lemma 2.3.1 implies the double inequality

$$K\big(\mathbf{\Omega}_{a_1} \cap V_{\delta_1}\big) - \delta \le K(\mathbf{\Omega}_a) \le K\big(K(\mathbf{\Omega}_a) \cap V_{\delta_1}\big), \qquad (2.3.14)$$

where $a_1 = a - \alpha(\delta)$ and $\delta_1 = \delta/2$.

We shall calculate now $K\big(\mathbf{\Omega}_{a_1} \cap V_{\delta_1}\big)$ and then shall pass to the limit in inequality (2.3.14) as $\delta \to 0$. In order to apply Theorem 1.8.1 let us enlarge the set V_{δ_1} up to $V'_{\delta_1/2}$. It will be proved that any extremal belongs to both sets for sufficiently small δ so the value of our extremal problem does not change.

One more auxiliary statement is connected with the properties of the space $\overset{\circ}{\mathbf{W}}_{p,m}[0,1]$. We recall that it contains the functions $x(t)$ that are absolutely continuous together with their derivatives up to order $m-1$ on the interval $[0, 1]$ and satisfy the boundary conditions $x(0) = x(1) = 0$ and, in addition, $x^{(m)} \in L^p[0,1], 0 < p < \infty$.

The norm in this space is defined as

$$\| x \|_{\overset{\circ}{\mathbf{W}}_{p,m}} := \left[\int_0^1 |x^{(m)}(t)|^p \, dt \right]^{1/p}.$$

We will also use the space $\overset{\circ}{\mathbf{W}}_{\infty,1}[0,1]$ with the norm

$$\| x \|_{\overset{\circ}{\mathbf{W}}_{\infty,1}} := \operatorname*{ess\,sup}_{0 \le t \le 1} |x'(t)|.$$

Lemma 2.3.2 *For any* $x \in \overset{\circ}{\mathbf{W}}_{1,2}[0,1]$

$$\sup_{0 \le t \le 1} |x(t)| \le \sup_{0 \le t \le 1} |x'(t)| \le \| x \|_{\overset{\circ}{\mathbf{W}}_{1,2}}.$$

Proof Using the boundary conditions we obtain

$$|x(t)| = \left| \int_0^t x'(s) \, ds \right| \le \int_0^1 |x'(s)| \, ds \le \sup_{0 \le t \le 1} |x'(t)|.$$

By Rolle's theorem there exists a point $t_0 \in [0, 1]$ such that $x'(t_0) = 0$. Therefore for all $t \in [0, 1]$

$$| x'(t) | = \left| \int_{t_0}^{t} x''(s)\, ds \right| \le \int_{0}^{1} | x''(s) |\, ds = \| x \|_{\overset{\circ}{\mathbf{W}}_{1,2}} .$$

Taking the supremum over all t we establish the conclusion of the lemma.

$\square$

It will be convenient to move subsequently from the class $\mathcal{F}$ to the class of absolutely continuous functions $x(t) = F(t) - t$, $F \in \mathcal{F}$. It is required to find the lower bound of the functional

$$f_0(x) := \int_{0}^{1} \big(x'(t) + 1 \big) \ln \big(x'(t) + 1 \big)\, dt , \qquad (2.3.15)$$

defined on an open subset $\mathcal{U}$ of the Banach space $\mathcal{X} = \overset{\circ}{\mathbf{W}}_{\infty,1}[0,1]$ containing elements x such that

$$x'(t) + 1 > \tfrac{1}{2}\delta_1 \qquad \text{for almost all } t \in [0, 1] \qquad (2.3.16)$$

under the supplementary constraint

$$f_1(x) := \int_{0}^{1} x^k(t) q(t)\, dt - a_1 \ge 0 . \qquad (2.3.17)$$

The inequality (2.3.16) yields that the functions f_0 and f_1 are strictly differentiable in $\mathcal{U}$, which is easily verified with the aid of Theorem 1.8.2.

Under such a formalization of the considered problem Theorem 1.8.1 guarantees the existence of Lagrange multipliers $\hat{\lambda}_0$ and $\hat{\lambda}_1$, $\hat{\lambda}_0^2 + \hat{\lambda}_1^2 > 0$, such that the function $\hat{x}(t)$, giving a local extremum in problem (2.3.15)–(2.3.17), satisfies the equation

$$\mathcal{L}_x \big(\hat{x},\, \hat{\lambda}_0,\, \hat{\lambda}_1 \big) = 0, \qquad (2.3.18)$$

where

$$\mathcal{L}\big(x,\, \hat{\lambda}_0,\, \hat{\lambda}_1 \big) = \hat{\lambda}_0\, f_0(x) + \hat{\lambda}_1\, f_1(x) ,$$

as well as the supplementary condition

$$\hat{\lambda}_1 f_1(\hat{x}) = 0 . \qquad (2.3.19)$$

Calculating (2.3.18) explicitly we get the equation

$$\int\limits_0^1 \Big[\hat{\lambda}_0\big(\ln\big(\hat{x}'(t)+1\big)+1\big)h'(t)+k\,\hat{\lambda}_1\,\hat{x}^{k-1}(t)\,q(t)\,h(t)\Big]\,dt=0\,,$$

which is valid for any $h(t)\in\overset{\circ}{\mathbf{W}}_{\infty,1}[0,1]$.

Integrating by parts, we reduce it to the form

$$\int_0^1\Big[\hat{\lambda}_0\big(\ln\big(\hat{x}'(t)+1\big)+1\big)+k\,\hat{\lambda}_1\int\limits_t^1\hat{x}^{k-1}(s)\,q(s)\,ds\Big]h'(t)\,dt=0\,.$$

Choosing in the capacity of $h(t)$ the elements of some ample system of functions, we obtain by a "basic lemma" of variational calculus (see, e.g., Young (1969), Sec. 7) that for almost all t

$$\hat{\lambda}_0\big(\ln\big(\hat{x}'(t)+1\big)+1\big)+k\,\hat{\lambda}_1\int\limits_t^1\hat{x}^{k-1}(s)\,q(s)\,ds=\text{const}\,.\qquad(2.3.20)$$

We may assume that (2.3.20) is valid for all t as this assumption does not change the value of functional (2.3.15). It follows from (2.3.20) that the function $\ln\big(\hat{x}'(t)+1\big)$ is absolutely continuous. Taking derivatives we obtain for the extremal $\hat{x}(t)$ the equation

$$\hat{\lambda}_0\,\hat{x}''(t)-k\,\hat{\lambda}_1\,\hat{x}^{k-1}(t)\big(\hat{x}'(t)+1\big)\,q(t)=0\,,\qquad(2.3.21)$$

and one can assume again that (2.3.21) holds for all $t\in[0,1]$. The condition $\hat{\lambda}_0=0$ contradicts (2.3.16)–(2.3.17). Hence $\hat{\lambda}_0\neq0$ and we come to the Euler–Lagrange equation

$$\hat{x}''(t)-\mu\,\hat{x}^{k-1}(t)\big(\hat{x}'(t)+1\big)\,q(t)=0\,,\qquad(2.3.22)$$

which should be considered together with the normalization condition (being a consequence of (2.3.17))

$$\int\limits_0^1\hat{x}^k(t)\,q(t)\,dt=a_1\qquad(2.3.23)$$

and the boundary conditions

$$\hat{x}(0)=\hat{x}(1)=0\,.\qquad(2.3.24)$$

2.4 Large Deviations of Integral Statistics. Constructing Solutions of the Euler–Lagrange Equations

We proceed to construct the solutions of problem (2.3.22)–(2.3.24). It will be convenient to introduce the following notation:

$$\varepsilon = a_1^{1/k}, \qquad \hat{x} = g\,\varepsilon, \qquad \nu = \mu\,\varepsilon^{k-2}. \qquad (2.4.1)$$

The equation (2.3.22) becomes

$$g'' - \nu\,g^{k-1}q - \nu\,\varepsilon\,g^{k-1}g'q = 0 \qquad (2.4.2)$$

and the normalization condition looks like

$$\int_0^1 g^k q\,dt = 1. \qquad (2.4.3)$$

The boundary conditions are unchanged:

$$g(0) = g(1) = 0. \qquad (2.4.4)$$

The left-hand side of equation (2.4.2) may be considered as an analytic operator $A(g,\nu,\varepsilon)$ from the space $\overset{\mathrm{o}}{\mathbf{W}}_{1,2}[0,1] \times \mathbf{R}^2$ into $L^1[0,1]$. It is natural to expect that the solution of problem (2.4.2)–(2.4.4) depends analytically on parameter $\varepsilon > 0$ for sufficiently small values of it. We shall prove subsequently that this is actually true.

Let us consider first the auxiliary problem coming from problem (2.4.2)–(2.4.4) for $\varepsilon = 0$:

$$g'' = \nu\,g^{k-1}\,q, \qquad g(0) = g(1) = 0, \qquad \int_0^1 g^k q\,dt = 1. \qquad (2.4.5)$$

The solutions of (2.4.5) give the main contribution to the solution of the initial problem.

Variational considerations imply that the solution of this problem always exists but is not obliged to be unique. Let us denote by $\lambda_0 = \lambda_0(q;k)$ the smallest in absolute value ν for which the solution exists (all such ν are negative) and by x_0 the solution corresponding to λ_0. Later the pair (x_0, λ_0) will be called the "principal" solution of problem (2.4.5).

In the easiest case $k = 1$ the solution is unique and can be easily obtained with the aid of the Green function $\mathcal{G}(t,s)$ as follows

$$\lambda_0(q;1) = -\left(\int_0^1 \int_0^1 \mathcal{G}(t,s)\, q(t)\, q(s)\, dt\, ds \right)^{-1}, \qquad (2.4.6)$$

$$x_0(t) = -\lambda_0 \int_0^1 \mathcal{G}(t,s)\, q(s)\, ds, \qquad (2.4.7)$$

where

$$\mathcal{G}(t,s) := \min(t,s) - ts, \qquad 0 \le t, s \le 1. \qquad (2.4.8)$$

If $k = 2$, we have the classical Sturm–Liouville problem with a discrete spectrum. Under the conditions imposed on q all the eigenvalues of this problem are simple and the first eigenfunction x_0 has no zeros within the interval $[0, 1]$ (see, e.g., Atkinson (1964), Chapter 8).

For $k > 2$ the structure of solutions of (2.4.5) is not well explored. The Caratheodory's theorem (see, e.g., Kamke (1959), Sec. 5.3) implies that the solution of the Cauchy problem for the equation

$$g'' = \lambda_0\, g^{k-1}\, q$$

is unique. Consequently, there does not exist $t_0 \in (0, 1)$ such that $x_0(t_0) = 0$ and $x_0'(t_0) = 0$. Combining this observation with variational arguments (for even k) and reasons for the convexity (for odd k), it is easy to prove that the solutions of (2.4.5), corresponding to λ_0, preserve their sign on $(0, 1)$. But we cannot rule out a priori the existence of multiple solutions (Krasnoselskii (1964), Chapter 7; Krasnoselskii and Zabreiko (1975), Chapter 6). At the same time it is well known that for $q \equiv 1$ the nontrivial nonnegative solution of (2.4.5) is unique (Krasnoselskii (1964), p. 309). This fact will be used in Chapter 6.

We initiate the construction of the solution of problem (2.4.2)–(2.4.4) with the easiest case $k = 1$. Let us make the substitution

$$x = g - x_0, \qquad\qquad \lambda = \nu - \lambda_0. \qquad (2.4.9)$$

Then equation (2.4.2) takes now the form

$$x'' - \lambda q - (\lambda + \lambda_0)\, \varepsilon\, (x' + x_0')\, q = 0 \qquad (2.4.10)$$

and the normalization condition will be transformed into

$$\int_0^1 (x + x_0)\, q\, dt = 1, \qquad (2.4.11)$$

whereas the boundary conditions remain unchanged:

$$x(0) = x(1) = 0 \, . \qquad (2.4.12)$$

The equation (2.4.10) together with conditions (2.4.12) may be treated as an implicit analytic operator $A(x, \varepsilon, \lambda)$: $\overset{\circ}{\mathbf{W}}_{1,2}[0,1] \times \mathbf{R}^2 \mapsto L^1[0,1]$. The Fréchet derivative of this operator for $\varepsilon = \lambda = 0$ appears as follows:

$$B\,x = A_x(x,0,0) = x'', \qquad x(0) = x(1) = 0 \, .$$

The operator B possesses a bounded inverse which may be easily constructed using the Green function (2.4.8). Hence we can apply Theorem 1.8.4 on implicit analytic operators and obtain the solution of (2.4.10) in the form

$$x = \sum_{i+j \geq 1} x_{ij}\, \varepsilon^i \lambda^j, \qquad (2.4.13)$$

where the series is absolutely convergent, that is, with respect to the norm of the space $\overset{\circ}{\mathbf{W}}_{1,2}[0,1]$ for sufficiently small $\varepsilon > 0$ and $|\lambda|$. Substituting (2.4.13) in (2.4.10) and comparing the coefficients of the same powers of ε and λ, one can calculate any number of coefficients x_{ij}, specifically

$$x_{01}(t) = \int\limits_0^1 \mathcal{G}(t,s)\, q(s)\, ds \, .$$

Now we put (2.4.13) in (2.4.11) and integrate termwise. The resulting equation is

$$-\frac{\lambda}{\lambda_0} + c_{10}\,\varepsilon + \sum_{i+j \geq 2} c_{ij}\, \varepsilon^i \lambda^j = 0, \qquad (2.4.14)$$

where c_{ij} are numerical coefficients and the series converges absolutely again for sufficiently small $\varepsilon > 0$ and $|\lambda|$. It follows from (2.4.14) that λ is an analytic function of ε in a neighborhood of the origin, hence we obtain from (2.4.13) that

$$x(t) = \sum_{j=1}^{\infty} x_j(t)\, \varepsilon^j,$$

where the series on the right-hand side converges for sufficiently small

$\varepsilon > 0$. It follows from (2.4.9) and (2.4.1) that

$$\hat{x}(t) = x_0\, a_1 + \sum_{j=1}^{\infty} x_j(t)\, a_1^{j+1}, \qquad (2.4.15)$$

$$\mu = \lambda_0\, a_1 + \sum_{j=1}^{\infty} \mu_j\, a_1^{j+1}. \qquad (2.4.16)$$

Thus we have constructed the unique solution of problem (2.3.22)–(2.3.24) for $k = 1$.

The case $k = 2$ turns out to be significantly more difficult and requires a different technique. After substitution (2.4.9), equation (2.4.2) takes the form

$$B\,x - R(x, \varepsilon, \lambda) = 0, \qquad (2.4.17)$$

where

$$B\,x = x'' - \lambda_0\, x\, q, \qquad (2.4.18)$$

$$R(x, \varepsilon, \lambda) = \lambda\, x\, q + \lambda\, x_0\, q + \varepsilon\, \lambda\, x\, x'q + \lambda_0\, \varepsilon\, x\, x'q + \varepsilon\, \lambda\, (x_0\, x)'q$$
$$+ \lambda_0\, \varepsilon\, x_0\, x_0'q + \lambda_0\, \varepsilon\, (x_0\, x)'q + \varepsilon\, \lambda\, x_0\, x_0'q. \qquad (2.4.19)$$

The normalization condition is now

$$\int_0^1 (x + x_0)^2 q\, dt = 1, \qquad (2.4.20)$$

and boundary conditions (2.4.12) are unchanged.

In contrast to the preceding case the operator $B\colon \overset{\circ}{\mathbf{W}}_{1,2}[0,1] \mapsto L^1[0,1]$ determined by (2.4.18) is noninvertible. Therefore the implicit operator theorem is inapplicable, and equation (2.4.17) may have multiple solutions which need not be analytic. To find these solutions we need to set up the Lyapunov–Schmidt branching equation.

Let us denote by $\mathbf{E}_1$ the space $\overset{\circ}{\mathbf{W}}_{1,2}[0,1]$, by $\mathbf{E}_2$ the space $L^1[0,1]$, and by $\mathbf{E}_i^*$, $i = 1, 2$, the spaces conjugate to $\mathbf{E}_i$. Let $B^*\colon \mathbf{E}_2^* \mapsto \mathbf{E}_1^*$ be the adjoint operator of B and the value of the functional α on an element x be denoted by (x, α). We denote by γ an element of $\mathbf{E}_1^*$ possessing the property $(x_0, \gamma) = 1$. The operator B in question is a Fredholm operator with the number of zeros equal to 1. The operator B^* then possesses the same properties. Let $\psi \in \mathbf{E}_2^*$ be an element generating the kernel of the operator B^* and z be an element of $\mathbf{E}_2$ such that $(z, \psi) = 1$. Since

B is a self-adjoint operator, these elements are easily determined on the basis of x_0 and q.

We now denote by $\widetilde{B}$ the operator related to B by the equality

$$\widetilde{B} x = B x + \xi z, \qquad \text{where } \xi = (x, \gamma).$$

By the generalized Schmidt lemma (see Vainberg and Trenogin (1974), Sec. 21.3) the operator $\widetilde{B}$ is invertible, its inverse operator $\Gamma = \widetilde{B}^{-1}$ is bounded, and

$$\Gamma z = x_0, \qquad\qquad \Gamma^* \gamma = \psi.$$

Rewriting equation (2.4.17) in the form

$$\widetilde{B} x = R(x, \varepsilon, \lambda) + \xi\, z, \qquad\qquad (2.4.21)$$

we see that the theorem on implicit operators is now applicable. It guarantees that the solution of (2.4.21) can be represented in the form

$$x = \sum_{i,\, j,\, k=0}^{\infty} x_{ijk}\, \xi^i\, \varepsilon^j \lambda^k \qquad\qquad (2.4.22)$$

and series (2.4.22) converges absolutely, that is, with respect to the norm of the space $\mathbf{E}_1$ for sufficiently small $|\xi|$, ε, and $|\lambda|$.

We now substitute series (2.4.22) into the equality $\xi = (x, \gamma)$. Then, setting $L_{ijk} := (x_{ijk}, \gamma)$, we obtain the so-called Lyapunov–Schmidt branching equation (see Vainberg and Trenogin (1974), Sec. 23.3):

$$\sum_{i=2}^{\infty} L_{i00}\, \xi^i + \sum_{i=0}^{\infty} \xi^i \sum_{j+k \geq 1} L_{ijk}\, \varepsilon^j \lambda^k = 0. \qquad\qquad (2.4.23)$$

It is well known that if the branching equation has small solutions, then they can be represented in the form of convergent series in integral or fractional powers of parameters ε and λ; these small solutions are in one-to-one correspondence with the small solutions of the original equation. In order to find the coefficients x_{ijk} of series (2.4.22) we equate the coefficients of like powers of ξ, ε, and λ in (2.4.21) after the substitution of (2.4.22). We get then

$$\widetilde{B}\,x_{100} = z\,, \qquad\qquad x_{100} = \Gamma\,z = x_0\,;$$

$$\widetilde{B}\,x_{i00} = 0\,, \qquad\qquad x_{i00} = 0\,,\ i \geq 2\,;$$

$$\widetilde{B}\,x_{001} = x_0\,q\,, \qquad\qquad x_{001} = \Gamma(x_0\,q)\,;$$

$$\widetilde{B}\,x_{010} = \lambda_0\,x_0\,x_0'\,q\,, \qquad\qquad x_{010} = \lambda_0\,\Gamma(x_0 x_0' q)\,;$$

$$\widetilde{B}\,x_{101} = x_{100}\,q\,, \qquad\qquad x_{101} = \Gamma(x_0 q)\,;$$

$$\widetilde{B}\,x_{110} = \lambda_0(x_0\,x_{100})'q\,, \qquad x_{110} = 2\,\lambda_0\,\Gamma(x_0 x_0' q)\,;$$

and so on.

Using the formula $L_{ijk} = (x_{ijk}, \gamma)$ we further obtain

$$L_{200} = L_{300} = \cdots = L_{k00} = \cdots = 0\,,$$

$$L_{001} = \big(\Gamma(x_0 q), \gamma\big) = (x_0 q, \psi) = 1\,,$$

$$L_{010} = \lambda_0\big(\Gamma(x_0 x_0' q), \gamma\big)\,,$$

$$L_{101} = \big(\Gamma(x_0 q), \gamma\big) = (x_0 q, \psi) = 1\,,$$

$$L_{110} = 2\,\lambda_0\big(\Gamma(x_0 x_0' q), \gamma\big) = 2\,L_{010}\,,$$

and so on.

Thus, if we restrict attention to first terms, the branching equation has the form

$$F_1(\xi, \varepsilon, \lambda) \equiv \lambda + \xi\,\lambda + L_{010}\,(\varepsilon + 2\,\xi\,\varepsilon) + L_{020}\,\varepsilon^2 + \cdots = 0\,. \quad (2.4.24)$$

We now use normalization condition (2.4.20) and Lemma 2.3.2. Since series (2.4.22) for x converges absolutely for sufficiently small $|\xi|$, ε, and $|\lambda|$, the same is true for x^2, and the series for x^2 may be integrated termwise with respect to the measure $q\,dt$. As a result we obtain

$$F_2(\xi, \varepsilon, \lambda) \equiv 2\,\xi + 2\,\lambda + a_{010}\,\varepsilon + a_{020}\,\varepsilon^2 + \cdots = 0, \qquad (2.4.25)$$

where $a_{010}, a_{020}, \ldots$ are numerical coefficients.

We consider now (2.4.24) and (2.4.25) simultaneously and use the classic variant of the implicit function theorem (Theorem 1.8.3). As the Jacobian

$$\frac{\partial(F_1, F_2)}{\partial(\xi, \lambda)}\,(0, 0) = \begin{vmatrix} 0 & 1 \\ 2 & 2 \end{vmatrix} \neq 0\,,$$

we obtain that

$$\xi = \sum_{k=1}^{\infty} c_k \, \varepsilon^k , \qquad \lambda = \sum_{k=1}^{\infty} d_k \, \varepsilon^k , \qquad (2.4.26)$$

and both these series converge for sufficiently small $\varepsilon > 0$.

Substituting these series in (2.4.22) and using (2.4.9) and (2.4.1), we find the "principal" solution of problem (2.3.22)–(2.3.24) for $k = 2$ in the form of absolutely convergent series

$$\hat{x}(t) = x_0 \, a_1^{1/2} + \sum_{j=1}^{\infty} x_j(t) \, a_1^{(j+1)/2} , \qquad (2.4.27)$$

$$\mu = \lambda_0 + \sum_{j=1}^{\infty} \mu_j \, a_1^{j/2} . \qquad (2.4.28)$$

The coefficients x_j and μ_j in formulas (2.4.15)–(2.4.16) and (2.4.27)–(2.4.28) are certainly different. We use the same notation in order to underline the similarity of these formulas.

In the nonlinear case $k > 2$ the arguments are based on the regular case $k = 1$ and the branching case $k = 2$ we have already considered. Using notation (2.4.9) we transform again equation (2.4.2) into the form (2.4.17) where

$$B \, x = x'' - \lambda_0 \, (k - 1) \, x_0^{k-2} \, x \, q , \qquad (2.4.29)$$

and the operator $R(x, \varepsilon, \lambda)$ contains the terms with ε, λ, or x in powers higher than one.

Now we consider the following two possibilities.

$\mathbf{I}^0$. *The operator B:* $\overset{\circ}{\mathbf{W}}_{1,2}[0,1] \mapsto L^1[0,1]$ *given by (2.4.29) has a bounded inverse (the regular case).*

Then we apply the theorem on the implicit analytic operator as in the case $k = 1$. From it we obtain the solution of problem (2.3.22)–(2.3.24) in the form

$$\hat{x}(t) = x_0 \, a_1^{1/k} + \sum_{j=1}^{\infty} x_j(t) \, a_1^{(j+1)/k} , \qquad (2.4.30)$$

$$\mu = \lambda_0 \, a_1^{(2/k-1)} + \sum_{j=1}^{\infty} \mu_j \, a_1^{(j-k+2)/k} , \qquad (2.4.31)$$

where the series converge for sufficiently small $a_1 > 0$.

II⁰. *The operator B is noninvertible (the case of branching).*

Since B is the Sturm–Liouville operator and therefore is a Fredholm operator, we can use for the determination of the solution the branching equation simultaneously with the norming condition. Most substantial is the fact that since B has only single zeros, it rules out the difficult case of multidimensional branching. The calculations will become now more lengthy and some formulas will change (e.g., the first two terms in (2.4.25) will be $k\,\xi + k\,\lambda$ instead of $2\,\xi + 2\,\lambda$). But all these changes are not basic and do not change the plan of the proof. Finally we discover that in the branching case also the solution of (2.3.22)–(2.3.24) has the form (2.4.30)–(2.4.31).

Thus we have constructed the "principal" solution of our extremal problem for any positive integer k. One should also emphasize that for $k \geq 2$ there can exist several extremals of the form (2.4.30) as even the solution of problem (2.4.5) is not necessary unique.

The extremal $\hat{x}(t)$ given by (2.4.30) admits termwise differentiation and the derivative's series converges uniformly due to Lemma 2.3.2. Therefore

$$\hat{x}'(t) = x_0'(t)\, a_1^{1/k} + \sum_{j=1}^{\infty} x_j'(t)\, a_1^{(j+1)/k}. \qquad (2.4.32)$$

It is clear that for sufficiently small a_1 and δ_1 the inequality $\hat{x}'(t)+1 > \delta_1$ holds so the extremal $\hat{x}(t)$ belongs to the set $\boldsymbol{\Omega}_{a_1} \cap V_{\delta_1/2}'$ as well as to $\boldsymbol{\Omega}_{a_1} \cap V_{\delta_1}$.

Substituting (2.4.32) in the functional $f_0(x)$ and integrating termwise we obtain

$$\int_0^1 \big(\hat{x}'(t)+1\big)\ln\big(\hat{x}'(t)+1\big)\,dt = -\frac{\lambda_0}{2}\, a_1^{2/k} + \sum_{j=3}^{\infty} b_j\, a_1^{j/k}, \qquad (2.4.33)$$

where $b_j = b_j(q;k)$ are numerical coefficients, $\lambda_0 = \lambda_0(q;k)$ is the "principal" eigenvalue of problem (2.4.5), and the series converges for all sufficiently small $a_1 > 0$. To obtain (2.4.33) we have used the equality

$$\int_0^1 {x_0'}^2(t)\,dt = -\int_0^1 x_0''(t)\,x_0(t)\,dt = -\lambda_0.$$

We recall now that $a_1 = a + \alpha(\delta) \to a$ as $\delta \to 0$. Passing to the limit in

(2.3.14) we have

$$K(\mathbf{\Omega}_a) \geq -\frac{\lambda_0}{2}\, a^{2/k} + \sum_{j=3}^{\infty} b_j\, a^{j/k}. \qquad (2.4.34)$$

On the other hand the value of $K\big(\mathbf{\Omega}_a \cap V_{\delta_1}\big)$ is computed analogously to the value of $K\big(\mathbf{\Omega}_{a_1} \cap V_{\delta_1}\big)$ and the corresponding extremals have the form (2.4.30) for $a_1 = a$. Therefore it follows again from (2.3.14) that

$$K(\mathbf{\Omega}_a) \leq -\frac{\lambda_0}{2}\, a^{2/k} + \sum_{j=3}^{\infty} b_j\, a^{j/k}. \qquad (2.4.35)$$

Now (2.4.34) and (2.4.35) imply (2.3.8) and hence Theorem 2.3.1 is completely proved. $\square$

Note that in principle one can calculate any number of coefficients b_j but it is necessary to know all the solutions of (2.4.5) corresponding to the principal eigenvalue λ_0. If there exist several such solutions one has also several extremals (2.4.30) and one must determine which of them gives the infimum of $f_0(x)$.

Theorem 2.3.2 in the case $k = 1$ may be obtained much easier by reducing the problem to the study of sums of independent random variables. Let us introduce the notation

$$Q(t) = \int_0^t q(s)\, ds\,, \qquad k = \int_0^1 Q(t)\, dt\,, \qquad \sigma^2 = \int_0^1 Q^2(t)\, dt - k^2\,.$$

Integrating by parts we have

$$\omega_{n,\,q}^1 = \int_0^1 \big[F_n(t) - t\big]\, q(t)\, dt = k - n^{-1} \sum_{j=1}^{n} Q(X_j)\,.$$

The Chernoff theorem (Theorem 1.6.2) now gives

$$\lim_{n\to\infty} n^{-1} \ln P\big(\omega_{n,\,q}^1 \geq a\big) = \ln \inf_{t>0} E \exp\big\{t\,[k - a - Q(X_1)]\big\}\,. \qquad (2.4.36)$$

The infimum is attained at the point t_a, a root of the equation

$$\int_0^1 [k - Q(x)]\, e^{-t\,Q(x)}\, dx = a \int_0^1 e^{-t\,Q(x)}\, dx\,.$$

For sufficiently small $a > 0$ one can guarantee that the root t_a can be represented as an analytic function of a: $t_a = \sum_{k\geq 1} c_k a^k$ (see, e.g., De

Bruijn (1961), Sec. 2.2). The calculations show that $c_1 = \sigma^{-2}$. Substituting t_a in (2.4.36) we have

$$\lim_{n \to \infty} n^{-1} \ln P\big(\omega_{n,q}^1 \geq a\big) = -\frac{a^2}{2\sigma^2} + \sum_{k \geq 3} \beta_k \, a^k$$

and the series on the right-hand side converges for sufficiently small $a > 0$. So we have again proved (2.3.5) for $k = 1$ and $\lambda_0(q; 1) = -1/\sigma^2$.

Integrating by parts one can easily verify that (2.4.6) gives the same result. Note that if $q \equiv 1$ one has $Q(x) = x$ and we get the well-known value

$$\lambda_0(1; 1) = -12. \tag{2.4.37}$$

Another interesting statistic is the so-called first component of the statistic ω^2 (see Durbin and Knott (1972, 1975) and Stephens (1974)), corresponding to the weight function $\tilde{q}(t) = \sin \pi t$:

$$\kappa_n = \int_{-\infty}^{\infty} \big[F_n(t) - F_0(t) \big] \sin\big\{ \pi F_0(t) \big\} \, dF_0(t). \tag{2.4.38}$$

In this case we have $\lambda_0(\tilde{q}; 1) = -2\pi^2$.

In concluding this section we note that the first coefficient of expansion (2.3.8) has been calculated by Jeurnink and Kallenberg (1990) for weighted quadratic statistics from a somewhat wider class by using a different method based on the approximation by sums of finite-dimensional random vectors.

2.5 Large Deviations of Integral Anderson–Darling, Watson, and Khmaladze–Aki Statistics

The statistics A_n^2, U_n^2, and L_n^2 introduced in (2.1.10), (2.1.11), and (2.1.13) are also frequently used in the case of goodness-of-fit testing. Though they are very close to the statistics $\omega_{n,q}^k$ studied in Sections 2.3–2.4, we cannot obtain their large-deviation asymptotics from Theorem 2.3.2. Therefore we devote the present section to their large deviations under H_0. One can again assume that $F_0(t) = t$, $0 \leq t \leq 1$.

The main difficulty in the study of A_n^2 lies in the fact that the weight $w(t) = \big[t(1 - t)\big]^{-1}$ is nonsummable on [0, 1]. So we must apply two auxiliary inequalities. We denote here $\ln^+ f = \max(\ln f, 0)$.

Lemma 2.5.1 (Hardy, Littlewood, and Polya 1934, Th. 241)
Let F be an absolutely continuous d.f. on [0, 1], $F(t) = \int_0^t f(u)\, du$,

$t \in [0, 1]$, such that $\int_0^1 f \ln^+ f \, dt < +\infty$. Then there exists an absolute constant C such that

$$\int_0^1 \frac{F(t)}{t} \, dt < C \left(\int_0^1 f \ln^+ f \, dt + 1 \right).$$

Corollary Under the conditions of Lemma 2.5.1 there exists an absolute constant C' such that

$$\int_0^1 \frac{|F(t) - t|}{t(1-t)} \, dt < C' \left(\int_0^1 f \ln f \, dt + 1 \right). \qquad (2.5.1)$$

Proof It is evident that

$$\int_0^1 \frac{|F(t) - t|}{t(1-t)} \, dt \leq 2 \int_0^{1/2} \frac{|F(t) - t|}{t} \, dt + 2 \int_{1/2}^1 \frac{|F(t) - t|}{(1-t)} \, dt$$
$$= 2 \left(J_1 + J_2 \right).$$

Now we estimate J_1 and J_2 using Lemma 2.5.1:

$$J_1 \leq \int_0^1 \frac{F(t)}{t} \, dt + \frac{1}{2} \leq C \left(\int_0^1 f \ln^+ f \, dt + 1 \right) + \frac{1}{2},$$

$$J_2 \leq \int_0^1 \frac{|1 - F(1-t)|}{t} \, dt + \frac{1}{2} \leq C \left(\int_0^1 f \ln^+ f \, dt + 1 \right) + \frac{1}{2}.$$

Finally we note that for any density f

$$\int_0^1 f \ln^+ f \, dt \leq \int_0^1 f \ln f \, dt + e^{-1}. \qquad \square$$

Now let an $a > 0$ be given. Denote by $\mathcal{A}_a$ the set of d.f.s F on $[0, 1]$ satisfying the inequality

$$\int_0^1 \left(F(t) - t \right)^2 w(t) \, dt \geq a. \qquad (2.5.2)$$

Further we put analogously to (1.2.10) and (2.3.7)

$$K(\mathcal{A}_a) := \inf \left\{ K(F) : F \in \mathcal{A}_a \right\}.$$

The following result is proved in the sequel.

Theorem 2.5.1 (Nikitin 1980) *For sufficiently small $a > 0$ the quantity $K(\mathcal{A}_a)$ admits the representation in the form of a convergent series with numerical coefficients*

$$K(\mathcal{A}_a) = a + \sum_{j \geq 4} c_j \, a^{j/2} \, . \qquad (2.5.3)$$

Relying on Theorem 2.5.1 we now prove the result analogous to (2.3.9).

Theorem 2.5.2 (Nikitin 1980) *For sufficiently small $a > 0$ under the hypothesis H_0*

$$\lim_{n \to \infty} n^{-1} \ln P\big(A_n^2 \geq a\big) = -K(\mathcal{A}_a), \qquad (2.5.4)$$

where $K(\mathcal{A}_a)$ admits representation (2.5.3).

Proof of Theorem 2.5.2 Let

$$\Gamma := \Big\{ F \in \mathcal{F} \colon \int_0^1 f \ln f \, dt < +\infty \Big\} \, .$$

We establish now that the functional

$$\chi_w(F) := \int_0^1 \big(F(t) - t \big)^2 w(t) \, dt$$

is continuous on Γ in the ρ-topology. For F_1, $F_2 \in \Gamma$ such that

$$\sup_{0 \leq t \leq 1} \big| F_1(t) - F_2(t) \big| \longrightarrow 0$$

we obtain, according to (2.5.1), that

$$\begin{aligned}
\Big| \chi_w(F_1) - \chi_w(F_2) \Big| &\leq \sup_{0 \leq t \leq 1} \big| F_1(t) - F_2(t) \big| \\
&\times \int_0^1 \bigg(\sum_{j=1}^2 \big| F_j(t) - t \big| \, w(t) \bigg) \, dt \\
&\leq C \sup_{0 \leq t \leq 1} \big| F_1(t) - F_2(t) \big| \bigg(\int_0^1 \big(f_1 \ln f_1 + f_2 \ln f_2 \big) \, dt + 2 \bigg) \longrightarrow 0 \, .
\end{aligned}$$

The continuity of χ_w in the ρ-topology implies its continuity in the τ-topology on Γ. The functional $a \mapsto K(\mathcal{A}_a)$ is continuous for sufficiently small a due to (2.5.3). So the result follows from Theorem 1.6.7 as all the conditions required there are fulfilled. $\square$

Proof of Theorem 2.5.1 The plan of the proof is very close to that of the proof of Theorem 2.3.1 and differs from it only in some details. So we dwell upon the main features of the proof only. Consider a Banach space $\mathcal{E}_w$ consisting of absolutely continuous functions $x(t)$ on $[0, 1]$, satisfying zero boundary conditions $x(0) = x(1) = 0$, for which

$$\operatorname*{ess\,sup}_{0 \le t \le 1} |x'(t)| < \infty$$

and, moreover,

$$\int_0^1 |x(t)|\, w(t)\, dt < \infty. \tag{2.5.5}$$

The norm in this space is given by

$$\|x\|_{\mathcal{E}_w} := \int_0^1 |x(t)|\, w(t)\, dt + \operatorname*{ess\,sup}_{0 \le t \le 1} |x'(t)|.$$

It follows from (2.5.1) that in calculating $K(\mathcal{A}_a)$ we can consider only the subspace of $\overset{\circ}{\mathbf{W}}_{\infty,1}$ specified by (2.5.5).

In the space $\mathcal{E}_w$ we consider an open set $\mathcal{U}$ given by the inequality (2.3.16), where δ_1 is a small positive number. We must find the infimum of the functional

$$f_0(x) = \int_0^1 \big(x'(t) + 1 \big)\, \ln\big(x'(t) + 1 \big)\, dt,$$

defined on the set $\mathcal{U} \subset \mathcal{E}_w$ under the additional condition

$$f_1(x) = \int_0^1 x^2(t)\, w(t)\, dt - a \ge 0.$$

Under such a formalization of the problem the functions $f_0(x)$ and $f_1(x)$ are strictly differentiable on $\mathcal{U}$. We can apply the Lagrange principle (Theorem 1.8.2) and obtain the following Euler–Lagrange equation for the extremal $\hat{x}$:

$$h'' - \mu h w - \mu \varepsilon h h' w = 0, \tag{2.5.6}$$

where $\varepsilon = \sqrt{a}$ and $h = \varepsilon^{-1}\hat{x}$. It should be considered simultaneously with the normalization condition

$$\int_0^1 h^2 w\, dt = 1 \tag{2.5.7}$$

and the boundary conditions $h(0) = h(1) = 0$.

It follows from (2.5.6) that $h'' \in \mathbf{L}^1[0,1]$ so h is an element of the space $\mathcal{S}_w$ consisting of elements x of the space $\overset{\circ}{\mathbf{W}}_{1,2}[0,1]$ that satisfy (2.5.5). The space $\mathcal{S}_w$ with the norm

$$\|x\|_{\mathcal{S}_w} := \int\limits_0^1 \big(|x|\, w + |x'| + |x''| \big)\, dt$$

is a Banach space.

We may interpret equation (2.5.6) as an implicit analytic operator from $\mathcal{S}_w \times \mathbf{R}^2$ into $L^1[0,1]$. Therefore the solution of problem (2.5.6)–(2.5.7) can be constructed with the aid of the branching equation in the same manner as in Section 2.4. The calculations show that the "principal" solutions of (2.5.6) have the form

$$h(t) = \pm\sqrt{6}\, t\, (1-t) + 3\, t\, (1-t)\, (1-2t)\, \varepsilon + \sum_{j\geq 2} s_j(t)\, \varepsilon^j, \qquad (2.5.8)$$

where $s_j \in \mathcal{S}_w$ and the series on the right-hand side converges absolutely for sufficient small $\varepsilon > 0$.

Substituting now $\hat{x} = h\,\varepsilon$ into the functional f_0 and integrating termwise, we get representation (2.5.3) for $K(\mathcal{A}_a)$ (the term with $a^{3/2}$ vanishes on integration). $\qquad\square$

It should be noted that Groeneboom and Shorack (1981) have obtained the weaker result

$$\lim_{n\to\infty} n^{-1} \ln P\big(A_n^2 \geq a\big) = -a + o(a) \qquad \text{as } a \to 0. \qquad (2.5.9)$$

The paper by Groeneboom and Shorack (1981) was published substantially later than the paper by Nikitin (1980) with the proof of Theorems 2.5.1 and 2.5.2. The earlier result (2.5.9) had been announced by Nikitin (1976). Another attempt at the proof of (2.5.9) was described by Inglot and Ledwina (1993), where Theorem 1.6.10 was generalized to the case of weighted empirical processes. The result

$$K(\mathcal{A}_a) = a + o(a) \qquad \text{as } a \to 0$$

was also obtained by Jeurnink and Kallenberg (1990) with a generalization for a larger class of weights.

We pass now to the study of large deviations of the Watson statistic U_n^2 given by (2.1.11).

Theorem 2.5.3 (Nikitin 1988) *If the hypothesis H_0 is true then*

$$\lim_{n\to\infty} n^{-1} \ln P\big(U_n^2 \geq a\big) = -2\,\pi^2 a + o(a) \qquad \text{as } a \to 0. \qquad (2.5.10)$$

Proof Theorem 2.5.3 may be proved basically in the same way as Theorems 2.3.1 and 2.5.2. But the operator B from (2.4.18) will map $\overset{\circ}{C}{}^2[0,1]$ into $C[0,1]$ by the means of the formula

$$B\,x = x'' - \lambda_0 \left(x - \int_0^1 x(s)\,ds \right), \qquad x(0) = x(1) = 0.$$

The equation $B\,x = 0$ has two linearly independent solutions corresponding to the principal eigenvalue $\lambda_0 = -4\pi^2$:

$$x_1(t) = \sin 2\pi t, \qquad x_2(t) = 1 - \cos 2\pi t.$$

Therefore when studying U_n^2 we face the case of two-dimensional branching when the subspace of zeros of the operator B has a dimension greater than 1 (Vainberg and Trenogin 1974). The construction of the branching equations becomes much more involved and their analysis requires a set of supplementary ideas. It is possible to overcome these difficulties (see, e.g., Abbakumov (1987) where the counterpart of the Watson statistic for symmetry testing – the Hill–Rao statistic (Hill and Rao 1977) – is considered), but we confine our attention to obtaining only the main term of the expansion of large-deviation probabilities that may be obtained via Theorem 1.6.3.

Let $X_1, X_2, \ldots$ be independent random variables with the uniform distribution on $[0,1]$. Consider a new sequence $\{Y_j\}$ of random variables taking on values in the space $\mathcal{Y} = L^2[0,1]$:

$$Y_j(t,\omega) := \mathbf{1}_{\{X_j < t\}}(\omega) - t + X_j(\omega) - \tfrac{1}{2}.$$

The r.v.s Y_j are independent and identically distributed. The space $\mathcal{Y}_1^*$ consists of functionals f^* on $L^2[0,1]$ acting on each element $y \in \mathcal{Y}$ by

$$f^*\big(y(t)\big) = \int_0^1 y(t)\,f(t)\,dt,$$

where $\|f\|_{L^2} = 1$. Denote by $\mathcal{Y}_1$ a unit ball in the space $\mathcal{Y}$.

It is evident that for any $f^* \in \mathcal{Y}_1^*$

$$Ef^*(Y_j) = E \int_0^1 Y_j(t,\omega)\,f(t)\,dt = \int_0^1 EY_j(t,\omega)\,f(t)\,dt = 0.$$

Also, one has $\|Y_j\| = (12)^{-1/2}$, and therefore for all $z \in \mathbf{R}^1$

$$E \exp\{z\,\|Y_j\|\} < \infty.$$

We see that the conditions of Theorem 1.6.3 are fulfilled. As

$$n^{-1}\,\|Y_1 + \cdots + Y_n\|_{L^2} = \left(U_n^2\right)^{1/2},$$

the conclusions of Theorem 1.6.3 give the required asymptotics. It remains only to calculate σ^2. But

$$\sigma^2 = \sup \left\{ \operatorname{Var} f^*(Y_1) : \ f^* \in \mathcal{Y}_1^* \right\}$$

$$= \sup \left\{ E \left(\int\limits_0^1 f(t) \, Y_1(t,\omega) \, dt \right)^2 : f \in \mathcal{Y}_1 \right\}$$

$$= \sup \left\{ \int\limits_0^1 \int\limits_0^1 f(s) \, f(t) \, E\big(Y_1(s,\omega) Y_1(t,\omega)\big) \, ds \, dt \ : \ f \in \mathcal{Y}_1 \right\}$$

$$= \sup \left\{ \int\limits_0^1 \int\limits_0^1 \mathcal{K}(t,s) \, f(t) \, f(s) \, dt \, ds \ : \ f \in \mathcal{Y}_1 \right\},$$

where

$$\mathcal{K}(t,s) = \min(t,\, s) - t\,s + \tfrac{1}{2}\,(t^2 + s^2) - \tfrac{1}{2}\,(t+s) + \tfrac{1}{12}\,.$$

By the extremal principle of Courant and Weyl (see, e.g., Zabreiko et al. (1968), Ch. 3, Sec. 4) one has

$$\sup \left\{ \int\limits_0^1 \int\limits_0^1 \mathcal{K}(t,s) \, f(t) \, f(s) \, dt \, ds \ : \ f \in \mathcal{Y}_1 \right\} = \mu_0,$$

where μ_0 is a principal eigenvalue of the integral operator with the kernel $\mathcal{K}(t,s)$. It follows from Watson (1961), where the spectrum of this operator has been found, that

$$\sigma^2 = \mu_0 = (4\,\pi^2)^{-1}\,.$$

This gives the conclusion of Theorem 2.5.3.   □

The arguments used for the proof of Theorem 2.5.3 may also be used for the analysis of large deviations of the statistic L_n^2.

Theorem 2.5.4 (Podkorytova 1993) *If the hypothesis H_0 is true then*

$$\lim_{n \to \infty} n^{-1} \ln P\big(L_n^2 \geq a\big) = -\frac{\pi^2}{8}\, a + o(a) \qquad \text{as } a \to 0. \qquad (2.5.11)$$

As a proof one has to take into account that in this case

$$Y_j(t,\omega) = \begin{cases} 1 + \ln\left(1 - X_j(\omega)\right) & \text{if } X_j(\omega) < t, \\ \ln\left(1 - t\right) & \text{if } X_j(\omega) \geq t, \end{cases}$$

$$\mathcal{K}(t,s) = \min\left(t,\,s\right), \qquad\qquad \sigma^2 = 4/\pi^2. \qquad\qquad \square$$

One should emphasize, however, that results (2.5.10)–(2.5.11) obtained on the basis of Sethuraman's theorem (Theorem 1.6.3) give substantially less information on the large-deviation probabilities than their full expansions obtained by our "variational" method in Sections 2.2–2.4. Specifically, such results are usually insufficient for proving the asymptotic log normality of P-values and calculating the corresponding asymptotic variances of test statistics in the sense of Lambert and Hall (1982). The "variational" method has also a clear advantage over the method based on the strong approximation (see Theorem 1.6.10) as the latter requires the existence of the large deviation limit that is difficult to prove. Another argument for the "variational" method is in the possibility of studying large-deviation probabilities under the alternative whereas the possibilities of other methods are at present not entirely clear.

2.6 Bahadur Efficiency of Goodness-of-Fit Tests

The results of Sections 2.2–2.5 concerning the large-deviation asymptotics of various goodness-of-fit statistics permit one to calculate the principal parts of their Bahadur exact slopes and consequently their local Bahadur efficiency.

In order to use results of Sections 1.1 and 1.2 and, particularly, Theorem 1.2.2, we denote by Θ the set of all continuous distribution functions $\theta(\cdot)$ on the real line with the uniform metric

$$\rho(\theta_1, \theta_2) = \sup_t |\theta_1(t) - \theta_2(t)|,$$

and by P_θ the distribution corresponding to $\theta(\cdot)$. We assume that the parametric set Θ_0 consists of a single distribution function F_0 and denote

$$\Theta_1 := \Theta \setminus \Theta_0.$$

The method of the computation of exact slopes (or their principal parts as $\rho(\theta, F_0) \to 0$) proposed by Theorem 1.2.2 is applicable to all the sequences of statistics that we have considered earlier in Section 2.1. It is only required to reduce them to the "standard" form, extracting

the root of the corresponding power and normalizing in such a way that
Theorem 1.2.2 becomes applicable. In the case of the statistic A_n^2 we
additionally assume that the alternative distribution function θ is such
that

$$\int\limits_{-\infty}^{\infty} \left(\theta - F_0 \right)^2 w(F_0)\, dF_0 \longrightarrow 0 \qquad \text{as } \rho(\theta, F_0) \to 0.$$

When considering the Khmaladze–Aki statistics K_n and L_n^2 we suppose
(see Aki (1986)) that the following two conditions are fulfilled.

(A) *There exist distribution function S_θ and real sequences $\{a_n\}$ and
$\{b_n\}$ such that $\sqrt{n}\, a_n \to \infty$, $b_n/(\sqrt{n}\, a_n) \to 0$, and*

$$R_\theta^n\big((z - b_n)/a_n\big) \longrightarrow S_\theta(z),$$

where

$$R_\theta(z) = \begin{cases} \theta\big(F_0^{-1}\big(1 - e^{-z}\big)\big) & \text{if } z \geq 0, \\[2mm] 0 & \text{if } z < 0; \end{cases}$$

(B)
$$\int\limits_{-\infty}^{\infty} \frac{1 - \theta(t)}{1 - F_0(t)}\, dF_0(t) < \infty.$$

Taking into account all these remarks we may easily find the expressions for the function $b(\theta)$ from Theorem 1.2.2 with the aid of the
Glivenko–Cantelli theorem (see Table 1).

As to the function $f(t)$ from Theorem 1.2.2, its continuity and representation in the form of a power series in some power of t convergent in
a neighborhood of zero or in the form of a principal part of this series
plus remainder are just the main results of Sections 2.2–2.5.

Knowing the functions $b(\theta)$ and the principal parts of functions $f(t)$
as $t \to 0$ we may calculate the principal parts of the exact slopes as
$\rho(\theta, F_0) \to 0$. These principal parts are further called the *local exact
slopes*. They enable us to find the *local Bahadur ARE*. The latter may
be defined in a natural way for any two sequences of statistics $\{T_n\}$ and
$\{V_n\}$ as the limit (see (1.2.5))

$$\Lambda_{V,T}^{B} := \lim_{\rho(\theta, F_0) \to 0} e_{V,T}^{B}(\theta) = \lim_{\rho(\theta, F_0) \to 0} \frac{c_V(\theta)}{c_T(\theta)}, \qquad (2.6.1)$$

assuming that this limit exists.

One may hope that in the most important case of close alternatives (it
corresponds to small $b(\theta)$ and t) the local ARE gives a sufficiently precise

Table 1. *The form of functions $b(\theta)$*

Statistic	Function $b(\theta)$
D_n	$\sup\limits_{-\infty<t<\infty} \lvert \theta(t) - F_0(t) \rvert$
D_n^+	$\sup\limits_{-\infty<t<\infty} [\theta(t) - F_0(t)]$
G_n	$\sup\limits_{-\infty<t<\infty} \left\lvert \theta(t) - F_0(t) - \int\limits_{-\infty}^{\infty} \big(\theta(s) - F_0(s)\big)\, dF_0(s) \right\rvert$
K_n	$\sup\limits_{-\infty<t<\infty} \left\lvert \theta(t) - F_0(t) - \int\limits_{-\infty}^{t} \dfrac{1 - \theta(s)}{1 - F_0(s)}\, dF_0(s) \right\rvert$
V_n	$\sup\limits_{-\infty<t<\infty} [\theta(t) - F_0(t)] - \inf\limits_{-\infty<t<\infty} [\theta(t) - F_0(t)]$
$\omega_{n,q}^{k}$	$\left[\int\limits_{-\infty}^{\infty} \big(\theta(t) - F_0(t)\big)^{k} q\big(F_0(t)\big)\, dF_0(t) \right]^{1/k}$
A_n^2	$\left[\int\limits_{-\infty}^{\infty} \big(\theta(t) - F_0(t)\big)^{2} w\big(F_0(t)\big)\, dF_0(t) \right]^{1/2}$
U_n^2	$\left[\int\limits_{-\infty}^{\infty} \left(\theta(t) - F_0(t) - \int\limits_{-\infty}^{\infty} \big(\theta(s) - F_0(s)\big)\, dF_0(s) \right)^{2} dF_0(t) \right]^{1/2}$
L_n^2	$\left[\int\limits_{-\infty}^{\infty} \left(\theta(t) - F_0(t) - \int\limits_{-\infty}^{t} \dfrac{1 - \theta(s)}{1 - F_0(s)}\, dF_0(s) \right)^{2} dF_0(t) \right]^{1/2}$

approximation to the exact value of the ARE. We give subsequently the local exact slopes of all statistics under consideration, assuming that they are consistent against the alternative θ (see Table 2).

Local exact slopes of the one-sided statistics $D_n^{\pm}$, $G_n^{\pm}$, and $K_n^{\pm}$ are omitted in Table 2 since the formulas for them are connected with obvious changes.

Before calculating the local Bahadur ARE we shall shortly discuss the question of calculating the approximate local Bahadur ARE based on the notion of the approximate slope or its principal part, the

Table 2. *Local exact slopes*

Statistic	Local exact slope
D_n	$4\left(\sup|\theta - F_0|\right)^2$
V_n	$4\left(\sup(\theta - F_0) - \inf(\theta - F_0)\right)^2$
G_n	$12\left(\sup\left|\theta - F_0 - \int\limits_{-\infty}^{\infty}(\theta - F_0)\,dF_0\right|\right)^2$
K_n	$\sup\left|\theta(t) - F_0(t) - \int\limits_{-\infty}^{t}\dfrac{1 - \theta(s)}{1 - F_0(s)}\,dF_0(s)\right|^2$
$\omega_{n,q}^k$	$-\lambda_0(q;k)\left[\int\limits_{-\infty}^{\infty}(\theta - F_0)^k\,q(F_0)\,dF_0\right]^2$
A_n^2	$2\left[\int\limits_{-\infty}^{\infty}(\theta - F_0)^2\left[F_0(1 - F_0)\right]^{-1}dF_0\right]$
U_n^2	$4\pi^2\left[\int\limits_{-\infty}^{\infty}(\theta - F_0)^2\,dF_0 - \left(\int\limits_{-\infty}^{\infty}(\theta - F_0)\,dF_0\right)^2\right]$
L_n^2	$\dfrac{\pi^2}{4}\left[\int\limits_{-\infty}^{\infty}\left(\theta(t) - F_0(t) - \int\limits_{-\infty}^{t}\dfrac{1 - \theta(s)}{1 - F_0(s)}\,dF_0(s)\right)^2 dF_0(s)\right]$

local approximate slope (see (1.2.17)). Note that we must find for it the asymptotics of the tail of the limiting distribution for the sequence of statistics under consideration instead of the large-deviation asymptotics. The latter problem is much simpler.

In this connection the following question is of great interest: When do the exact and approximate slopes locally coincide? This question has already been propounded by Bahadur (1960b) and studied by a number of authors, for example, Mikulski (1976) and Kallenberg and Ledwina (1987).

We shall show now that this coincidence actually occurs for all sequences of statistics considered earlier and represented in Tables 1 and 2.

Since the function $b(\theta)$ is the same for the exact and approximate slopes, we must show that the coefficient a from (1.2.18) coincides with the coefficient of $b(\theta)$ or some power of it from Table 2. For the statistics D_n, $D_n^{\pm}$, and V_n this follows from Abrahamson (1967) and for K_n, $K_n^{\pm}$, and L_n^2 from Aki (1986). The fact that the coefficient a for the Watson statistic U_n^2 equals $4\pi^2$ follows from the explicit formula for the limiting distribution in Watson (1961). For the statistics G_n and $G_n^{\pm}$ this coefficient equals 12, which results from the inequality for the distribution of the supremum of the limiting Gaussian process given by Fernique (1971) and Marcus and Shepp (1972) (see also Lemma 4.8.1). The limiting distribution of the statistic A_n^2 has a very complicated form (Anderson and Darling 1952, 1954). Therefore, one should use the asymptotic representation for the distribution of A_n^2 of the form $\sum_{j=1}^{\infty} \xi_j^2/(j(j+1))$, where $\{\xi_j\}$ is a sequence of standard independent Gaussian variables, as given by Durbin and Knott (1972). Applying the well-known result of Zolotarev (1961) we obtain that the coefficient a equals 2.

Now let us show that the indicated coincidence occurs also for the integral statistics $\omega_{n,q}^k$. Since $\omega_{n,q}^k$ may be considered as a continuous functional of the empirical process $\xi_n(t) = \sqrt{n}\,(F_n(t) - t)$ converging weakly in the Skorokhod space $D[0,1]$ to the Brownian bridge $\xi(t)$ (see, e.g., Billingsley (1968)), we have for any real x

$$\lim_{n\to\infty} P\Big([\omega_{n,q}^k]^{1/k} \geq x \Big) = P\Big(\Big[\int_0^1 \xi^k(t)\, q(t)\, dt \Big]^{1/k} \geq x \Big).$$

Denote now

$$a(q;k) := -\lim_{x\to\infty} \frac{2}{x^2}\, \ln P\Big(\Big[\int_0^1 \xi^k(t)\, q(t)\, dt \Big]^{1/k} \geq x \Big). \qquad (2.6.2)$$

It follows from Kallianpur and Oodaira (1978) that $a(q;k)$ is the value of the following extremal problem

$$a(q;k) := \inf\Big\{ \int_0^1 g'^2(t)\, dt\colon\ g \in A(q;k) \Big\},$$

where $A(q;k)$ is the set of absolutely continuous functions $g(t)$ on $[0, 1]$ such that $g(0) = g(1) = 0$ and $\int_0^1 g^k q\, dt \geq 1$.

This extremal problem may be solved in the same manner as the problems considered in Sections 2.3–2.4, but the calculations turn out to be much easier since the functional $\int_0^1 g'^2(t)\, dt$ belongs to the standard functionals of variational calculus in contrast to the "nonregular" functional $\int_0^1 g'\ln g'\, dt$. The Euler–Lagrange equation is now simply equation

(2.4.5), which must be taken simultaneously with the boundary conditions and the normalization condition. Therefore, $a(q; k) = -\lambda_0(q; k)$.

The constants $a(q; k)$ for $q \equiv 1$ may be found explicitly. Strassen (1964) has found the value of the extremal problem

$$\mathcal{C}(k) := \inf \left\{ \int_0^1 |g|^k \, dt : g \in \mathcal{B} \right\},$$

where $\mathcal{B}$ is the set of absolutely continuous functions $g(t)$ on $[0, 1]$ such that $g(0) = 0$ and $\int_0^1 g'^{\,2}(t) \, dt \le 1$. We cite only the final result omitting the calculations made by Strassen

$$\mathcal{C}(k) = 2 \, (k + 2)^{k/2 - 1} \, k^{-k/2} \left[\int_0^1 (1 - t^k)^{-1/2} \, dt \right]^{-k}.$$

The tables by Gradshteyn and Ryzhik (1971) (formula (3.249.7)) give

$$\int_0^1 (1 - t^k)^{-1/2} \, dt = k^{-1} \, B \left(\tfrac{1}{k}, \tfrac{1}{2} \right),$$

where $B(x, y) = \int_0^1 t^{x-1} (1 - t)^{y-1} \, dt$ is the Euler beta function.

Arguing similarly to Strassen (1964) one may obtain the value of $a(1; k)$ (it has been done by Kallianpur and Oodaira (1978)). Namely,

$$a(1; k) = 4 \, [\mathcal{C}(k)]^{-2/k} = 2^{\,2-2/k} \, B^2 \left(\tfrac{1}{k}, \tfrac{1}{2} \right) k^{-1} \, (k + 2)^{2/k - 1}. \qquad (2.6.3)$$

Particularly,

$$a(1; 1) = 12 \qquad\qquad \text{(cf. (2.4.37))},$$

$$a(1; 2) = \pi^2 \approx 9.87 \,,$$

$$a(1; 3) = \frac{2\pi}{3} \left(\frac{2}{5} \right)^{1/3} \left[\Gamma\!\left(\tfrac{1}{3}\right) \Big/ \Gamma\!\left(\tfrac{5}{6}\right) \right]^2 \approx 8.69 \,,$$

$$a(1; 4) = \frac{\pi \sqrt{3}}{6} \left[\Gamma\!\left(\tfrac{1}{4}\right) \Big/ \Gamma\!\left(\tfrac{3}{4}\right) \right]^2 \approx 7.94 \,.$$

Now we shall present some examples of the calculation of the local Bahadur efficiency (as we have just seen, the exact and approximate efficiencies do coincide locally for all statistics we are considering). As these examples have purely illustrative character, we shall confine ourselves to the location alternative and three standard distributions: Gaussian with the density $\varphi(x) = \frac{1}{\sqrt{2\pi}} \exp\left[-\tfrac{1}{2} x^2\right]$, logistic with the density $e^x \left[1 + e^x\right]^{-2}$, and Cauchy with the density $[\pi(1 + x^2)]^{-1}$. In the family of statistics $\omega_{n, q}^k$ we consider only the cases $q \equiv 1$ and $k = 1, 2, 3, 4$.

In the Gaussian case, that is, for

$$F_0(x) = \Phi(x) = \int\limits_{-\infty}^{x} \varphi(u)\, du\,, \qquad \theta(x) = \Phi(x + \theta)\,,$$

we obtain, for example, that

$$c_D(\theta) \sim 4 \sup_{x} \left| \Phi(x + \theta) - \Phi(x) \right|^2 \sim \frac{2}{\pi} \cdot \theta^2\,, \qquad \theta \to 0\,;$$

$$c_{\omega^2}(\theta) \sim \pi^2 \int\limits_{-\infty}^{\infty} \left[\Phi(x + \theta) - \Phi(x) \right]^2 d\Phi(x)$$

$$\sim \pi^2 \int\limits_{-\infty}^{\infty} \varphi^3(x)\, dx \cdot \theta^2 \sim \frac{\pi\sqrt{3}}{6} \cdot \theta^2\,, \qquad \theta \to 0\,.$$

In a completely similar way one computes the principal parts of exact slopes of the remaining statistics. Some technical difficulties connected with the Khmaladze–Aki statistics K_n and L_n^2 were overcome by Podkorytova (1990, 1993). Since all these principal parts have the form const $\cdot\, \theta^2$ as $\theta \to 0$, it is sufficient to compare the coefficients of θ^2, which are conveniently called the *local indices*.

Below, in Table 3, we have collected the local indices of the sequences of statistics considered in Chapter 2. In addition, the last row refers to the likelihood ratio statistic λ_n, which in the Bahadur theory is asymptotically optimal and has the largest exact slope and local index.

In a number of cases the local indices have been computed from tables by Gradshteyn and Ryzhik (1971) and in other cases numerical integration was used. As a rule we have restricted ourselves to an accuracy of 0.01 and we increase the accuracy only in order to distinguish the close values of indices.

Using the data of Table 3, we easily can compute the local Bahadur ARE of any pair of the considered statistics: It is sufficient to find the ratio of the corresponding local indices.

It is interesting that among nonparametric statistics not intended specifically for verifying the hypothesis on the distribution with the location alternative, there exist asymptotically optimal representatives. In the case of the logistic distribution this is $\omega_{n,1}^1$, G_n, and A_n^2, whereas in the case of the Cauchy distribution it is U_n^2.

For the Gaussian distribution no asymptotically optimal statistic was found among the statistics under consideration. It is highly probable that

Table 3. *Bahadur local indices for location alternatives*

Statistics	Gaussian	Logistic	Cauchy
D_n	0.64	0.25	0.407
G_n	0.955	0.333	0.304
G_n^+	0.16	0.08	0.304
K_n	0.81	0.25	0.326
U_n^2	0.49	0.22	0.500
A_n^2	0.96	0.333	0.33
L_n^2	0.94	0.329	0.299
$\omega_{n,1}^1$	0.955	0.333	0.304
$\omega_{n,1}^2$	0.91	0.329	0.38
$\omega_{n,1}^3$	0.87	0.32	0.41
$\omega_{n,1}^4$	0.56	0.30	0.42
λ_n	1.00	0.333	0.500

the statistic $\omega_{n,q}^2$ has this property under a special choice of the weight q, namely, for $\hat{q}(x) = \left\{ \varphi[\Phi^{-1}(x)] \right\}^{-2}$. This weight was considered for the first time, apparently, by de Wet and Venter (1973) who proved that $\lambda_0(\hat{q};2) = -1$ so the local index of the approximate slope of the de Wet and Venter statistic is indeed equal to 1. The author does not know whether it is true in the case of exact slopes. The fact is that the weight $\hat{q}(x)$ is nonsummable on $[0, 1]$ which prevents us from finding the large-deviation asymptotics.

Some results on local indices for scale alternatives have been given by Bahadur (1971) and by Groeneboom and Shorack (1981). It is curious that in one of the first places on any list of statistics appears the Kuiper statistic V_n, which is never worse and often four times more efficient than the Kolmogorov statistic whereas in the location case they are equivalent.

It should be noted that the Bahadur exact slopes turn out to be useful not only for calculating the ARE, but also for formulating the so-called Érdös–Rényi laws of large numbers. See the papers by S. Csörgö (1979a, 1979b).

In conclusion it is necessary to mention that the limiting (as $\alpha \to 0$) Pitman ARE (see Section 1.4) for any pair of statistics under consideration coincides with the local approximate and consequently with the local exact Bahadur ARE. For the statistics D_n, $D_n^\pm$, V_n, $\omega_{n,1}^2$, and U_n^2 it has been proved by Wieand (1976) by the direct verification of condition (1.4.3). Gregory (1980) proved it for the statistic A_n^2. Other statistics may be treated similarly (taking into account the supplementary regularity conditions for K_n and L_n^2).

An interesting problem is constructing optimal *combinations* of independent and dependent test statistics in the sense of the Pitman and Bahadur efficiencies. It has been discussed by a number of authors, for example, by Oosterhoff (1969); Littell and Folks (1971, 1973); Berk and Jones (1978); Jones (1978); Berk and Cohen (1979); Cohen, Marden, and Singh (1982); Pesarin and Zucchetto (1986); and Pallini (1990).

2.7 Hodges–Lehmann ARE of Goodness-of-Fit Tests

In this section we compare goodness-of-fit tests based on empirical d.f.s using the Hodges–Lehmann ARE, considered in Section 1.3. Our main result, published already in Nikitin (1986c, 1987a, 1987c) is the following: *The two-sided variants of the Kolmogorov and ω^2 statistics are asymptotically optimal (AO) in the Hodges–Lehmann sense.*

As we have noted in the introduction, this fact seems to be somewhat unexpected because the asymptotic optimality of these statistics in the Bahadur and Pitman senses may occur only locally and in exceptional cases (see Section 2.6 and Chapter 6).

On the other hand, Tusnády (1977, p. 391) when considering a particular example stated without proof that the majority of tests are optimal in the sense of the Hodges–Lehmann efficiency, so the results of this section may serve as a justification and extension of this presumption.

Using the results on large deviations given in Section 1.7 we prove at once the Hodges–Lehmann optimality for the multisample goodness-of-fit tests.

Let $X_{11}, \dots, X_{1,n_1}; \dots; X_{c,1}, \dots, X_{c,n_c}$ be $c \geq 1$ independent samples of independent observations taking on values in $\mathbf{R}^1$ and having distributions $P_1, \dots, P_c$. It is assumed that sample sizes n_i satisfy the relations

$$\lim_{N \to \infty} \frac{n_i}{N} = \rho_i > 0, \qquad (2.7.1)$$

where $N = n_1 + \cdots + n_c$. We are testing the hypothesis H_0^c: $P_1 = \cdots = P_c = P_0$ for some given distribution P_0 with the density p_0. As an alternative we consider the hypothesis A_0^c: $P_1 = R_1, \ldots, P_c = R_c$, where $\mathbf{R} = (R_1, \ldots, R_c)$ is a given element of Λ^c and its components R_i are absolutely continuous with the corresponding densities r_i.

Suppose that the test for this problem is based on some sequence of statistics $\{T_N\}$, their large values being significant. Let $\beta_N(\alpha, \mathbf{R})$ be the power of this test, corresponding to the significance level $\alpha \in (0, 1)$. According to (1.3.1) we call the *Hodges–Lehmann index* of a sequence $\{T_N\}$ a real function $d_T(\mathbf{R})$ of the alternative $\mathbf{R}$ such that

$$\lim_{N \to \infty} N^{-1} \ln \left[1 - \beta_N(\alpha, \mathbf{R}) \right] = -\tfrac{1}{2} d_T(\mathbf{R}), \qquad 0 < d_T(\mathbf{R}) < \infty .$$
$$(2.7.2)$$

We start out by finding an upper bound for the index of multisample goodness-of-fit statistics.

Theorem 2.7.1 (Nikitin 1986c, 1987a) *The following inequality holds:*

$$\varliminf_{N \to \infty} N^{-1} \ln \left[1 - \beta_N(\alpha, \mathbf{R}) \right] \geq - \sum_{i=1}^{c} \rho_i K(P_0, R_i) \qquad (2.7.3)$$

(the Kullback–Leibler information $K(P_0, R_i)$ has been defined in (1.2.9)).

Proof We follow the arguments used by Rao (1962) in the proof of his Lemma 4.1 with necessary changes. Without loss of generality one may assume that $K(P_0, R_i) < \infty$ for any integer i, $1 \leq i \leq c$. Denote by Ξ_N the acceptance region of the hypothesis H_0^c in the sample space $\mathbf{R}^N$ and by λ_N the Lebesgue measure on it. For the reasons of brevity let $\mathbf{x}^{(n_i)}$, $i = 1, \ldots, c$, denote the group of variables $(x_{i1}, \ldots, x_{i, n_i})$ corresponding to the ith sample.

Using the Jensen inequality we obtain

$$N^{-1} \ln \frac{1 - \beta_N(\alpha, \mathbf{R})}{1 - \alpha} = N^{-1} \ln \int_{\Xi_N} \prod_{i=1}^{c} r_i(\mathbf{x}^{(n_i)}) \frac{d\lambda_N}{1 - \alpha}$$

$$\geq N^{-1} \int_{\Xi_N} \ln \prod_{i=1}^{c} \frac{r_i(\mathbf{x}^{(n_i)})}{p_0(\mathbf{x}^{(n_i)})} \prod_{i=1}^{c} p_0(\mathbf{x}^{(n_i)}) \frac{d\lambda_N}{1 - \alpha}$$

$$= - \sum_{i=1}^{c} \rho_i K(P_0, R_i) + \gamma_N,$$

where

$$\gamma_N := \int_{\Xi_N} \left[\sum_{i=1}^{c} \rho_i \, K(P_0, R_i) - N^{-1} \sum_{i=1}^{c} \ln \frac{p_0\big(\mathbf{x}^{(n_i)}\big)}{r_i\big(\mathbf{x}^{(n_i)}\big)} \right]$$

$$\times \prod_{i=1}^{c} p_0\left(\mathbf{x}^{(n_i)}\right) \frac{d\lambda_N}{1-\alpha} \, .$$

It remains to pass to the limit as $N \to \infty$ taking in mind that by the ergodic theorem in the mean $\gamma_N \to 0$ (see, e.g., Shiryaev (1984), Chapter V).   □

Now we put in agreement with the notation of Section 1.7

$$J_\rho(P_0, \mathbf{R}) := \sum_{i=1}^{c} \rho_i \, K(P_0, R_i) \, . \tag{2.7.4}$$

If the index $d_T(\mathbf{R})$ of the sequence of statistics $\{T_N\}$ exists, estimate (2.7.3) implies

$$d_T(\mathbf{R}) \le 2 J_\rho(P_0, \mathbf{R}) \, , \tag{2.7.5}$$

which is the c-sample analog of (1.3.5). In the cases when the equality occurs in (2.7.5) for any set of alternative distributions $\mathbf{R}$, the sequence $\{T_N\}$ is said to be *Hodges–Lehmann optimal*.

Consider now some particular sequences of statistics based on empirical d.f.s that may be used for verifying H_0^c against A_0^c. In order to use the notation of Section 1.7 put for any $x \in \mathbf{R}^1$ and $j = 1, 2, \ldots, c$

$$F_{j, \, n_j}(x) = P_{j, \, n_j}\big\{ (-\infty, x) \big\} \, ,$$

$$F_j(x) = R_j\big\{ (-\infty, x) \big\} \, ,$$

$$F_0(x) = P_0\big\{ (-\infty, x) \big\} \, .$$

We deal with the following statistics ($\kappa \geq 1$):

$$D_N^\kappa = \sup_{-\infty < x < \infty} \sum_{j=1}^{c} \rho_j^{\kappa/2} \left| F_{j,\,n_j}(x) - F_0(x) \right|^\kappa,$$

$$G_N^\kappa = \sup_{-\infty < x < \infty} \sum_{j=1}^{c} \rho_j^{\kappa/2} \left| F_{j,\,n_j}(x) - F_0(x) \right.$$

$$\left. - \int_{-\infty}^{\infty} \left[F_{j,\,n_j}(y) - F_0(y) \right] dF_0(y) \right|^\kappa,$$

$$W_{N,\,q}^\kappa = \sum_{j=1}^{c} \rho_j^{\kappa/2} \int_{-\infty}^{\infty} \left| F_{j,\,n_j}(x) - F_0(x) \right|^\kappa q_j(F_0(x))\, dF_0(x),$$

$$U_N^\kappa = \sum_{j=1}^{c} \rho_j^{\kappa/2} \int_{-\infty}^{\infty} \left| F_{j,\,n_j}(x) - F_0(x) \right.$$

$$\left. - \int_{-\infty}^{\infty} \left[F_{j,\,n_j}(y) - F_0(y) \right] dF_0(y) \right|^\kappa dF_0(x),$$

where the weight functions q_j, $j = 1, 2, \ldots, c$, in the definition of $W_{N,\,q}^\kappa$ are assumed to be positive and summable on $(0, 1)$.

The particular cases of these statistics are well known and have been extensively studied. For instance, the statistic D_N^κ for $\kappa = 1$ and $c = 1$ is the classic Kolmogorov statistic (2.1.1), its c-sample variant for $c > 1$ and $\kappa = 2$ has been investigated by Kiefer (1959). The statistic $W_{N,\,q}^\kappa$ for $\kappa = 2$, $c = 1$, and $q \equiv 1$ coincides with the classic ω^2 statistic of Cramér, von Mises, and Smirnov (2.1.8). Its c-sample variant has been considered also by Kiefer (1959). The statistic G_N^κ for $\kappa = 1$ and $c = 1$ becomes the Watson–Darling statistic (2.1.5). As for the statistic U_N^κ, it has been introduced for $\kappa = 2$ and $c = 1$ by Watson (1961) (see (2.1.11)) and for $c > 1$ by Maag (1966).

Theorem 2.7.2 (Nikitin 1986c, 1987a) *The sequences of statistics* $\{D_N^\kappa\}$, $\{G_N^\kappa\}$, $\{W_{N,\,q}^\kappa\}$, *and* $\{U_N^\kappa\}$ *are Hodges–Lehmann optimal.*

Proof All statistics under consideration are τ-continuous functionals of empirical d.f.s $F_{j,\,n_j}$ and consequently of empirical distributions $\mathcal{P}_{j,\,n_j}$. We may use Theorem 1.7.2.

Suppose that the particular sequence of statistics $\{T_N\}$ is representable in the form $T_N := T(P_N)$ where $P_N := \left(\mathcal{P}_{1,n_1}, \ldots, \mathcal{P}_{c,n_c} \right)$, and denote for any $\delta \geq 0$

$$\Delta_\delta^T := \left\{ Q \in \Lambda^c \colon T(Q) \leq \delta \right\}.$$

By Lemma 1.7.6 for any $\varepsilon > 0$ there exists $\delta > 0$ such that

$$- J_\rho \left(\Delta_\delta^T, \mathbf{R} \right) \leq -J_\rho \left(\Delta_0^T, \mathbf{R} \right) + \varepsilon. \tag{2.7.6}$$

The probability of error of the second kind for any sequence $\{T_N\}$ has the form

$$1 - \beta_N(\alpha, \mathbf{R}) = P\left(T_N \leq \sigma_{N,\alpha} \right),$$

where $\sigma_{N,\alpha}$ is the critical value. Considering that $\sigma_{N,\alpha} \to 0$ as $N \to \infty$ for all statistics listed in the statement of Theorem 2.7.2 and using Theorem 1.7.2, we obtain

$$\varlimsup_{N \to \infty} N^{-1} \ln \left(1 - \beta_N(\alpha, \mathbf{R}) \right) \leq -J_\rho \left(\Delta_\delta^T, \mathbf{R} \right).$$

Due to estimate (2.7.6), we have after passing to the limit as $\varepsilon \to 0$

$$\varlimsup_{N \to \infty} N^{-1} \ln \left(1 - \beta_N(\alpha, \mathbf{R}) \right) \leq -J_\rho \left(\Delta_0^T, \mathbf{R} \right). \tag{2.7.7}$$

Note now that a feature of all the statistics we study is that the set Δ_0^T is reduced to the unique element $(P_0, \ldots, P_0) \in \Lambda^c$. It remains to compare the two inequalities (2.7.7) and (2.7.3) to establish the conclusion of the theorem.   □

An analysis of the proof of Theorem 2.7.2 shows that we have not used the concrete form of the statistics under consideration except for their τ-continuity as functionals of empirical distributions and the particular structure of the set Δ_0^T. Another important factor is due to the shrinking of the acceptance regions to the null set in a "coarse" way. All this enables one to formulate a slight generalization of Theorem 2.7.2. See Kallenberg and Kourouklis (1992), who used this type of argument to show the Hodges–Lehmann optimality of the chi-square test as well as the likelihood ratio test in the case of regular exponential families.

One should also note that the Hodges–Lehmann optimality of the Kolmogorov statistic (2.1.1) follows from Theorem 13 of Sanov (1957) and from equality (12) of Borovkov (1984, Ch. 3, Sec. 3), though these authors do not mention the notion of the Hodges–Lehmann efficiency.

Hodges–Lehmann optimality does not hold for the statistics $\omega_{n,q}^k$ introduced by (2.1.9) or their c-sample variants in the case of odd k. We will show it in the simplest case $c = 1$, $k = 1$ by calculating the corresponding Hodges–Lehmann local index explicitly.

Let $X_1, X_2, \ldots, X_n$ be a sample with continuous d.f. $G(x; \theta)$, $\theta \geq 0$, such that $G(x; \theta) \neq G(x; 0)$ for $\theta \neq 0$. Suppose we are testing H_0: $\theta = 0$ against the alternative A_0: $\theta > 0$ and have the sequence of statistics

$$\omega_{n,q}^1 = \int\limits_{-\infty}^{\infty} \left[F_n(x) - G(x; 0) \right] q\big(G(x; 0)\big) \, dG(x; 0),$$

assuming for $\theta > 0$

$$\int\limits_{-\infty}^{\infty} \left[G(x; \theta) - G(x; 0) \right] q\big(G(x; 0)\big) \, dG(x; 0) > 0 \tag{2.7.8}$$

(this condition guarantees the consistency of the test based on $\omega_{n,q}^1$).

As for the regularity properties of the family $\{G(x; \theta)\}$ we suppose that there exists the derivative $G'_\theta(x; \theta)$ in θ and that for almost all x and θ

$$\left| G'_\theta(x; \theta) \right| \leq L, \tag{2.7.9}$$

L being a positive constant. Another requirement is that all d.f.s $G(x; \theta)$ under consideration are absolutely continuous in x for almost all θ and that the corresponding densities $g(x; \theta)$ for almost all x and θ satisfy the inequality

$$g(x; \theta) > 0. \tag{2.7.10}$$

Under these conditions there exist strictly monotone inverse d.f.s $G^{-1}(x; \theta)$ and we denote for brevity

$$H(x; \theta) := G\big(G^{-1}(x; \theta); 0 \big), \tag{2.7.11}$$

$$\Psi(x) := G'_\theta\big(G^{-1}(x; \theta); \theta \big) \Big|_{\theta=0}. \tag{2.7.12}$$

Theorem 2.7.3 *Let a weight function q be summable and positive on $(0, 1)$ and the family of distribution functions $\{G(x; \theta)\}$ satisfy conditions (2.7.8)–(2.7.10). Then*

$$d_{\omega^1}(\theta) \sim \frac{\left(\int\limits_0^1 \psi(u) q(u) \, du \right)^2 \cdot \theta^2}{\int\limits_0^1 \int\limits_0^1 \left[\min(t, s) - t\,s \right] q(t) q(s) \, dt \, ds} \qquad \text{as } \theta \to 0. \tag{2.7.13}$$

Proof Introduce

$$Q(t) = \int\limits_0^t q(s)\,ds.$$

The condition of Theorem 2.7.3 ensures that Q is a strictly monotone function and that there exists the inverse function Q^{-1} with the same properties.

Let us represent the statistics $\omega_{n,\,q}^1$ under the alternative A_0 as

$$
\begin{aligned}
\omega_{n,\,q}^1 &= \int\limits_0^1 \Big[F_n(t) - H(t;\theta) \Big]\, dQ\big(H(t;\theta) \big) \\
&= n^{-1} \sum_{i=1}^n \int\limits_{X_i'}^1 dQ\big(H(t;\theta) \big) - \int\limits_0^1 s\,dQ(s) \\
&= n^{-1} \sum_{i=1}^n \left[\int\limits_0^1 Q(s)\,ds - Q\big(H(X_i';\theta)\big) \right] \equiv n^{-1} \sum_{i=1}^n (-Z_i),
\end{aligned}
$$

where $X_1', X_2', \ldots, X_n'$ is a sample from the uniform distribution on $[0, 1]$ and $F_n(t)$ is the corresponding empirical d.f. The random variables Z_i, $i = 1, \ldots, n$, are independent and identically distributed and, moreover,

$$
\begin{aligned}
P\big(Z_i > 0\big) &= P\left(Q(H(X_i';\theta)) > \int\limits_0^1 Q(s)\,ds \right) \\
&= P\left(H(X_i';\theta) > Q^{-1}\left(\int\limits_0^1 Q(s)\,ds \right) \right) \\
&= 1 - G\left(G^{-1}\left(Q^{-1}\left(\int\limits_0^1 Q(s)\,ds \right);0 \right);\theta \right) > 0,
\end{aligned}
$$

as

$$\int\limits_0^1 Q(s)\,ds < Q(1)$$

and all the functions in superposition are strictly monotone.

Therefore, when calculating the Hodges–Lehmann index of the sequence of statistics $\{\omega_{n,\,q}^1\}$ we can apply the Chernoff theorem (Theorem

1.6.2). Denoting by $\sigma_{n,\alpha}$ the critical value with an $0 < \alpha < 1$ one obtains under the alternative A_0

$$d_{\omega^1}(\theta) = \lim_{n \to \infty} n^{-1} \ln P\left(\omega_{n,q}^1 \leq \sigma_{n,\alpha}\right)$$

$$= \lim_{n \to \infty} n^{-1} \ln P\left(n^{-1} \sum_{i=1}^{n} Z_i \geq -\sigma_{n,\alpha}\right)$$

$$= \ln \inf_{t>0} E \exp\{t Z_i\}. \tag{2.7.14}$$

It remains to calculate $\inf_{t>0} E \exp\{t Z_i\}$. This infimum is attained at the point t^*, a root of the equation

$$\varphi(t,\theta) \equiv E\left[Q(H(X_i';\theta)) - \bar{Q}\right] \exp\left[t Q(H(X_i';\theta))\right] = 0, \tag{2.7.15}$$

where we denote for brevity $\bar{Q} = \int_0^1 Q(s)\, ds$. It is clear that $\varphi(0,0) = 0$ and

$$\varphi_t'(0,0) = \int_0^1 Q^2(s)\, ds - \bar{Q}^2 > 0.$$

Now we apply the implicit function theorem to equation (2.7.15). Since the function φ is differentiable in t and θ, the implicit function $t^*(\theta)$ is also differentiable in θ and simple calculations show that

$$t^*(\theta) = \frac{\left(\int_0^1 \psi(t)\, q(t)\, dt\right) \cdot \theta}{\int_0^1 Q^2(s)\, ds - \left(\int_0^1 Q(s)\, ds\right)^2} + o(\theta) \qquad \text{as } \theta \to 0.$$

Substituting this value in (2.7.14) we establish the conclusion of the theorem. The identity

$$\int_0^1 \int_0^1 \left[\min(t,s) - t s\right] q(t)\, q(s)\, dt\, ds = \int_0^1 Q^2(s)\, ds - \left(\int_0^1 Q(s)\, ds\right)^2$$

may be proved using integration by parts. □

It follows from (2.7.13) that even locally the Hodges–Lehmann index is not equal to the principal part of its upper bound given by the right-hand side of (2.7.5) (for $c = 1$). The equality may occur only in specific cases that will be discussed in Chapter 6.

2.8 Hodges–Lehmann and Chernoff Efficiencies of the Smirnov Statistic

We have shown in the previous section that the Kolmogorov statistic (2.1.1) possesses the property of Hodges–Lehmann optimality. Now we consider the one-sided counterparts of this statistic, the Smirnov statistics (2.1.2)–(2.1.3), and show that under appropriate conditions they are also Hodges–Lehmann optimal. In the general case this is not, however, true.

As in Section 2.7, let $X_1, X_2, \ldots, X_n$ be a sample with continuous d.f. $G(x; \theta)$, $\theta \geq 0$, such that $G(x; \theta) \neq G(x; 0)$ for $\theta \neq 0$. The testing problem is again H_0: $\theta = 0$ and the alternative A_0: $\theta > 0$. As usual we denote by F_n the empirical d.f. based on $X_1, X_2, \ldots, X_n$ and consider the one-sided Smirnov statistics (cf. (2.1.2), (2.1.3))

$$D_n^+ = \sup_{-\infty < x < \infty} \left[F_n(x) - G(x; 0) \right], \quad D_n^- = \sup_{-\infty < x < \infty} \left[G(x; 0) - F_n(x) \right].$$

We study only the first of these statistics, the second may be considered quite analogously.

Let us introduce some restrictions on the family $\{G(x; \theta)\}$. First we assume that for all $\theta > 0$

$$\sup_x \left[G(x; \theta) - G(x; 0) \right] > 0 \qquad (2.8.1)$$

(this ensures the consistency of the test based on $\{D_n^+\}$). Moreover, we suppose that for all x and θ the function $G(x; \theta)$ is continuously differentiable in x, the corresponding density $g(x; \theta)$ being strictly positive. Then there exists the differentiable inverse d.f. $G^{-1}(x; \theta)$ such that for all θ

$$\frac{\partial}{\partial x} G^{-1}(x; \theta) = \left[g\big(G^{-1}(x; \theta); \theta \big) \right]^{-1}, \qquad 0 \leq x \leq 1.$$

These conditions guarantee that $H(x; \theta)$ introduced earlier in (2.7.11) is continuously differentiable in x for all θ.

Theorem 2.8.1 (Nikitin 1985) *Suppose that the previous conditions imposed on the family $\{G(x; \theta)\}$ are fulfilled and, in addition, the distribution function $H(x; \theta)$ is convex in x on $[0, 1]$ for all θ. Then the sequence $\{D_n^+\}$ is Hodges–Lehmann optimal.*

Proof Let θ be fixed. Denote by $\beta_n^+(\theta)$ the power of the test based on $\{D_n^+\}$. For arbitrary $\varepsilon > 0$ and sufficiently large n we have

$$1 - \beta_n^+(\theta) \leq P\big(\bar{F}_n(t) \leq H(t; \theta) + \varepsilon, \ 0 \leq t \leq 1 \big), \qquad (2.8.2)$$

where the empirical d.f. $\bar{F}_n$ is already based on a sample from the uniform distribution on $[0, 1]$. It is well known (see, e.g., Borovkov (1984), Ch. 1), that $\bar{F}_n(t)$ has the same distribution as $n^{-1}\eta(n\,t)$ under the condition $\eta(n) = n$, where $\eta(t)$ is the standard Poisson process with parameter 1.

Denote by $\boldsymbol{\vartheta}_\varepsilon(t)$ the shortest path joining the points $(0, 0)$ and $(1, 1)$ and not touching the set

$$\mathbf{V}_\varepsilon := \big\{(t, y)\colon 0 \le t \le 1,\ y(t) > H(t; \theta) + \varepsilon\big\}.$$

The path $\boldsymbol{\vartheta}_\varepsilon(t)$ may also be viewed (see Borovkov and Sycheva (1970)) as the infimum of the convex hull of the set $\mathbf{V}_\varepsilon$ with the adjoined points $(0, 0)$ and $(1, 1)$.

From the results of Borovkov (1967) and Borovkov and Sycheva (1970) concerning the logarithmic asymptotics of large-deviation probabilities for random walks in "the second boundary value problem with fixed end" and for Poisson processes it follows that

$$\lim_{n\to\infty} n^{-1} \ln P\big(\bar{F}_n(t) \le H(t;\theta)+\varepsilon,\ 0 \le t \le 1\big) = -\int_0^1 \boldsymbol{\vartheta}'_\varepsilon(t)\, \ln \boldsymbol{\vartheta}'_\varepsilon(t)\, dt.$$

$$(2.8.3)$$

From the convexity of $H(t; \theta)$ it follows that the graph of the function $\boldsymbol{\vartheta}_\varepsilon(t)$ consists of two segments tangent to $H(t; \theta) + \varepsilon$ drawn from the points $(0, 0)$ and $(1, 1)$, respectively, and of the part of the graph of the function $H(t; \theta) + \varepsilon$ between the tangency points. Since

$$\boldsymbol{\vartheta}'_\varepsilon(t) \longrightarrow H'(t; \theta) \qquad \text{as } \varepsilon \to 0$$

for each t and the derivative $\boldsymbol{\vartheta}'_\varepsilon(t)$ is uniformly bounded with respect to t and ε, we obtain

$$\lim_{\varepsilon\to0} \int_0^1 \boldsymbol{\vartheta}'_\varepsilon(t)\, \ln \boldsymbol{\vartheta}'_\varepsilon(t)\, dt = \int_0^1 H'(t; \theta)\, \ln H'(t; \theta)\, dt. \qquad (2.8.4)$$

But it is true that

$$\int_0^1 H'(t; \theta)\, \ln H'(t; \theta)\, dt = K(0, \theta), \qquad (2.8.5)$$

where $K(0, \theta)$ is the Kullback–Leibler information (1.2.9) corresponding to d.f.s $G(x; 0)$ and $G(x; \theta)$. From (2.8.2)–(2.8.5) one obtains

$$\varlimsup_{n\to\infty} n^{-1} \ln \big[1 - \beta_n^+(\theta)\big] \le -K(0, \theta).$$

In the particular case $c = 1$ Theorem 2.7.1 implies the opposite inequality

$$\lim_{n \to \infty} n^{-1} \ln\left[1 - \beta_n^+(\theta)\right] \geq -K(0, \theta). \qquad \square$$

The assumption about the convexity of the function $H(t; \theta)$ is essential since, otherwise, ϑ_ε will converge for $\varepsilon \to 0$ not to H but to the largest convex minorant of H, that is, to its second Young–Fenchel conjugate, which coincides with H only under the condition of convexity (see, e.g., Alexeev et al. (1979), Sec. 2.6.3).

In order to understand how restrictive the condition of convexity is we assume that the function $H(t; \theta)$ is twice differentiable in t. Computations show that

$$H''(t; \theta) = \frac{g(x; 0)}{g^2(x; \theta)} \left[\frac{g'(x; 0)}{g(x; 0)} - \frac{g'(x; \theta)}{g(x; \theta)} \right] \Bigg|_{x = G^{-1}(t; \theta)}.$$

If now we assume that $G(x; \theta)$ is a location family, that is, $G(x; \theta) = G(x + \theta)$, $\theta \geq 0$, then nonnegativity of H'' is equivalent to convexity of H and at the same time to nondecrease of the function $-(\ln g)' = -g'/g$. The densities g having this property are *strongly unimodal* in the Ibragimov (1956) sense. (See also Hájek and Šidák (1967), Ch. I, Sec. 2.2.) In particular, the Gaussian, Laplace, logistic, and many other standard densities are of such a kind.

The statistic D_n^- may be considered similarly; it turns out to be Hodges–Lehmann optimal under the condition of the concavity of the function H.

The Chernoff efficiency of the statistics considered in Chapter 2 is almost uninvestigated. One may suppose that the Chernoff indices of the Kolmogorov–Smirnov and integral statistics have complicated expressions depending on the structure of the function $H(x; \theta)$.

In a subsequent extension to Theorem 2.8.1 we prove the Chernoff local optimality of the Smirnov statistics for a particular parametric family: the family of the Laplace distributions with a location parameter. As we have seen in this section, the Smirnov statistics are in this case Hodges–Lehmann optimal. Simple arguments based on results of Section 2.6 show that they are also Bahadur locally optimal (and in Chapter 6 it will be proved that it may hold only in the Laplace case). Therefore, the fact of the Chernoff local optimality appears quite natural.

Direct calculations show that for the Laplace d.f. $G(x;\theta) = G_1(x+\theta)$ with a location parameter θ, where

$$G_1(x) := \begin{cases} \frac{1}{2}\,e^x & \text{for } x \le 0, \\[2mm] 1 - \frac{1}{2}\,e^{-x} & \text{for } x > 0, \end{cases}$$

the function $H(x;\theta)$ has the form

$$H(x;\theta) := \begin{cases} x\,e^{-\theta} & \text{for } 0 \le x \le \frac{1}{2}, \\[2mm] \frac{1}{4}\,(1-x)^{-1}\,e^{-\theta} & \text{for } \frac{1}{2} \le x \le 1 - \frac{1}{2}\,e^{-\theta}, \\[2mm] 1 - (1-x)\,e^{\theta} & \text{for } 1 - \frac{1}{2}\,e^{-\theta} \le x \le 1. \end{cases} \qquad (2.8.6)$$

As function (2.8.6) is convex in x and piecewise linear (except an interval of small length) we can find explicitly the expression for the path ϑ_ε. One may assume that ε is sufficiently small, say $\varepsilon < \theta/2$. Denote for brevity

$$\xi = (1+4\,\varepsilon\,e^{\theta})^{1/2}, \qquad \tau_1 = 1-(1+\xi)^{-1}, \qquad \tau_2 = 1 - \frac{1}{2}\,(1-\varepsilon)^{-1}e^{-\theta}.$$

In this notation we have

$$\vartheta_\varepsilon(t) = \begin{cases} \varepsilon\,(\xi+1)\,(\xi-1)^{-1}\,t & \text{for } 0 \le t \le \tau_1, \\[2mm] \frac{1}{4}\,(1-t)^{-1}\,e^{-\theta}+\varepsilon & \text{for } \tau_1 \le t \le \tau_2, \\[2mm] (1-\varepsilon)^2\,(t-1)\,e^{\theta}+1 & \text{for } \tau_2 \le t \le 1. \end{cases}$$

It is possible now to compute $\int_0^1 \vartheta_\varepsilon'(t)\,\ln\vartheta_\varepsilon'(t)\,dt$ explicitly. Note first of all that

$$\int_{\tau_1}^{\tau_2} \frac{e^{-\theta}}{4\,(1-t)^2}\,\ln\frac{e^{-\theta}}{4\,(1-t)^2}\,dt = O(\theta^3), \qquad \theta \to 0.$$

The integrals taken over the remaining part of the interval $[0, 1]$ are easy to compute and a minor manipulation yields

$$\int_0^1 \vartheta_\varepsilon'(t)\,\ln\vartheta_\varepsilon'(t)\,dt = \frac{1}{2}\,(\theta-2\,\varepsilon)^2 + o(\theta^2+\varepsilon^2) \qquad \text{as } \theta^2+\varepsilon^2 \to 0.$$

Substituting this in (2.8.3) we get the large-deviation asymptotics of $\{D_n^+\}$ under the alternative. Comparing it with Theorem 2.2.1 where such asymptotics has been found under H_0 and using Theorem 1.5.1 we obtain for the Chernoff index $\rho_{D^+}(\theta)$ of the sequence $\{D_n^+\}$ that

$$\rho_{D^+}(\theta) = \frac{1}{8}\,\theta^2\,(1+o(1)) \qquad \text{as } \theta \to 0. \qquad (2.8.7)$$

The upper bound for the Chernoff indices given by inequality (1.5.4) is in our case

$$\rho_{D^+}(\theta) \leq -\inf\{\psi(t)\colon 0 \leq t \leq 1\}, \tag{2.8.8}$$

where

$$\psi(t) := \ln \int_{-\infty}^{\infty} \exp\left\{-t\left(|x+\theta| - |x|\right)\right\} \tfrac{1}{2} e^{-|x|}\, dx.$$

Elementary calculations, taking into account only terms up to order θ^2, show that the infimum of $\psi(t)$ is attained at the point $t = t^* = \tfrac{1}{2} + o(1)$, since we have

$$\int_{-\infty}^{\infty} \left(|x+\theta| - |x|\right) e^{-|x|}\, dx = \theta^2 \left(1 + o(1)\right),$$

$$\int_{-\infty}^{\infty} \left(|x+\theta| - |x|\right)^2 e^{-|x|}\, dx = 2\,\theta^2 \left(1 + o(1)\right), \qquad \theta \to 0.$$

Expanding the exponent in a series one gets after simple manipulations that

$$\psi(t^*) = \inf\{\psi(t)\colon 0 \leq t \leq 1\} = -\tfrac{1}{8}\theta^2 \left(1 + o(1)\right), \qquad \theta \to 0.$$

Therefore (2.8.8) yields

$$\rho_{D^+}(\theta) \leq \tfrac{1}{8}\theta^2(1 + o(1)) \qquad \text{as } \theta \to 0.$$

Comparing this inequality with (2.8.7) we see that the sequence $\{D_n^+\}$ is actually Chernoff locally optimal. If one has under the alternative $G(x;\theta) = G(x-\theta)$, $\theta \geq 0$, then this property passes to the sequence $\{D_n^-\}$.

In the case of families with a scale parameter on the half-line $G(x;\theta) = G(x\,e^\theta)$, $x \geq 0$, the part of the Laplace distribution functions G_1 is played by the distribution function

$$G_2(x) = \begin{cases} \tfrac{1}{2}x & \text{for } 0 \leq x \leq 1, \\ 1 - (2\,x)^{-1} & \text{for } x > 1, \end{cases}$$

which is related to G_1 by the change of variable

$$G_2(x) = G_1(\ln x), \qquad x > 0.$$

3

Asymptotic Efficiency of Nonparametric Homogeneity Tests

3.1 Problem Statement and Tests under Consideration

Let $X_1, X_2, \ldots, X_m$ and $Y_1, Y_2, \ldots, Y_n$ be two independent samples with continuous distribution functions F_1 and F_2. The problem is to test the hypothesis about the equality of these functions

$$H_1: F_1 \equiv F_2. \qquad (3.1.1)$$

The hypothesis H_1 is often called the *hypothesis of homogeneity*. Assume that if the alternative hypothesis A_1 is true then $F_1(t) = R_1(t)$ and $F_2(t) = R_2(t)$, where R_1 and R_2 are some continuous distribution functions on the real line and $R_1 \not\equiv R_2$. Kendall and Stuart (1967, Chapter 31) emphasized that just such simple parametric hypotheses are most natural and reasonable alternatives to nonparametric ones.

Denote by F_m and G_n the empirical distribution functions based on initial samples. For the problem of testing H_1 against A_1 there exist many statistics based on the difference between F_m and G_n analogous to goodness-of-fit statistics considered in Section 2.1.

Denote $N = m + n$ and suppose that the sample sizes m and n tend to infinity in such a way that

$$\lim_{m,\,n\to\infty} \frac{m}{n+m} = \rho_1, \qquad \lim_{m,\,n\to\infty} \frac{n}{n+m} = \rho_2, \qquad (3.1.2)$$

where ρ_1 and ρ_2 are some positive constants.

In our case the following statistics

$$D_{m,n} = \frac{\sqrt{mn}}{m+n} \sup_{-\infty < t < \infty} \left| F_m(t) - G_n(t) \right|, \qquad (3.1.3)$$

$$D_{m,\,n}^+ = \frac{\sqrt{mn}}{m+n} \sup_{-\infty < t < \infty} \left[F_m(t) - G_n(t) \right], \qquad (3.1.4)$$

94

$$D^-_{m,n} = \frac{\sqrt{mn}}{m+n} \sup_{-\infty < t < \infty} \left[G_n(t) - F_m(t) \right], \qquad (3.1.5)$$

proposed by Smirnov (1939), are analogs of the Kolmogorov–Smirnov statistics (2.1.1)–(2.1.3) for goodness-of-fit testing. An analog of the Kuiper statistic (2.1.4) is the statistic

$$V_{m,n} = \frac{\sqrt{mn}}{m+n} \left\{ \sup_{-\infty < t < \infty} \left[F_m(t) - G_n(t) \right] - \inf_{-\infty < t < \infty} \left[F_m(t) - G_n(t) \right] \right\},$$

$$(3.1.6)$$

which was studied by Maag and Stephens (1968), and by Gnedenko (1954) in the case $m = n$. It is possible to consider two-sample variants of the Watson–Darling statistics (2.1.5)–(2.1.7) and Khmaladze–Aki statistic (2.1.12), but they have not been investigated in depth. We shall not touch upon them as well as upon two-sample counterparts of the Watson statistic (2.1.11) and the Anderson–Darling statistic (2.1.10) introduced by Watson (1962) and Pettitt (1976).

The best integral-type statistic for testing H_1 against A_1 is the so-called Lehmann–Rosenblatt statistic (see Lehmann (1951), Rosenblatt (1952b), and Pettitt (1979)):

$$W^2_{m,n,1} = \frac{mn}{(m+n)^2} \int_{-\infty}^{\infty} \left(F_m(t) - G_n(t) \right)^2 dH_{m+n}(t),$$

where

$$H_{m+n}(t) = \frac{m}{m+n} F_m(t) + \frac{n}{m+n} G_n(t)$$

is the empirical distribution function corresponding to the pooled sample. We consider also its generalization in the sense of (2.1.9), namely,

$$W^k_{m,n,q} = \left(\frac{mn}{m+n} \right)^{k/2} \int_{-\infty}^{\infty} \left(F_m(t) - G_n(t) \right)^k q(H_{m+n}(t)) \, dH_{m+n}(t),$$

$$(3.1.7)$$

where the weight function q is positive and differentiable on $(0, 1)$, and

$$\operatorname*{ess\,sup}_{0 \le x \le 1} |q'(x)| \le L < \infty \qquad (3.1.8)$$

and k is a positive integer. For $k = 2$ and $q \equiv 1$ one obtains the classical statistic $W^2_{m,n,1}$.

In the problem of homogeneity testing a natural and very popular competitor of the Kolmogorov–Smirnov and ω^2-type statistics is the class of linear rank statistics. Denote by R_i, $i = 1, \ldots, m$, the rank of an observation X_i in the variational series corresponding to the pooled sample. We call a *simple linear rank statistic* the statistic having the form (Hájek and Šidak 1967, Lehmann 1975):

$$S_N = N^{-1} \sum_{i=1}^{m} a_N\big(R_i/(N+1)\big), \qquad (3.1.9)$$

where the function $a_N(u)$ is constant on all intervals of the form $\big\{(i-1)/N \le u < i/N\big\}$, $i = 1, \ldots, N$. We assume, moreover, that

$$a_N(u) \xrightarrow{L^2} J(u), \qquad N \to \infty, \qquad (3.1.10)$$

where J is a nonconstant function on $[0, 1]$ called a *score function*. Statistics of the form (3.1.9) possess property A introduced in Section 1.6 (see Woodworth (1970)).

Without loss of generality we may suppose

$$\int_0^1 J(u)\, du = 0, \qquad \int_0^1 J^2(u)\, du = 1. \qquad (3.1.11)$$

In the case $J(u) = \sqrt{12}\,(u - \frac{1}{2})$ one obtains the famous Wilcoxon rank statistic and in the case $J(u) = \Phi^{-1}(u)$ the Fisher–Yates–Terry–Hoeffding normal scores statistic or, asymptotically equivalent to it, van der Waerden statistic (see Hájek and Šidak (1967)). Statistics (3.1.9) are also functionals of empirical distribution functions in view of the well-known representation

$$S_N = \frac{m}{N} \int_{-\infty}^{\infty} a_N\big(\big(N H_N(t) + 1\big)/(N+1)\big)\, dF_m(t), \qquad (3.1.12)$$

belonging to Chernoff and Savage (1958) (see also Govindarajulu, Le Cam, and Raghavachari (1967)).

As we have already mentioned repeatedly, the comparison of all sequences of statistics based on the Bahadur, Hodges–Lehmann and Chernoff AREs requires the knowledge of their large-deviation asymptotics under the null hypothesis as well as under the alternative.

Large deviations of statistics (3.1.3)–(3.1.6) under the hypothesis H_1 may be obtained in the same way as it has been done for statistics (2.1.1)–(2.1.4) in Theorem 2.2.1.

Theorem 3.1.1 (Abrahamson 1967) *Let $\{T_{m,n}\}$ be any of statistics $\{D_{m,n}\}$, $\{D_{m,n}^{\pm}\}$, or $\{V_{m,n}\}$. Then under the hypothesis H_1 we have*

$$\lim_{m,n\to\infty} (m+n)^{-1} \ln P\left(T_{n,m} \geq a\right) = -g_T(a), \quad 0 < a < 1, \quad (3.1.13)$$

where the function g_T is continuous for sufficiently small positive a and, moreover,

$$g_T(a) = 2\,a^2\left(1 + o(1)\right) \qquad as \ \ a \to 0.$$

It is possible to derive large-deviation asymptotics of sequence (3.1.9) from Theorems 1.6.11 and 1.6.12. We get asymptotically equivalent statistics by putting in (1.6.26) the expression

$$J(u,v) = J(u) \cdot \mathbf{1}_{\{v < \rho_1\}}.$$

It follows that

$$\widetilde{J}(u,v) = J(u) \cdot \mathbf{1}_{\{v < \rho_1\}} - J(u), \qquad b^2 = \rho_1\,\rho_2.$$

This yields the following result.

Theorem 3.1.2 *Let H_1 be true, condition (3.1.11) be fulfilled, and*

$$\int_0^1 \exp\{r\,J(u)\}\,du < \infty \qquad for \ some \ r > 0. \qquad (3.1.14)$$

Then

$$\lim_{N\to\infty} N^{-1} \ln P\left(S_N \geq a\right) = -\frac{a^2}{2\,\rho_1\rho_2} + o(a^2) \quad as \ a \to 0. \qquad (3.1.15)$$

3.2 Large Deviations of Integral Two-Sample Statistics

The large-deviation analysis of the statistics $W_{m,n,q}^k$ under the hypothesis H_1 may be carried out by methods used in Sections 2.3–2.4 for the statistics $\omega_{n,q}^k$. For this purpose we present it here in a condensed form.

Without loss of generality one may assume that the empirical distribution functions F_m and G_n are based on samples from the uniform distribution on $[0, 1]$. Denote for arbitrary d.f.s F and G on $[0, 1]$

$$\chi_q(F,G) := \int_0^1 \left[F(t) - G(t)\right]^k q\big(\rho_1\,F(t) + \rho_2\,G(t)\big)\,d\big(\rho_1\,F(t) + \rho_2\,G(t)\big),$$

$$\rho(F,G) := \sup_{0 \leq t \leq 1} \big|F(t) - G(t)\big|,$$

as in Section 2.6. Denote, moreover,

$$\xi(m,n) := \max \left(\left| \frac{m}{N} - \rho_1 \right|, \left| \frac{n}{N} - \rho_2 \right| \right).$$

Let C_i, $i = 1, 2$, be some constants depending on k, q, and L. It is easy to check, using (3.1.2), that

$$\left| W^k_{m,n,q} - (\rho_1 \rho_2)^{k/2} \chi_q(F_m, G_n) \right| \leq C_1 \xi(m,n) \longrightarrow 0, \qquad N \to \infty, \tag{3.2.1}$$

and

$$\left| \chi_q(F_1, G_1) - \chi_q(F_2, G_2) \right| \leq C_2 \max \left(\rho(F_1, F_2), \rho(G_1, G_2) \right). \tag{3.2.2}$$

The estimate (3.2.2) yields the continuity of functional χ_q in the uniform topology of the space $\Lambda^2 [0,1]$ of pairs of probability measures on $[0,1]$.

Consider now for any $a > 0$ the set Ω_a of all pairs (f,g) of distribution functions on $[0,1]$ satisfying the inequality

$$(\rho_1 \rho_2)^{k/2} \chi_q(f,g) \geq a. \tag{3.2.3}$$

For any pair (f,g) of absolutely continuous distribution functions on $[0,1]$ we introduce the Kullback–Leibler information by the formula, analogous to (1.7.1),

$$J_{\rho_1, \rho_2}(f,g) := \rho_1 \int_0^1 f' \ln f' \, dt + \rho_2 \int_0^1 g' \ln g' \, dt,$$

and in accordance with (1.7.2) put

$$J_{\rho_1, \rho_2}(\Omega_a) := \inf \left\{ J_{\rho_1, \rho_2}(f,g) : (f,g) \in \Omega_a \right\}.$$

The following statement will be proved in the sequel.

Theorem 3.2.1 (Nikitin 1980) *For sufficiently small $a > 0$ the function*

$$a \mapsto J_{\rho_1, \rho_2}(\Omega_a)$$

can be represented in the form of a convergent series with numerical coefficients

$$J_{\rho_1, \rho_2}(\Omega_a) = -\sum_{j=2}^{\infty} \alpha_j \, a^{j/k}, \tag{3.2.4}$$

where $\alpha_2 = \frac{1}{2} \lambda_0(q; k)$ and $\lambda_0(q; k)$ is the principal eigenvalue of problem
(2.4.5).

It is clear that (3.2.4) implies the continuity of $J_{\rho_1, \rho_2}(\Omega_a)$ as a function of a in a neighborhood of zero. Due to inequality (3.2.2) we may apply Theorem 1.7.1 to the statistics $W^k_{m, n, q}$. As a result we obtain the following statement.

Theorem 3.2.2 *For sufficiently small $a > 0$ under H_1*

$$\lim_{N \to \infty} N^{-1} \ln P \left(W^k_{m, n, q} \ge a \right) = -J_{\rho_1, \rho_2}(\Omega_a),$$

where $J_{\rho_1, \rho_2}(\Omega_a)$ may be calculated by formula (3.2.4).

Proof of Theorem 3.2.1 Lemma 1.7.1 and Lemma 1.7.5 enable one to conclude that the functional J_{ρ_1, ρ_2} attains its lower bound on Ω_a. To calculate the variation of this functional let us first restrict the class of admissible elements to a subset where the variation exists. Denote by $\mathcal{F} \times \mathcal{F}$ the set of all pairs (f, g) of absolutely continuous distribution functions on $[0, 1]$. For any δ put

$$V_\delta := \left\{ (f, g) \in \mathcal{F} \times \mathcal{F} \colon f' \ge \delta. \ g' \ge \delta \quad \text{a.e on } [0, 1] \right\},$$

$$V'_\delta := \left\{ (f, g) \in \mathcal{F} \times \mathcal{F} \colon f' > \delta, \ g' > \delta \quad \text{a.e on } [0, 1] \right\}.$$

Similarly to Lemma 2.3.1 one proves the double inequality

$$J_{\rho_1, \rho_2}(\Omega_{a_1} \cap V_{\delta/2}) - \delta \le J_{\rho_1, \rho_2}(\Omega_a) \le J_{\rho_1, \rho_2}(\Omega_a \cap V_{\delta/2}), \quad (3.2.5)$$

where $a_1 = a_1(\delta) \to a$ as $\delta \to 0$. Hence we need to calculate $J_{\rho_1, \rho_2}(\Omega_{a_1} \cap V_{\delta/2})$ and then pass to the limit in (3.2.5) as $\delta \to 0$.

As in Section 2.3, let us enlarge the set $\Omega_{a_1} \cap V_{\delta/2}$ up to the set $\Omega_{a_1} \cap V'_{\delta/4}$. Then we may use the Lagrange principle (Theorem 1.8.1) and must find the infimum of the functional

$$\int_0^1 \left\{ \rho_1 \big(x'(t) + 1 \big) \ln \big(x'(t) + 1 \big) + \rho_2 (y'(t) + 1) \ln \big(y'(t) + 1 \big) \right\} dt,$$

$$(3.2.6)$$

defined on an open subset $\mathcal{U}$ of the Banach space

$$\overset{\circ}{\mathbf{W}}_{\infty, 1} [0, 1] \times \overset{\circ}{\mathbf{W}}_{\infty, 1} [0, 1],$$

consisting of elements (x, y) such that a.e.

$$x'(t) + 1 > \tfrac{1}{4} \delta, \qquad\qquad y'(t) + 1 > \tfrac{1}{4} \delta \qquad (3.2.7)$$

under the complementary restriction

$$(\rho_1\rho_2)^{k/2} \chi_q \left(x(t) + t, y(t) + t \right) \geq a_1 . \tag{3.2.8}$$

Theorem 1.8.1 guarantees that there exist Lagrange multipliers $\hat{\lambda}_0$ and $\hat{\lambda}_1$, $\hat{\lambda}_0^2 + \hat{\lambda}_1^2 > 0$, such that the pair of functions $(\hat{x}, \hat{y})$ on which the local extremum in problem (3.2.6)–(3.2.8) is attained satisfies the following system of Euler–Lagrange equations:

$$\begin{cases} \hat{\lambda}_0 \, \rho_1 \, \hat{x}'' - k \, \hat{\lambda}_1 \, (\rho_1\rho_2)^{k/2} \, (\hat{x} - \hat{y})^{k-1} \, q(\hat{z} + t) \, (\hat{x}' + 1) \, (\hat{y}' + 1) = 0 , \\ \hat{\lambda}_0 \, \rho_2 \, \hat{y}'' + k \, \hat{\lambda}_1 \, (\rho_1\rho_2)^{k/2} \, (\hat{x} - \hat{y})^{k-1} \, q(\hat{z} + t) \, (\hat{x}' + 1) \, (\hat{y}' + 1) = 0, \end{cases}$$
$$\tag{3.2.9}$$

where $\hat{z} = \rho_1\hat{x} + \rho_2\hat{y}$.

This system should be considered together with the boundary conditions

$$\hat{x}(0) = \hat{x}(1) = 0 , \qquad \hat{y}(0) = \hat{y}(1) = 0 \tag{3.2.10}$$

and the normalization condition

$$(\rho_1\rho_2)^{k/2} \int_0^1 \left(\hat{x} - \hat{y} \right)^{k/2} q(\hat{z} + t) \, (\hat{z}' + 1) \, dt = a_1 . \tag{3.2.11}$$

The assumption $\hat{\lambda}_0 = 0$ leads to a contradiction. Hence one may put $\hat{\lambda}_0 = 1$. Adding together equations (3.2.9) we obtain, due to (3.2.10), that

$$\rho_1 \, \hat{x} + \rho_2 \, \hat{y} = \hat{z} \equiv 0 . \tag{3.2.12}$$

Now let us make the substitutions $a_1^{1/k} = \varepsilon$ and $\hat{x} - \hat{y} = u\varepsilon/\sqrt{\rho_1\rho_2}$. Then the function u must satisfy the nonlinear differential equation

$$u'' - \mu u^{k-1} q \left(1 + u' \varepsilon \, \frac{\rho_2 - \rho_1}{\sqrt{\rho_1\rho_2}} - u'^2 \, \varepsilon^2 \right) = 0 , \tag{3.2.13}$$

the boundary conditions

$$u(0) = u(1) = 0, \tag{3.2.14}$$

and the normalization condition

$$\int_0^1 u^k q \, dt = 1 . \tag{3.2.15}$$

The problem (3.2.13)–(3.2.15) is of the same type as problem (2.3.22)–(2.3.24), considered earlier in Section 2.3. Its solution may be constructed in the same way as in Section 2.4. Finally we obtain

$$u(t) = x_0(t) + \sum_{j=1}^{\infty} u_j(t)\, \varepsilon^j\,,$$

$$\mu = \lambda_0 + \sum_{j=1}^{\infty} \mu_j\, \varepsilon^j\,,$$

(3.2.16)

where (x_0, λ_0) is a "principal" (possibly not unique) solution of problem (2.4.5) and the first series in (3.2.16) converges absolutely, that is, with respect to the norm of the space $\overset{\circ}{\mathbf{W}}_{1,2}[0,1]$.

From the equations

$$\hat{x} - \hat{y} = u\varepsilon \big/ \sqrt{\rho_1 \rho_2}\,, \qquad \rho_1\,\hat{x} + \rho_2\,\hat{y} = 0\,,$$

it follows that

$$\hat{x} = u\varepsilon\,\sqrt{\rho_2/\rho_1}\,, \qquad \hat{y} = -u\varepsilon\,\sqrt{\rho_1/\rho_2}\,.$$

(3.2.17)

It is clear that inequalities (3.2.7) are valid for sufficiently small ε so the pair $(\hat{x}, \hat{y})$ of extremals belongs to $\Omega_{a_1} \cap V'_{\delta/4}$ as well as to $\Omega_{a_1} \cap V_{\delta/2}$. Substituting (3.2.16) in (3.2.17) and then in (3.2.6), and passing to the limit as $\delta \to 0$, we prove Theorem 3.2.1.   □

We recall that for the weight $q \equiv 1$ one has $\lambda_0 = -12$ for $k = 1$ and $\lambda_0 = -\pi^2$ for $k = 2$. Therefore Theorems 3.2.1 and 3.2.2 imply

$$\lim_{N \to \infty} N^{-1} \ln P\,(W_{m,n,1}^1 \geq a) = -6\,a^2 + \sum_{j=3}^{\infty} \alpha_j'\, a^j$$

and

$$\lim_{N \to \infty} N^{-1} \ln P\,(W_{m,n,1}^2 \geq a) = -\tfrac{1}{2}\, \pi^2 a + \sum_{j=3}^{\infty} \alpha_j''\, a^j$$

for sufficiently small $a > 0$.

3.3 Bahadur Efficiency of Homogeneity Tests

Taking as a basis large-deviation asymptotics of the integral statistics $W_{m,n,q}^k$ obtained in Theorems 3.2.1 and 3.2.2, we may now calculate the principal parts of the Bahadur exact slopes of these statistics which enables one to compare them by the local Bahadur ARE with other statistics used for testing homogeneity.

To use the results of Sections 1.1–1.2 we suppose Θ is the set of all pairs of continuous distribution functions on $\mathbf{R}^1$ with the uniform metric ρ and Θ_0 is the subset of such pairs from Θ with identical components. Let P_θ be the distribution corresponding to the pair $\boldsymbol{\theta} = (\theta_1, \theta_2)$ and

$$\Theta_1 = \Theta \setminus \Theta_0 \,.$$

The method of calculating the exact slopes proposed by Theorem 1.2.2 is applicable to the sequence of statistics $\{W^k_{m,n,q}\}$. Using the Glivenko–Cantelli theorem and the first term of the expansion given by Theorem 3.2.1, we obtain that the principal part of the exact slope $c_{W^k}(\boldsymbol{\theta})$ for these statistics has the form

$$c_{W^k}(\boldsymbol{\theta}) \sim -\rho_1 \rho_2 \, \lambda_0(q; k) \left[\int\limits_{-\infty}^{\infty} (\theta_1 - \theta_2)^k \, q(\psi) \, d\psi \right]^{2/k}, \quad \rho(\boldsymbol{\theta}, \Theta_0) \to 0,$$

$$(3.3.1)$$

denoting for brevity $\psi = \rho_1 \theta_1 + \rho_2 \theta_2$.

The exact slopes (or their principal parts) of other statistics for homogeneity testing considered in Section 3.1 may be easily found. It was proved by Abrahamson (1967) that, as $\rho(\boldsymbol{\theta}, \Theta_0) \to 0$,

$$c_D(\boldsymbol{\theta}) \sim 4\,\rho_1\,\rho_2 \left\{ \sup_{-\infty < t < \infty} \left| \theta_1(t) - \theta_2(t) \right| \right\}^2, \qquad (3.3.2)$$

$$c_{D^+}(\boldsymbol{\theta}) \sim 4\,\rho_1\,\rho_2 \left\{ \sup_{-\infty < t < \infty} \left[\theta_1(t) - \theta_2(t) \right] \right\}^2, \qquad (3.3.3)$$

$$c_{D^-}(\boldsymbol{\theta}) \sim 4\,\rho_1\,\rho_2 \left\{ \sup_{-\infty < t < \infty} \left[\theta_2(t) - \theta_1(t) \right] \right\}^2, \qquad (3.3.4)$$

$$c_V(\boldsymbol{\theta}) \sim 4\,\rho_1\,\rho_2 \left\{ \sup_{-\infty < t < \infty} \left[\theta_1(t) - \theta_2(t) \right] \right.$$

$$\left. - \inf_{-\infty < t < \infty} \left[\theta_1(t) - \theta_2(t) \right] \right\}^2 \qquad (3.3.5)$$

under the natural premise that all quantities in braces are positive.

The problem of the validity of the law of large numbers for linear rank statistics (3.1.9) under the alternative is well explored. If (3.1.10) and property A (see Section 1.6) are fulfilled, it is possible to approximate the function a_N by appropriate bounded continuous function a and to pass then to the score function J that simplifies the calculation of the limit (for details refer to Woodworth (1970), Hájek (1974), and Müller-Funk (1983)). Finally one obtains that $\mathbf{P}_\theta$-almost surely

$$S_N \longrightarrow \rho_1 \int_0^1 J(\psi)\, d\theta_1 := b_J(\boldsymbol{\theta})\,. \tag{3.3.6}$$

It is natural to assume here that for all $\boldsymbol{\theta} \in \Theta_1$ we have $b_J(\boldsymbol{\theta}) > 0$ which guarantees the consistency of the test based on $\{S_N\}$. The subsequent simplification of the function $b_J(\boldsymbol{\theta})$ depends on the distribution of observations under the alternative and may be carried out under additional assumptions about smoothness of the score function J.

Using the information concerning large deviations of linear rank statistics under H_1 given by Theorem 3.1.2, we can find their local exact slopes.

Now let us discuss the question of the maximal value of the exact slope of any sequence of statistics $\{T_N\}$ for testing homogeneity. Analogously to the proof of Theorem 1.2.3 we make sure that

$$c_T(\boldsymbol{\theta}) \leq 2\, \inf\left\{ \rho_1\, K(\theta_1, \theta_0) + \rho_2\, K(\theta_2, \theta_0)\right\},$$

where the infimum is taken over the set of all absolutely continuous distribution functions θ_0 on $\mathbf{R}^1$. It is easy to check that the "least favorable" distribution function θ_0 has the form

$$\theta_0 = \psi = \rho_1\, \theta_1 + \rho_2\, \theta_2$$

and, hence, one can find (see Hájek (1974))

$$c_T(\boldsymbol{\theta}) \leq 2\left[\rho_1\, K(\theta_1, \psi) + \rho_2\, K(\theta_2, \psi)\right] \equiv 2\, K^*(\boldsymbol{\theta}, \Theta_0)\,. \tag{3.3.7}$$

It is interesting to note that the quantity on the right-hand side of inequality (3.3.7) coincides with the so-called *informational radius,* playing an important role in problems of mathematical taxonomy and cluster analysis due to Jardine and Sibson (see, e.g., Jardine and Sibson (1971), Rao (1977)).

Hájek (1974) proved the property of "asymptotic sufficiency" of the vector of ranks in the Bahadur sense. In other words the equality holds in (3.3.7) for a certain rank statistic for which the score function may be determined by the alternative.

To construct some examples of the comparison of tests by consideration of their local Bahadur efficiency we restrict ourselves, as in Section 2.6, to the case of a location parameter and put

$$\theta_1(t) := F(t)\,, \qquad\qquad \theta_2(t) := F(t - \theta)\,,$$

where F is a known and regular enough d.f. on the real line and $\theta \geq 0$. In this case all the sequences of statistics we have studied (except $D_{m,n}^-$) have local exact slopes of the form

$$\rho_1\,\rho_2\,l(F)\cdot\theta^2\left(1+o(1)\right), \qquad \theta\to 0.$$

As to the right-hand side of inequality (3.3.7) when d.f. F has sufficiently regular density f and

$$h(x) = \rho_1\,f(x) + \rho_2\,f(x-\theta),$$

we deduce

$$2\,K^*(\theta,\,\Theta_0) = 2\int\limits_{-\infty}^{\infty}\left\{\rho_1\,\ln\frac{f(x)}{h(x)}\,f(x) + \rho_2\,\ln\frac{f(x-\theta)}{h(x)}\,f(x-\theta)\right\}dx$$

$$\sim \rho_1\,\rho_2\int\limits_{-\infty}^{\infty}\frac{f'^2(x)}{f(x)}\,dx\cdot\theta^2$$

$$= \rho_1\,\rho_2\,I\left(f\right)\cdot\theta^2 \qquad \text{as } \theta\to 0,$$

where $I\left(f\right)$ is again the Fisher information, $0 < I\left(f\right) < \infty$.

Therefore for the local comparison in the Bahadur sense of tests considered previously it is sufficient to compare only corresponding constants $l(F)$, taking into account that their maximal value (which is attained for the likelihood ratio test) is the Fisher information $I\left(f\right)$. The constants under comparison differ from the local Bahadur indices of corresponding goodness-of-fit tests (see Section 2.6) only by the factor $\rho_1\rho_2$.

We choose the Wilcoxon test with $J(u) = \sqrt{12}\,(u - \frac{1}{2})$ and van der Waerden test with $J(u) = \Phi^{-1}(u)$ as representatives of the class of linear rank tests and take again, as in Section 2.6, Gaussian, logistic, and Cauchy d.f.s as examples of F.

Part of the data is borrowed from Table 3 of Section 2.6. To calculate the local exact slopes of linear rank tests we used tables by Gradshteyn and Ryzhik (1971) or numerical integration.

As in Chapter 2, we restrict ourselves to an accuracy of 0.01 increasing it only when it is necessary to distinguish the close values of indices.

As the local index of the Wilcoxon rank test coincides with the local index of the test based on $\omega_{n,1}^1$, it is no wonder that corresponding rows in Tables 3 and 4 coincide. The value 0.95 for the Gaussian distribution is the famous constant $3/\pi$, first found by Pitman (1949) in the case of the Pitman efficiency. Comparatively small values of the local indices of van der Waerden and Wilcoxon tests in the case of the Cauchy distribution

Table 4. *Local indices $L(F)$ for the location alternative*

Statistic	Gaussian	Logistic	Cauchy
$D_{m,n}$	0.64	0.25	0.41
$W^2_{m,n,1}$	0.91	0.329	0.38
$W^4_{m,n,1}$	0.56	0.30	0.42
Wilcoxon	0.95	0.333	0.30
van der Waerden	1.00	0.32	0.16
Fisher information	1.00	0.333	0.50

may be explained by the fact that these tests are locally most powerful rank tests for Gaussian and logistic distributions correspondingly and are not obliged to behave well for distributions with heavy tails.

Note also that the approximate Bahadur ARE coincides locally, as $\theta \to 0$, with the exact Bahadur ARE. This conclusion follows from the analysis of limiting distributions of these statistics under H_1. It is well known (see Doob (1949), Rosenblatt (1952b), and Korolyuk and Borovskikh (1984)) that under condition (3.1.1) the limiting distributions of the statistics $W^k_{m,n,q}$, $D_{m,n}$, and $D^{\pm}_{m,n}$ coincide with the limiting distributions of the usual one-sample statistics $\omega^k_{n,q}$, D_n and $D^{\pm}_n$ so their approximate slopes differ only by factors $(\rho_1\rho_2)^{k/2}$ or $\rho_1\rho_2$ from approximate slopes of the one-sample statistics calculated in Section 2.6. As to rank statistics their asymptotic normality follows from the famous Chernoff–Savage theorem (see Chernoff and Savage (1958) or Hájek and Šidak (1967)) which enables one to calculate easily the approximate slopes locally coinciding with their exact counterparts (cf. Kremer (1979a,b, 1982, 1983)).

The limiting Pitman ARE is also the same as the local Bahadur ARE. For asymptotically normal statistics such as linear rank statistics it is a consequence of Bahadur (1960b); for the remaining ones it is sufficient to verify Wieand's condition (1.4.3) from Section 1.4 which does not require new ideas in addition to goodness-of-fit statistics.

3.4 Hodges–Lehmann Efficiency of Homogeneity Tests

In Section 2.7 we studied the Hodges–Lehmann ARE of goodness-of-fit tests. The main result obtained there is the conclusion that two-sided tests of the Kolmogorov–Smirnov and ω^2 type are asymptotically optimal in the sense of the Hodges–Lehmann efficiency. We will now show that this situation also remains valid for homogeneity tests.

Again let X_1, X_2, …, X_m and Y_1, Y_2, …, Y_n be two independent samples with absolutely continuous distributions P_1 and P_2. Suppose that condition (3.1.2) is fulfilled. We are testing the composite hypothesis H_1: $P_1 = P_2$ against the alternative A_1: $P_1 = R_1$, $P_2 = R_2$ for given distributions R_1 and R_2 on the real line with distinct positive densities r_1 and r_2. In order to apply to this problem the general framework of Section 1.3 we shall consider, as a parametric set, the set of pairs $\boldsymbol{\theta} = (P_1, P_2)$ of absolutely continuous probability measures on $\mathbf{R}^1$. The set of such pairs with equal components plays the role of Θ_0 and the set of pairs with distinct components having positive densities plays the role of Θ_1. Then we may assume, as in Section 1.3, that we are testing the hypothesis H_1: $\boldsymbol{\theta} \in \Theta_0$ against the simple alternative A_1: $\boldsymbol{\theta} = \mathbf{R} = (R_1, R_2) \in \Theta_1$.

If $\{T_{m,n}\}$ is a sequence of statistics designed for testing H_1 against A_1, we suppose that

$$P_{\theta_0}(T_{m,n} < t) = V_{m,n}(t) \qquad (3.4.1)$$

for all $\theta_0 \in \Theta_0$ and all t, so that the distribution of $T_{m,n}$ does not depend on $\theta_0 \in \Theta_0$. This condition is fulfilled for all statistics considered in this section.

We start out by finding an upper bound for the Hodges–Lehmann indices of such statistics.

Theorem 3.4.1 (Nikitin 1986c, 1987a) *Let $\beta_{1N}(\alpha; R_1, R_2)$ be the power function of a test at a level $\alpha \in (0, 1)$ based on the sequence $\{T_{m,n}\}$ satisfying (3.4.1). Then the following inequality holds:*

$$\varliminf_{N \to \infty} N^{-1} \ln \left[1 - \beta_{1N}(\alpha; R_1, R_2) \right] \geq \ln \int_{-\infty}^{\infty} r_1^{\rho_1}(x)\, r_2^{\rho_2}(x)\, dx. \quad (3.4.2)$$

The quantity on the right-hand side of (3.4.2) multiplied by -1 was introduced by Rényi (1961). According to his terminology it is the information of order ρ_1.

Proof Let P_0 be an unknown common distribution of samples under the hypothesis H_1 (with density p_0) and $\gamma_N(\alpha; R_1, R_2, P_0)$ be the probability of error of the second kind for the best test at a level α of the simple hypothesis H'_1: $P_1 = P_2 = P_0$ against A_1. Property (3.4.1) implies that $T_{m,n}$ may be viewed also as a test statistic at a level α of H'_1 against A_1, but

$$1 - \beta_{1N}(\alpha; R_1, R_2) \geq \gamma_N(\alpha; R_1, R_2, P_0) \,.$$

Applying Theorem 2.7.1 to the lower estimate of γ_N with $c = 2$, we have

$$\lim_{N \to \infty} N^{-1} \ln \left[1 - \beta_{1N}(\alpha; R_1, R_2) \right] \geq - \inf_{P_0} \sum_{i=1}^{2} \rho_i \, K(P_0, R_i) \,. \quad (3.4.3)$$

It only remains to find the explicit expression for the right-hand side of (3.4.3). Using the language of variational calculus we have to look for the infimum of the functional

$$\int_{-\infty}^{\infty} \left[\rho_1 \, u' \, \ln \frac{u'}{r_1} + \rho_2 \, u' \, \ln \frac{u'}{r_2} \right] dt \quad (3.4.4)$$

in the class of all absolutely continuous distribution functions u on the real line. The solution exists on account of Lemma 1.7.5. On determining the variation we find that the extremal u may be calculated by the relation

$$u' = C \, r_1^{\rho_1} \, r_2^{\rho_2}, \quad \text{where } C = \left(\int_{-\infty}^{\infty} r_1^{\rho_1}(x) \, r_2^{\rho_2}(x) \, dx \right)^{-1} \,.$$

Substituting it into minimized functional (3.4.4), we establish the assertion of Theorem 3.4.1. $\quad\square$

Let a sequence of statistics $\{T_{m,n}\}$ have the Hodges–Lehmann index $d_T(R_1, R_2)$. Then from (3.4.2) it follows that

$$d_T(R_1, R_2) \leq -2 \ln \int_{-\infty}^{\infty} r_1^{\rho_1}(x) \, r_2^{\rho_2}(x) \, dx \,. \quad (3.4.5)$$

If in (3.4.5) the equality is valid, then again, as in Section 2.7, the sequence of statistics $\{T_{m,n}\}$ is said to be Hodges–Lehmann asymptotically optimal.

Now we consider several nonparametric statistics used to verify H_1 against A_1. Let us apply the notation of Section 3.1. Let $D_{m,n}$ and

$V_{m,n}$ be given by formulas (3.1.3) and (3.1.6). We introduce

$$\omega_{m,n,q}^{\kappa} = \left(\frac{mn}{(m+n)^2} \right)^{\kappa/2} \int\limits_{-\infty}^{\infty} |F_m(t) - G_n(t)|^{\kappa} \, q\big(H_N(t)\big) \, dH_N(t),$$

$$U_{m,n}^{\kappa} = \left(\frac{mn}{(m+n)^2} \right)^{\kappa/2} \int\limits_{-\infty}^{\infty} \left| F_m(t) - G_n(t) \right.$$

$$\left. - \int\limits_{-\infty}^{\infty} \big(F_m(y) - G_n(y)\big) \, dH_N(y) \right|^{\kappa} dH_N(t).$$

In these definitions it is assumed that $\kappa \geq 1$ and the weight function q is positive and has a bounded derivative on $(0, 1)$.

Observe that the statistic $\omega_{m,n,q}^{\kappa}$ for $\kappa = 2$ and $q \equiv 1$ is the classic Lehmann–Rosenblatt statistic already considered in Sections 3.1–3.2. For any even integer κ it coincides with the statistic $W_{m,n,q}^{\kappa}$ investigated there. Finally, the statistic $U_{m,n}^{\kappa}$ was studied by Watson (1962) for $\kappa = 2$.

Theorem 3.4.2 (Nikitin 1986c, 1987a) *The sequences of statistics* $\{D_{m,n}\}$, $\{V_{m,n}\}$, $\{\omega_{n,q}^{\kappa}\}$ *and* $\{U_{m,n}^{\kappa}\}$ *are Hodges–Lehmann asymptotically optimal.*

Proof All of these statistics are uniformly continuous functionals of pairs of empirical d.f.s F_m and G_n in the ρ-topology and of corresponding empirical distributions. Let T_N be any of the statistics under consideration, $T_N = T(F_m, G_n)$, and $1 - \beta_N(\alpha; R_1, R_2)$ be the probability of error of the second kind at a significance level $\alpha \in (0, 1)$. Applying Theorem 1.7.2 as in the proof of Theorem 2.7.2 for the sets having the form (1.7.5) we obtain

$$\varlimsup_{N \to \infty} N^{-1} \ln \left[1 - \beta_N(\alpha; R_1, R_2) \right]$$

$$\leq - \inf \left\{ \int\limits_{-\infty}^{\infty} \left(\rho_1 \, u' \ln \frac{u'}{r_1} + \rho_2 \, v' \ln \frac{v'}{r_2} \right) dt : (u, v) \in \Lambda_T^2 \right\}, \quad (3.4.6)$$

where Λ_T^2 is the class of all pairs (u, v) of distribution functions on the real line satisfying the condition

$$T(u, v) \leq 0. \qquad (3.4.7)$$

But for all considered statistics it follows from (3.4.7) that either $u \equiv v$ or u and v differ perhaps on just a set where the corresponding densities

u' and v' are simultaneously zero. Hence, the right-hand side of (3.4.6) becomes, up to the sign, functional (3.4.4), which has already been calculated in Theorem 3.4.1, and, hence, we obtain for the Hodges–Lehmann index $d_T(R_1, R_2)$ an estimate opposite to (3.4.5). □

This result may be slightly generalized along lines similar to those of the paper by Kallenberg and Kourouklis (1992).

The property of Hodges–Lehmann asymptotic optimality is not universal and may fail for the one-sided test statistics. We will show it for the example of linear rank statistics in the following section.

3.5 Hodges–Lehmann and Chernoff Efficiencies of Linear Rank Tests

Consider again the problem of testing the homogeneity hypothesis H_1 on the basis of two independent samples $X_1, X_2, \ldots, X_m$ and $Y_1, Y_2, \ldots, Y_n$. By way of the alternative A_1 we suppose that the first and the second samples have d.f.s $G(x; 0)$ and $G(x; \theta)$, for some $\theta > 0$, respectively. The family $\{ G(x; \theta), \theta > 0 \}$ will be the subject of some regularity conditions stated subsequently.

To test H_1 against A_1 we will use the statistics having the form

$$S_N = N^{-1} \sum_{i=1}^{m} J\big(R_i/(N+1)\big), \qquad (3.5.1)$$

which are asymptotically equivalent to statistics (3.1.9). It is assumed that the score function J satisfies (3.1.11) and sample sizes m and n obey (3.1.2).

In previous sections of this chapter we have seen that the problem of calculating the Pitman and Bahadur efficiencies of linear rank tests was explored by many authors. Already in Chernoff and Savage (1958) it was shown that under appropriate regularity conditions imposed on J and the family $\{G(x; \theta)\}$, the Pitman efficacy of a given sequence of statistics $\{S_N\}$ is

$$\mathrm{eff}\,(J; G) = \rho_1 \rho_2 \bigg(\int\limits_{-\infty}^{\infty} J'(G(x; 0))\, G'_\theta(x; 0)\, dG(x; 0) \bigg)^2, \qquad (3.5.2)$$

where

$$G'_\theta(x; 0) = \frac{d}{d\theta}\, G(x; \theta)\Big|_{\theta=0}.$$

Later these conditions were weakened by Hájek (1968), Lai (1975), and Denker (1985), among others. It follows from results quoted in Section 3.3 that the Bahadur exact slope $c_S(\theta)$ of sequence (3.5.1), again under the natural regularity conditions, satisfies the relationship

$$c_S(\theta) = \text{eff}\,(J;\,G) \cdot \theta^2 \left(1 + o(1)\right), \qquad \theta \to 0, \qquad (3.5.3)$$

hence the local Bahadur ARE of two such sequences coincides with the Pitman ARE.

It follows from the paper of Kallenberg (1983a) that the same is true (again under corresponding regularity conditions) for the intermediate efficiency (see Section 1.4).

But the question of the computation of the Hodges–Lehmann and Chernoff AREs of statistics (3.5.1) had been left unexplored until recently. An attempt at finding these types of ARE made by Hwang (1978) for the Wilcoxon statistic did not succeed in obtaining substantial results. The reason has to do with some technical difficulties connected with calculating large-deviation probabilities under the alternative when the rank statistics are no longer "distribution-free."

However, it turns out that the "variational" method used previously when studying large deviations under the null hypothesis may be applied under the alternative too. We prove that the principal parts of the Hodges–Lehmann and Chernoff indices (up to the constant factor) coincide with the right-hand side of (3.5.3). This implies the coincidence of the local Hodges–Lehmann, Chernoff, and Bahadur efficiencies with the intermediate and Pitman efficiencies for a broad class of linear rank statistics.

The problem of the equivalence of the Pitman efficiency with the local Hodges–Lehmann and Chernoff efficiencies was also addressed by Kourouklis (1989, 1990) in the context of testing the one-sided hypotheses about a location parameter using M-estimators.

Now let us give the conditions imposed on the score function and the distribution of initial observations. In the first place we suppose that the score function J is twice continuously differentiable on $[0, 1]$.

Henceforth it is assumed that d.f. $G(x;\,\theta)$ is defined for $x \in \mathbf{R}^1$ and $\theta \in [0,\,\theta_0]$, $\theta_0 > 0$, and $G(x;\,\theta) \neq G(x;\,\theta')$ for $\theta \neq \theta'$ and all $x \in \mathbf{R}^1$. In addition, assume $G(x;\,\theta)$ is absolutely continuous in x for all θ, and the corresponding density $g(x;\,\theta)$ is strongly positive. Denote by $G^{-1}(x;\,0)$ the inverse of $G(x;\,0)$ and introduce the auxiliary d.f. $r(x;\,\theta)$ specified on $[0, 1]$ by the relation

$$r(x; \theta) = G\big(G^{-1}(x; 0); \theta\big).$$

The principal restrictions on the family $\{G(x; \theta)\}$ are introduced subsequently in terms of $r(x; \theta)$. Suppose that for all $x \in [0, 1]$ and $\theta \in [0, \theta_0]$ the function $r(x; \theta)$ possesses all mixed derivatives up to second order in x and up to first order in θ that are jointly continuous in x and θ. The derivatives in x are denoted by r' and r'', and the derivative in θ and mixed derivatives by r_θ, r'_θ, and r''_θ, respectively.

We introduce just one more parameter

$$\tau := \rho_1 \rho_2 \int\limits_0^1 J'(t) \, r_\theta(t; 0) \, dt$$

and require that $\tau > 0$. This condition ensures the consistency of a test based on $\{S_N\}$ in the case of verifying H_1 against A_1 for small θ. Note also that

$$\tau^2 = \rho_1 \rho_2 \text{ eff } (J; G).$$

Let Λ^2 be the set of all pairs (f, g) of absolutely continuous d.f.s on $[0, 1]$. Consider for any $a \in \mathbf{R}^1$ the following set

$$\Delta_a^S := \Big\{ (f, g) \in \Lambda^2 \colon \rho_1 \int\limits_0^1 J(\rho_1 f + \rho_2 g) \, f' \, dt \le a \Big\}.$$

The Kullback–Leibler information is, as usual, defined by

$$I_\rho(f, g; r) := \rho_1 \int\limits_0^1 f' \ln f' \, dt + \rho_2 \int\limits_0^1 g' \ln \frac{g'}{r'} \, dt$$

and for any $\Omega \subset \Lambda^2$ we put

$$I_\rho(\Omega; r) := \inf \Big\{ I_\rho(f, g; r) \colon (f, g) \in \Omega \Big\}.$$

The following main result formulated already by Nikitin (1987c) will be proved in the sequel.

Theorem 3.5.1 (Nikitin 1990a, 1991) *Let the conditions stated for the score function J and the family of d.f.s $\{G(x; \theta)\}$ be fulfilled. If $|a|$ and θ are sufficiently small such that $\tau \theta > a$, then the function $a \mapsto I_\rho(\Delta_a^S; r)$ is continuous. Moreover,*

$$I_\rho(\Delta_a^S; r) = (2 \, \rho_1 \rho_2)^{-1} (\tau \theta - a)^2 + o\, (\theta^2 + a^2) \qquad (3.5.4)$$

as $a \to 0$ and $\theta \to 0$.

It is easy to deduce from this theorem another theorem directly connected with large deviations under the alternative.

Theorem 3.5.2 *Let the alternative hypothesis A_1 be true and the regularity conditions stated earlier be fulfilled. Then for any real sequence $\gamma_N \to 0$ and sufficiently small $|a|$ and θ such that $\tau \theta > a$, we have*

$$\lim_{N\to\infty} N^{-1} \ln P\left(S_N \le a + \gamma_N\right) = -I_\rho\left(\Delta_a^S; r\right).$$

Proof of Theorem 3.5.2 The distribution of statistic (3.5.1) under the alternative does not change after a Smirnov transform on the initial samples. As a result the elements of the first sample will be uniformly distributed on $[0, 1]$ whereas the elements of the second sample will have d.f. $r(x; \theta)$, $0 \le x \le 1$, with the density

$$r'(x; \theta) = \frac{g\left(G^{-1}(x; 0); \theta\right)}{g\left(G^{-1}(x; 0); 0\right)}. \tag{3.5.5}$$

Let F_m and G_n be the empirical d.f.s of transformed samples and put

$$H_N(t) := \frac{m}{m+n} F_m(t) + \frac{n}{m+n} G_n(t).$$

The well-known representation

$$S_N = \frac{m}{N} \int_0^1 J\left(\left(NH_N(t)+1\right)/(N+1)\right) dF_m(t)$$

belongs to Chernoff and Savage (1958) (cf. with (3.1.12)). It will be convenient to proceed further to the statistic

$$S_N' = \rho_1 \int_0^1 J\left(H_N(t)\right) dF_m(t).$$

Condition (3.1.2) together with the properties of J guarantee the existence of a sequence of real numbers $\{\gamma_N\}$ such that

$$\left| S_N - S_N' \right| \le \gamma_N \longrightarrow 0 \tag{3.5.6}$$

with probability 1 as $m, n \to \infty$.

The statistic S_N' may be viewed as a continuous functional of a pair of empirical d.f.s F_m and G_n with respect to the uniform topology in Λ^2. Hence we may apply the large-deviation results from Chapter 1, especially Theorem 1.6.7. Using (3.5.6) and Theorem 3.5.1, we establish the assertion of Theorem 3.5.2. □

Proof of Theorem 3.5.1 When minimizing the functional $I_\rho\left(\Delta_a^S; r\right)$ one can restrict the set of admissible elements to the set

$$\mathcal{M}_\delta := \left\{ (f,g) \in \Lambda^2 \colon f' \geq \delta,\, g' \geq \delta \quad \text{a.e. on } [0,\,1] \right\},$$

where δ is an arbitrary small positive number. Exactly in the same way as (2.3.14) one proves the inequality

$$I_\rho\left(\Delta_{a_1}^S \cap \mathcal{M}_{\delta/2};\, r\right) - \delta \;\leq\; I_\rho\left(\Delta_a^S;\, r\right) \;\leq\; I_\rho\left(\Delta_a^S \cap \mathcal{M}_{\delta/2};\, r\right),$$

$$(3.5.7)$$

where $a_1 = a + \alpha(\delta) \to a$ as $\delta \to 0$. (See also Nikitin (1991), pp. 314–15.)

In order to prove Theorem 3.5.1 it is sufficient to evaluate $I_\rho\left(\Delta_{a_1}^S \cap \mathcal{M}_{\delta/2};\, r\right)$ and then to pass to the limit as $\delta \to 0$ in inequality (3.5.7).

To formalize the problem of evaluating $I_\rho\left(\Delta_{a_1}^S \cap \mathcal{M}_{\delta/2};\, r\right)$ in the same way as in Section 3.2 we put

$$x(t) = f(t) - t, \qquad y(t) = g(t) - r(t;\theta).$$

Applying the Lagrange principle and calculating the variation we obtain, after some standard transforms, the following system of nonlinear differential equations for the extremals x, y and some nonnegative indeterminate multiplier μ:

$$\begin{cases} \rho_1\, x'' + \mu\, (x'+1)\,(y'+r')\, J'(z+h) = 0, \\[2mm] \rho_2\, y''\, r' - \mu\,(x'+1)\,(y'+r')\,J'(z+h)\,r' - \rho_2\, y'\, r'' = 0, \end{cases} \qquad (3.5.8)$$

where

$$z(t) := \rho_1\, x(t) + \rho_2\, y(t) \qquad \text{and} \quad h(t) := \rho_1\, t + \rho_2\, r(t;\theta).$$

The system (3.5.8) should be considered together with the inequalities

$$x' + 1 > \tfrac{1}{2}\,\delta, \qquad y' + r' > \tfrac{1}{2}\,\delta, \qquad (3.5.9)$$

the boundary conditions

$$x(0) = x(1) = 0, \qquad y(0) = y(1) = 0, \qquad (3.5.10)$$

and the normalization condition

$$\rho_1 \int_0^1 J(z+h)\,(x'+1)\,dt = a_1. \qquad (3.5.11)$$

There are two small parameters in our problem: θ and a_1. We begin by finding the solutions of this problem for the case $\theta = a_1 = 0$. Denote these solutions by x_0 and y_0.

Lemma 3.5.1 *The system (3.5.8)–(3.5.11) for* $\theta = a_1 = 0$ *has the unique solution* $x_0 = y_0 = 0$.

Proof of Lemma 3.5.1 The system (3.5.8) for $\theta = a_1 = 0$ becomes equal to

$$
\begin{cases}
\rho_1\, x_0'' = -\mu_0 \left(x_0' + 1\right)\left(y_0' + 1\right) J'(z_0 + t)\,, \\[2mm]
\rho_2\, y_0'' = \mu_0 \left(x_0' + 1\right)\left(y_0' + 1\right) J'(z_0 + t)\,,
\end{cases}
\tag{3.5.12}
$$

from which it follows immediately that $z_0 \equiv 0$. Normalization condition (3.5.11) turns into

$$
\int\limits_0^1 J(t)\,(x_0' + 1)\, dt = 0\,.
$$

Since $x_0' = -(\rho_2/\rho_1)\, y_0'$, we have

$$
\rho_2\, y_0'' = -\mu_0\,(y_0' + 1)\left(1 - \frac{\rho_2}{\rho_1}\, y_0'\right) J'(t)\,.
$$

Putting $v = y_0' + 1$ and $\nu = -(\rho_1 \rho_2)^{-1}\mu_0$, one obtains

$$
v' = \nu\, v\,(1 - \rho_2\, v)\, J'(t)\,.
$$

Integrating this equation we have

$$
\rho_2\, v(t) \;=\; \frac{\exp\{\nu\, J(t)\}}{c + \exp\{\nu\, J(t)\}} \qquad \text{with }\; c > 0\,.
$$

Since v is a probability density and the normalization condition should be fulfilled, we get the following system of equations for unknown constants c and ν:

$$
\int\limits_0^1 \frac{c}{c + \exp\{\nu\, J(t)\}}\, dt = \rho_1\,,
\tag{3.5.13}
$$

$$
\int\limits_0^1 \frac{J(t)}{c + \exp\{\nu\, J(t)\}}\, dt = 0\,.
\tag{3.5.14}
$$

It is clear that for every ν there exists only one positive solution $c = c(\nu)$ of equation (3.5.13). Equation (3.5.14) is valid when $\nu = 0$ (then $c = \rho_1\rho_2^{-1}$), but (3.5.14) cannot have other solutions as its left-hand side is a strictly decreasing function of ν. It follows that $\mu_0 = 0$, hence, due to boundary conditions (3.5.10), we have $x_0 = y_0 = 0$. $\qquad\square$

We shall construct now the "perturbation" of the zero solution for $\theta \neq 0$ and $a_1 \neq 0$. The equation (3.5.8) together with the boundary conditions can be viewed as an implicit operator

$$A(x, y;\, \theta, \mu) = 0$$

from a neighborhood $\mathcal{U}$ of zero of the space

$$\overset{\circ}{C}{}^2[0,1] \times \overset{\circ}{C}{}^2[0,1] \times \mathbf{R}^2$$

into the space

$$C[0,1] \times C[0,1].$$

Here $\overset{\circ}{C}{}^2[0,1]$ is the Banach space of twice continuously differentiable functions $x(t)$ on $[0,1]$ satisfying the boundary conditions $x(0) = x(1) = 0$ with the usual norm

$$\| x \| := \max_{0 \le t \le 1} |x''(t)|.$$

It is clear that $A(0,0;0,0) = 0$. In order to apply the implicit operator theorem (Theorem 1.8.5) we must verify that its conditions are fulfilled.

First of all we note that the Fréchet derivative

$$B = A'_{(x,\,y)}(0,0;0,0)$$

is a continuously invertible operator from the space $\mathbf{E}_1 = \overset{\circ}{C}{}^2[0,1] \times \overset{\circ}{C}{}^2[0,1]$ into the space $\mathbf{E}_2 = C[0,1] \times C[0,1]$. Indeed, the value of the mapping B acting on an arbitrary element $(h_1, h_2)^{\mathrm{T}} \in \mathbf{E}_1$ is

$$B(h_1, h_2)^{\mathrm{T}} = \left(\rho_1\, h_1'',\ \rho_2\, h_2'' \right)^{\mathrm{T}},$$

where ^{T} signifies transposition. Therefore the inverse operator exists and is continuous (it may be easily constructed with the aid of Green function (2.4.8)).

Secondly we note that the operator A belongs to the class $C^1(\mathcal{U})$. It is sufficient for this fact (Alexeev et al. (1979), Section 2.2.4)) that all partial derivatives in x, y, μ, and θ of components A_1 and A_2 of the operator A are continuous in the uniform operator topology. We write out these derivatives, denoting for brevity

$$w(t) := \rho_1\, x(t) + \rho_2\, y(t) + \rho_1\, t + \rho_2\, r(t;\, \theta).$$

For the derivatives in x and y we give here their values at an arbitrary element $h \in \mathbf{E}_1$:

$$A_{1x}(x, y; \mu, \theta)\, h = \rho_1\, h'' + \mu\, (y' + r')\, J'(w)\, h'$$
$$+ \mu\, \rho_1\, (x' + 1)\, (y' + r')\, J''(w)\, h\,,$$

$$A_{1y}(x, y; \mu, \theta)\, h = \mu\, (x' + 1)\, J'(w)\, h' + \mu\, \rho_2\, (x' + 1)\, (y' + r')\, J''(w)\, h\,,$$

$$A_{1\mu}(x, y; \mu, \theta) = (x' + 1)\, (y' + r')\, J'(w)\,,$$

$$A_{1\theta}(x, y; \mu, \theta) = \mu\, (x' + 1)\, J'(w)\, r'_\theta + \mu\, \rho_2\, (x' + 1)\, (y' + r')\, J''(w)\, r'_\theta\,,$$

$$A_{2x}(x, y; \mu, \theta)\, h = \mu\, (y' + r')\, J'(w)\, r'\, h$$
$$+ \mu\, \rho_1\, (x' + 1)\, (y' + r')J''(w)\, r'\, h\,,$$

$$A_{2y}(x, y; \mu, \theta)\, h = \rho_2\, r'\, h'' - \rho_2\, r''\, h' - \mu\, (x' + 1)\, J'(w)\, r'\, h'$$
$$- \mu\, \rho_2\, (x' + 1)\, (y' + r')\, J''(w)\, r'\, h\,,$$

$$A_{2\mu}(x, y; \mu, \theta) = (x' + 1)\, (y' + r')\, J'(w)\, r'\,,$$

$$A_{2\theta}(x, y; \mu, \theta) = \rho_2\, y''\, r'_\theta - \rho_2\, y'\, r''_\theta + \mu\, (x' + 1)\, (y' + r')\, J'(w)\, r'$$
$$+ \mu\, (x' + 1)\, J'(w)\, r'_\theta + \mu\, \rho_2\, (x' + 1)\, (y' + r')\, J''(w)\, r'\, r'_\theta\,.$$

Elementary arguments show that these partial derivatives are really continuous under the conditions of Theorem 3.5.1. Thus, by that conclusion of Theorem 1.8.5, there exist positive numbers $\varepsilon_1 > 0$ and $\delta_1 > 0$ and mappings

$$x \colon \mathcal{D}_{\delta_1}(0, \mathbf{R}^2) \mapsto \overset{\circ}{C}{}^2[0, 1]$$

and

$$y \colon \mathcal{D}_{\delta_1}(0, \mathbf{R}^2) \mapsto \overset{\circ}{C}{}^2[0, 1]$$

in the class $C^1\big(\mathcal{D}_{\delta_1}(0, \mathbf{R}^2)\big)$ such that

$$x(0,0) = 0\,, \qquad\qquad y(0,0) = 0\,, \qquad\qquad (3.5.15)$$

$$\| x(\mu, \theta) \| < \varepsilon_1\,, \qquad\qquad \| y(\mu, \theta) \| < \varepsilon_1 \qquad\qquad (3.5.16)$$

for $\mu^2 + \theta^2 < \delta_1^2$, and

$$A\big(x(\mu, \theta), y(\mu, \theta); \mu, \theta\big) = 0\,. \qquad\qquad (3.5.17)$$

Moreover, the mappings x and y are locally unique.

We may expand x and y in series in a neighborhood of zero and write

$$x(t; \mu, \theta) = x_{10}(t)\,\mu + x_{01}(t)\,\theta + o\left(\sqrt{\mu^2 + \theta^2}\right), \qquad (3.5.18)$$

$$y(t; \mu, \theta) = y_{10}(t)\,\mu + y_{01}(t)\,\theta + o\left(\sqrt{\mu^2 + \theta^2}\right), \qquad (3.5.19)$$

with

$$\left\| o\left(\sqrt{\mu^2 + \theta^2}\right) \right\| \longrightarrow 0 \qquad \text{as } \mu^2 + \theta^2 \longrightarrow 0\,.$$

The substitution of (3.5.18)–(3.5.19) in system (3.5.8) permits one to determine the coefficients $x_{10}, x_{01}, y_{10}, y_{01}$. We have

$$x''_{10}(t) = -\rho_1^{-1}\,J'(t)\,, \qquad\qquad y''_{10}(t) = \rho_2^{-1}\,J'(t)\,,$$

thus because of the boundary conditions

$$x_{10}(t) = \rho_1^{-1} \int\limits_0^1 \left[\min\,(t, s) - t\,s\right] J'(s)\,ds\,, 0 \qquad (3.5.20)$$

$$y_{10}(t) = -\rho_2^{-1} \int\limits_0^1 \left[\min\,(t, s) - t\,s\right] J'(s)\,ds\,, \qquad (3.5.21)$$

and, moreover,

$$x_{01}(t) = y_{01}(t) = 0\,.$$

Substituting the functions x and y in normalization condition (3.5.11) we obtain

$$F(\mu, \theta, a_1) \equiv \rho_1 \int\limits_0^1 J(\rho_1 x + \rho_2 y + \rho_1 t + \rho_2 r)\,(x' + 1)\,dt - a_1 = 0\,. \quad (3.5.22)$$

It is clear that $F(0, 0, 0) = 0$ and the function F is continuously differentiable in θ and a_1. Note that

$$F'_\mu(0, 0, 0) = \rho_1 \int\limits_0^1 J'(t)\,(\rho_1 x_{10}(t) + \rho_2 y_{10}(t))\,dt + \rho_1 \int\limits_0^1 J(t)\,x'_{10}(t)\,dt$$

$$= -\int\limits_0^1\!\!\int\limits_0^1 \left[\min\,(t, s) - t\,s\right] J'(t)\,J'(s)\,dt\,ds$$

$$= -\int\limits_0^1 J^2(t)\,dt = -1\,.$$

Now consider equation (3.5.22) from the point of view of the implicit function theorem. It is evident that for sufficiently small θ and a_1 there exists a continuously differentiable with respect to both arguments function $\mu(\,\cdot\,,\,\cdot\,)$ such that $\mu(0,0) = 0$ and $G\big(\mu(\theta, a_1), \theta, a_1\big) = 0$. Moreover,

$$\mu(\theta, a_1) = \mu_{10}\,\theta + \mu_{01}\,a_1 + o\left(\sqrt{\theta^2 + a_1^2}\right), \qquad \theta^2 + a_1^2 \longrightarrow 0. \quad (3.5.23)$$

The coefficients μ_{10} and μ_{01} may be found after the substitution of (3.5.23) in (3.5.22):

$$\mu_{10} = \tau = \rho_1\rho_2 \int\limits_0^1 J'(t)\,r_\theta(t;\,0)\,dt\,, \qquad \mu_{01} = -1\,. \quad (3.5.24)$$

Using (3.5.24) and (3.5.23) in representations (3.5.18)–(3.5.19) we obtain, passing to the functions f and g, that

$$f(t) \;=\; t + x_{10}(t)\,(\mu_{10}\,\theta - a_1) + o\left(\sqrt{\theta^2 + a_1^2}\right),$$

$$g(t) \;=\; r(t;\,\theta) + y_{10}(t)\,(\mu_{10}\,\theta - a_1) + o\left(\sqrt{\theta^2 + a_1^2}\right)$$

as $\theta^2 + a_1^2 \to 0$. Note that $\mu_{10}\theta - a_1 = \tau\,\theta - a_1 > 0$ for small θ and a_1 because the inequality $\mu > 0$ should be valid.

Since f and g are elements of the space $\overset{\circ}{C}{}^2[0, 1]$, we can write

$$f'(t) \;=\; 1 + x'_{10}(t)\,(\mu_{10}\,\theta - a_1) + o\left(\sqrt{\theta^2 + a_1^2}\right), \quad (3.5.25)$$

$$g'(t) \;=\; r'(t;\,\theta) + y'_{10}(t)\,(\mu_{10}\,\theta - a_1) + o\left(\sqrt{\theta^2 + a_1^2}\right). \quad (3.5.26)$$

It follows from these formulas that the conditions $f' > \delta/2$ and $g' > \delta/2$, equivalent to (3.5.9), are fulfilled for sufficiently small θ and a_1, so the extremals f and g belong not only to $\Delta_{a_1}^S$ but also to $\Delta_{a_1}^S \cap \mathcal{M}_{\delta/2}$. Moreover, it is clear from (3.5.18)–(3.5.19) that the remainder terms in (3.5.25)–(3.5.26) tend to zero uniformly in $t \in [0, 1]$.

Now put expressions (3.5.25) and (3.5.26) into the minimized functional

$$I_\rho\left(\Delta_{a_1}^S \cap \mathcal{M}_{\delta/2};\, r\right).$$

By properties of extremals the considered integrands are bounded and continuously depend on a_1. If $\delta \to 0$ then $a_1 \to a$ but the continuity in a is preserved. Passing to the limit in (3.5.7), as $\delta \to 0$, we obtain the continuity of the function $a \mapsto I_\rho\left(\Delta_a^S;\, r\right)$.

Taking into account (3.1.11) we omit elementary calculations and give only the final result

$$I_\rho \left(\mathbf{\Delta}_a^S; r \right) = \left(2 \rho_1 \rho_2 \right)^{-1} \left(\mu_{10} \theta - a \right)^2 + o \left(\sqrt{\theta^2 + a^2} \right). \qquad (3.5.27)$$

Due to (3.5.24) this implies the conclusion of Theorem 3.5.1. □

We turn now to discuss the regularity conditions introduced before the formulation of Theorem 3.5.1. The smoothness condition imposed on J is typical for a number of papers on the limiting behavior of linear rank statistics under the alternative. On the other hand the conditions on the distribution of observations are more restrictive. Apparently they are connected with the method of a proof. It is essential to know how regular is the structure of "comparison density" (3.5.5) (the term goes back to E. Parzen and is quoted, e.g., by Eubank, La Riccia, and Rosenstein (1987)). For instance, the Gaussian location family has the comparison density

$$r'(x; \theta) = \exp \left\{ \theta \ \Phi^{-1}(x) - \tfrac{1}{2} \theta^2 \right\}$$

and the indicated conditions are violated at the ends of the interval $[0, 1]$. Just the opposite is the case for the logistic location family which has a very simple comparison density

$$r'(x; \theta) = e^\theta \left(e^\theta + (1 - e^\theta) x \right)^{-2}$$

satisfying all aforementioned conditions. The same holds for the Cauchy location family with

$$r'(x; \theta) = (1 + \theta \sin 2\pi x + \theta^2 \sin^2 \pi x)^{-1}$$

and for many other well-known families of distributions.

Using Theorems 3.5.1–3.5.2 we may compute the Hodges–Lehmann and Chernoff local indices of the sequence of statistics (3.5.1). Clearly the critical value $\gamma_N(\alpha)$ corresponding to a level α is of order $O(N^{-1/2})$. Therefore, combining Theorems 3.5.1 and 3.5.2 with $a = 0$, we establish the following assertion.

Theorem 3.5.3 (Nikitin 1991) *Under the regularity conditions of Theorem 3.5.1 the Hodges–Lehmann index $d_S(\theta)$ of the sequence of statistics (3.5.1) satisfies the asymptotic relation*

$$d_S(\theta) = \text{eff}(J; G) \cdot \theta^2 \left(1 + o(1) \right), \qquad \theta \to 0. \qquad (3.5.28)$$

Comparing (3.5.28) with (3.5.3) we see that the Hodges–Lehmann indices and Bahadur exact slopes for statistics (3.5.1) are locally equivalent under general conditions.

Now we proceed to the calculation of the Chernoff local index. Combining Theorems 3.1.2 and 3.5.1 and choosing in the first of them

$$c^* = \tfrac{1}{2}\tau\theta\left(1 + o(1)\right), \qquad \theta \to 0,$$

we obtain the following result.

Theorem 3.5.4 (Nikitin 1991) *Under the regularity conditions of Theorem 3.5.1 the Chernoff index $\rho_S(\theta)$ of the sequence of statistics (3.5.1) satisfies the asymptotic relation*

$$\rho_S(\theta) = \tfrac{1}{8}\,\mathit{eff}(J;\,G)\cdot\theta^2\left(1 + o(1)\right), \qquad \theta \to 0. \qquad (3.5.29)$$

Therefore the local Chernoff index of the sequence of statistics (3.5.1) differs only up to the constant factor $\tfrac{1}{8}$ from the Hodges–Lehmann local index (3.5.28) and the Bahadur local exact slope (3.5.3). Hence for linear rank statistics satisfying the regularity conditions of Theorem 3.5.1 the local asymptotic ordering does not depend on the type of efficiency.

3.6 Asymptotic Comparison of Tests in Problems of Signal Detection in Noise of Unknown Level

One of the high priority problems of statistical radio physics is the two-sample problem of detection of weak signals in Gaussian noise of unknown level (see, e.g., Davenport and Root (1958), Prokof'yev (1973), Levin (1976)). Its mathematical formulation is as follows. We consider two independent samples $X_1, X_2, \ldots, X_n$ and $Y_1, Y_2, \ldots, Y_n$ of size n. It is required to test the null hypothesis H_1^* under which both samples have the Rayleigh distribution with the density

$$g(x;\,0,\sigma^2) = \begin{cases} \dfrac{x}{\sigma^2}\,\exp\left\{-\dfrac{x^2}{2\sigma^2}\right\} & \text{if } x \geq 0, \\[2mm] 0 & \text{if } x < 0. \end{cases} \qquad (3.6.1)$$

Under the alternative A_1^* the second sample has once again the Rayleigh distribution with density (3.6.1) and the first sample has the Rice distribution with the density

$$
g(x;\,\theta,\sigma^2) = \begin{cases} \dfrac{x}{\sigma^2}\,\exp\Big\{-\dfrac{x^2+\theta^2}{2\sigma^2}\Big\}\,I_0\Big(\dfrac{x\,\theta}{\sigma^2}\Big) & \text{if } x \ge 0\,, \\[2mm] 0 & \text{if } x < 0, \end{cases} \tag{3.6.2}
$$

where $I_0(\,\cdot\,)$ is the modified Bessel function of order zero. We emphasize that $\sigma^2 > 0$ is here an unknown nuisance parameter. The construction of tests for H_1^* against the alternative A_1^* not depending on σ^2 is often referred to in statistical radio physics and radio engineering as "overcoming a priori uncertainty."

Prokof'yev (1973) obtained the locally most powerful test invariant with respect to parameter σ and based on the statistic

$$
\Pi_{n,n} = \sum_{i=1}^{n} X_i^2 \Big/ \sum_{i=1}^{n} Y_i^2 \,. \tag{3.6.3}
$$

One often uses also invariant tests proposed for heuristic reasons. The so-called test of logarithmic contrast based on the statistic

$$
\Lambda_{n,n} = n^{-1} \sum_{i=1}^{n} \ln\big(X_i/Y_i\big), \tag{3.6.4}
$$

proposed by Artamonov and Shishkin (1972), may serve as a typical example.

On the other hand nonparametric tests based on the difference between empirical distribution functions as well as linear rank tests also may be used for testing H_1^* against A_1^* as all them are "distribution-free" under the null hypothesis. As representatives of these classes of tests we consider here (in the notation of Section 3.1) the Smirnov and Lehmann–Rosenblatt tests based on statistics

$$
D_{n,n} = \sup_{x \ge 0} \big| F_n(x) - G_n(x) \big|\,, \tag{3.6.5}
$$

$$
W_{n,n,1}^2 = \int_0^\infty \big(F_n(x) - G_n(x)\big)^2 \, d\big[(F_n(x) + G_n(x))/2\big] \tag{3.6.6}
$$

and the Wilcoxon–Mann–Whitney test based on the statistic

$$
Q_{n,n} = \Big[\int_0^\infty F_n(x)\,dG_n(x) - \tfrac{1}{2} \Big]. \tag{3.6.7}
$$

It is quite natural to use also the locally most powerful rank test for the problem under consideration. Such a test has been constructed by Nikitin and Filimonov (1981). It is based on the statistic

$$\Xi_{n,n} = \frac{1}{2n} \sum_{k=1}^{n} \left[\ln \left(1 - \frac{R_k}{2n+1} \right) + 1 \right], \qquad (3.6.8)$$

where R_k is the rank of the observation X_k from the pooled sample. Note that statistic (3.6.8) coincides with the well-known Savage linear rank statistic (see Savage (1956) and Hájek and Šidak (1967)).

In this section we effectuate the local comparison of statistics (3.6.3)–(3.6.8) on the basis of the Bahadur and Pitman AREs. As a result we obtain some ordering of these statistics (the same for both types of AREs), which permits us to give practical recommendations for the utilization of corresponding tests. Let us underline that the case of small θ we are considering just covers the most interesting case of weak signals in noise of level determined by parameter σ.

The calculation of the Hodges–Lehmann ARE is not so interesting here as it follows from Section 3.4 that statistics (3.6.5) and (3.6.6) are asymptotically optimal. The local Hodges–Lehmann indices of the statistics $\Pi_{n,n}$ and $\Lambda_{n,n}$ have been found by Nikitin and Filimonov (1989). As things turn out, the test based on $\Pi_{n,n}$ is locally asymptotically optimal and the test based on $\Lambda_{n,n}$ is less efficient by a factor of about 1.6.

As has been repeatedly noted already, the Bahadur efficiency of nonparametric tests is calculated usually for simple parametric alternatives. Therefore we restrict the alternative A_1^* to the simple hypothesis A_1^{**} under which the first sample has the density $g(x; \theta, \tau^2)$ and the second has the density $g(x; 0, \tau^2)$, where θ and τ^2 are given positive numbers.

Now we compute an upper bound for the Bahadur exact slopes in terms of the Kullback–Leibler information. As H_1^* is a composite hypothesis, we should, following Theorem 1.2.3, calculate

$$\inf_\sigma K(\theta, \tau^2; 0, \sigma^2) \equiv \inf_\sigma \left\{ \frac{1}{2} \left[\int_0^\infty \ln \frac{g(x; \theta, \tau^2)}{g(x; 0, \sigma^2)} \, g(x; \theta, \tau^2) \, dx \right. \right.$$

$$\left. \left. + \int_0^\infty \ln \frac{g(x; 0, \tau^2)}{g(x; 0, \sigma^2)} \, g(x; 0, \tau^2) \, dx \right] \right\}$$

$$= \inf_\sigma \frac{1}{2} \left[K_1 + K_2 \right].$$

Simple calculations show that, as $\theta \to 0$,

$$K_1 = \ln \frac{\sigma^2}{\tau^2} + \left(\frac{\tau^2}{\sigma^2} - 1 \right) \left(1 + \frac{\theta^2}{2\tau^2} + o(\theta^4) \right) + \frac{\theta^4}{8\sigma^4},$$

$$K_2 = \ln \frac{\sigma^2}{\tau^2} + \frac{\tau^2}{\sigma^2} - 1,$$

and the quantity $o(\theta^4)$ does not depend on σ. It is evident now that the infimum over σ of the expression $K(\theta, \tau^2; 0, \sigma^2)$ is attained at

$$\sigma^2 = \tau^2 \left(1 + \frac{\theta^2}{4\tau^2} \right)$$

and is equal to

$$\frac{\theta^4}{32\,\tau^4} \left(1 + o(1) \right).$$

Therefore the exact slope $c_T(\theta)$ of any sequence of statistics $\{T_{n,n}\}$ for testing H_1^* against A_1^{**} must satisfy the asymptotic inequality

$$c_T(\theta) \le \frac{\theta^4}{16\,\tau^4} \left(1 + o(1) \right), \qquad \theta \to 0. \qquad (3.6.9)$$

The computation of exact slopes of statistics (3.6.3)–(3.6.4) will be effectuated with the aid of Theorem 1.2.2. The large-deviation asymptotics of the sequence $\{\Pi_{n,n}\}$ under H_1^* may be found on the basis of Theorem 1.6.2. Standard calculations using the tables of Gradshteyn and Ryzhik (1971) demonstrate that for $\varepsilon > 0$

$$\lim_{n\to\infty} (2n)^{-1} \ln P\left(\Pi_{n,n} \ge \varepsilon \right) = \lim_{n\to\infty} (2n)^{-1} \ln P\left(\sum_{i=1}^{n} (X_i^2 - \varepsilon Y_i^2) \ge 0 \right)$$

$$= -\frac{1}{2} \ln \frac{4\varepsilon}{(1+\varepsilon)^2} \equiv -f_\Pi(\varepsilon).$$

To find the function $b_\Pi(\theta)$ we need the limit in probability of the ratio

$$\sum_{i=1}^{n} X_i^2 \Big/ \sum_{i=1}^{n} Y_i^2$$

under A_1^{**}. By the law of large numbers this limit is equal to

$$E X_1^2 \,/\, E Y_1^2 = 1 + \theta^2/(2\tau^2) + o(\theta^4). \qquad (3.6.10)$$

Substituting (3.6.10) in the expression obtained for $f_\Pi(\varepsilon)$ and expanding the result in a series we have

$$c_\Pi(\theta) = \frac{\theta^4}{16\,\tau^4} \left(1 + o(1) \right), \qquad \theta \to 0. \qquad (3.6.11)$$

Comparing with (3.6.9) we see that the locally most powerful test by Prokof'yev turns out to be also Bahadur locally optimal.

Large deviations of the sequence of statistics (3.6.4) may be found with the aid of Theorem 1.6.2 again. It is easy to verify that the density of the random variable $Z = X_1/Y_1$ has the form (under H_1^* and A_1^*, correspondingly)

$$p(z \mid H_1^*) = \frac{2\,z}{(1+z^2)^2}\,,$$

$$p(z \mid A_1^{**}) = \frac{2\,z}{(1+z^2)^3}\left(1+z^2\left(1+\frac{\theta^2}{2\,\tau^2}\right)\right)\exp\left\{-\frac{\theta^2}{2\,\tau^2\,(1+z^2)}\right\}.$$

We see that the moment generating function of $W = \ln Z$ is finite in a neighborhood of zero and the first moments of W under H_1^* are as follows

$$E\,W = 0\,. \qquad \operatorname{Var} W = \tfrac{1}{12}\,\pi^2\,.$$

If hypothesis H_1^* is true then by Theorem 1.6.2 we have

$$\lim_{n\to\infty}(2\,n)^{-1}\ln P\left(\Lambda_{n,n}\ge\varepsilon\right) = -\frac{6}{\pi^2}\,\varepsilon^2\left(1+o(1)\right),\qquad \varepsilon\to 0\,.$$

On the other hand, by the law of large numbers under hypothesis A_1^{**} the following convergence in probability takes place:

$$\Lambda_{n,n}\longrightarrow b_\Lambda(\theta) = \int\limits_0^\infty \ln z\,p(z\mid A_1^{**})\,dz \sim \frac{\theta^2}{4\,\tau^2}\,,\qquad \theta\to 0\,.$$

Therefore the local exact slope of the sequence of statistics $\{\Lambda_{n,n}\}$ is determined by the formulas

$$c_\Lambda(\theta) = \frac{3}{8\,\pi^2}\cdot\frac{\theta^4}{\tau^4}\left(1+o(1)\right),\qquad \theta\to 0\,. \tag{3.6.12}$$

We have already obtained in Sections 3.1–3.2 the expressions for the local exact slopes of Smirnov statistic (3.6.5) and Lehmann–Rosenblatt statistic (3.6.6). Choosing as θ_2 the d.f. corresponding to the density $g(x;\,0,\tau^2)$ and as θ_1 the d.f. corresponding to the density $g(x;\,\theta,\tau^2)$, we obtain after elementary calculations

$$c_D(\theta) = \frac{1}{4\,e^2}\cdot\frac{\theta^4}{\tau^4}\left(1+o(1)\right),\qquad \theta\to 0\,, \tag{3.6.13}$$

$$c_{W^2}(\theta) = \frac{\pi^2}{216}\cdot\frac{\theta^4}{\tau^4}\left(1+o(1)\right),\qquad \theta\to 0\,. \tag{3.6.14}$$

As to the Wilcoxon statistic (3.6.7) and the "Rayleigh scores" statistic (3.6.8), we may apply the results from Section 3.3, yielding

$$c_Q(\theta) = \frac{3}{64} \cdot \frac{\theta^4}{\tau^4} \left(1 + o(1)\right), \qquad\qquad \theta \to 0, \qquad (3.6.15)$$

$$c_\Xi(\theta) = \frac{1}{16} \cdot \frac{\theta^4}{\tau^4} \left(1 + o(1)\right), \qquad\qquad \theta \to 0. \qquad (3.6.16)$$

It follows from (3.6.11)–(3.6.16) that the tests under consideration admit the following local ordering with respect to the Bahadur efficiency. There is a tie for first place between the locally optimal tests of Prokof'yev and of Rayleigh scores. Then follow, in order of decreasing local Bahadur efficiency, the Wilcoxon test, the Lehmann–Rosenblatt test, and the test of logarithmic contrast, with the Smirnov test turning out to be the worst of them.

The local Bahadur ARE of any pair of tests we have studied may be found by the ratio of their local exact slopes. Let us compute its values for all considered tests with respect to the Prokof'yev test:

$$e^B_{\Xi,\Pi} = 1, \qquad e^B_{Q,\Pi} = \frac{3}{4} = 0.750, \qquad e^B_{W^2,\Pi} = \frac{2\pi^2}{27} \sim 0.731,$$

$$e^B_{\Lambda,\Pi} = \frac{6}{\pi^2} \sim 0.608, \qquad\qquad\qquad e^B_{D,\Pi} = \frac{4}{e^2} \sim 0.541.$$

As we have already noted in Section 1.4, the Pitman ARE (or its limiting value as $\alpha \to 0$) coincides with the local approximate Bahadur ARE if some additional conditions are fulfilled. These conditions are valid for the asymptotically normal statistics $\Pi_{n,n}$, $\Lambda_{n,n}$, $Q_{n,n}$, and $\Xi_{n,n}$. For the last two statistics $W^2_{n,n,1}$ and $D_{n,n}$ it is sufficient to verify the Wieand condition (1.4.3), which may be done in a standard way. It remains to prove the local coincidence of the exact and approximate Bahadur efficiencies. It is easy to effectuate it analyzing asymptotics of the tails of limiting distributions for the tests under consideration. (See more detail in Nikitin and Filimonov (1979).)

Therefore the ordering of tests in our problem on the basis of the Pitman ARE (or its limiting value as $\alpha \to 0$) is the same as for the local Bahadur ARE.

The previously discussed problem of discrimination between the Rayleigh and Rice distributions arises in the cases when the observed process is the narrow-band Gaussian noise or is a mixture of such noise with the effective signal. But in numerous applied problems the Gaussian model of the initial noise should be rejected. Examples of such situations were given, for example, by Akimov et al. (1978). There it was shown that in the general non-Gaussian case d.f. $F(t; \theta)$ of the envelope of the sum of the narrow-band noise and the signal with amplitude θ has the form

$$F(t; \theta) = \frac{1}{\pi} \int\limits_{|\theta - t|}^{|\theta + t|} \arccos \left[\frac{y^2 + \theta^2 - t^2}{2\,\theta\,y} \right] dG(y) + G(t - \theta), \quad (3.6.17)$$

where $G(t)$ is the d.f. of the envelope of the noise that is, as at the beginning of this section, known up to a scale parameter σ (it corresponds to the unknown noise level). It has been proved by Nikitin and Filimonov (1984) that for small θ the density $f(t; \theta)$ conforming to (3.6.17) satisfies the asymptotic relation

$$f(t; \theta) = g(t) + \frac{\theta^2}{4} \cdot \left(g''(t) - \left(\frac{g(t)}{t} \right)' \right) + o\,(\theta^2), \qquad \theta \to 0. \quad (3.6.18)$$

This formula is valid for sufficiently smooth densities g such that $g(0) = 0$. Using (3.6.18) Nikitin and Filimonov (1984) constructed the locally most powerful rank and invariant tests for the problem under consideration. A similar problem in the case of no amplitude but phase detection has been solved by Voshtshenko, Nikitin, and Filimonov (1990).

The nonstandard local structure of alternative (3.6.18) betokens interesting mathematical problems in the study of the efficiencies of various statistical tests in this model.

4

Asymptotic Efficiency of Nonparametric Symmetry Tests

4.1 Problem Statement and Tests under Consideration

Let $X_1, X_2, \ldots, X_n$ be a sample of size n with continuous d.f. F. We consider in this chapter the problem of testing the hypothesis H_2 about the *symmetry* of F relative to the origin, that is, the relation

$$H_2\colon\ 1 - F(x) - F(-x) = 0 \qquad \text{for all } x \in \mathbf{R}^1\,. \tag{4.1.1}$$

As the alternative A_2 we assume that the d.f. of observations X_i, $i = 1, \ldots, n$, is for some $\theta \in \Theta$ a known continuous d.f. $G(x; \theta)$, $x \in \mathbf{R}^1$, and Θ is a nonnegative neighborhood of zero in $\mathbf{R}^1$. We suppose also that

$$1 - G(x; \theta) - G(-x; \theta) = 0 \qquad \text{for all } x \in \mathbf{R}^1$$

only in the case $\theta = 0$.

The testing of hypothesis H_2 is one of the basic problems of nonparametric statistics. For instance, the well-known problem of *paired comparisons* when one must study the treatment effect if one subject of each pair is assigned to treatment and the other to control may be reduced to testing H_2 (see Lehmann (1975), Ch. 4). Hollander and Wolfe (1973), Lehmann (1975), and Hettmansperger (1984) have given numerous examples of practical situations connected with testing H_2.

There have been proposed a number of tests for testing the hypothesis of symmetry. The oldest and most well-known test is probably the *sign test* based on the statistic

$$E_n = n^{-1} \sum_{i=1}^{n} \left[\mathbf{1}_{\{X_i > 0\}} - \tfrac{1}{2} \right]. \tag{4.1.2}$$

As a matter of fact the sign test had been used already by Arbuthnott (1710).

In the sequel for any d.f. F on $\mathbf{R}^1$ we denote for brevity

$$\Delta F(x) := 1 - F(x) - F(-x),$$

and for any d.f. F defined on $[0, 1]$ we put

$$\Delta F(x) := 1 - F(x) - F(1 - x).$$

The statistical analog of the classic Kolmogorov statistic for goodness-of-fit testing is now

$$I_n = \sup_{-\infty < x < \infty} \left| \Delta F_n(x) \right|, \qquad (4.1.3)$$

introduced by Smirnov (1947) and studied later by Chatterjee and Sen (1973). It is possible to consider also the one-sided statistics I_n^+ and I_n^-, the definitions of which are evident.

In this situation the following statistics are the counterparts of the Watson–Darling statistics (2.1.5)–(2.1.7)

$$H_n = \sup_{-\infty < x < \infty} \left| \Delta F_n(x) - \int_{-\infty}^{\infty} \Delta F_n(y) \, dF_n(y) \right|, \qquad (4.1.4)$$

$$H_n^+ = \sup_{-\infty < x < \infty} \left[\Delta F_n(x) - \int_{-\infty}^{\infty} \Delta F_n(y) \, dF_n(y) \right], \qquad (4.1.5)$$

$$H_n^- = \sup_{-\infty < x < \infty} \left[\int_{-\infty}^{\infty} \Delta F_n(y) \, dF_n(y) - \Delta F_n(x) \right], \qquad (4.1.6)$$

considered for the first time by Abbakumov (1987) and later by Abbakumov and Nikitin (1993).

The ω^2-type statistic

$$R_n^2 = \int_{-\infty}^{\infty} \left(\Delta F_n(x) \right)^2 dF_n(x) \qquad (4.1.7)$$

for testing symmetry was proposed by Chentsov (1958). Later Orlov (1972), Rothmann and Woodroofe (1972), Srinivasan and Godio (1974), and Gregory (1977) as well as Koziol (1980) studied various properties of this statistic and its weighted variants. We shall consider the generalization of this statistic in the spirit of (2.1.9), namely,

$$R_{n, q}^k = \int_{-\infty}^{\infty} \left(\Delta F_n(x) \right)^k q\big(F_n(x)\big) \, dF_n(x), \qquad (4.1.8)$$

where k is a positive integer and q is a weight function. One supposes that q is a nonnegative function, symmetric with respect to the point $\frac{1}{2}$ and satisfying appropriate smoothness conditions.

By analogy with the Watson statistic (2.1.11) Hill and Rao (1977) proposed the statistic

$$N_n^2 = \int\limits_{-\infty}^{\infty} \left(\Delta\, F_n(x) - \int\limits_{-\infty}^{\infty} \Delta\, F_n(y)\, dF_n(y) \right)^2 dF_n(x), \qquad (4.1.9)$$

which may also be used for testing the hypothesis H_2.

In the determination of statistics (4.1.3)–(4.1.9) one may consider, by following Rothmann and Woodroofe (1972), the function

$$F_n^*(x) = \tfrac{1}{2} \left(F_n(x+0) + F_n(x-0) \right)$$

instead of $F_n(x)$. Then these statistics become invariant with respect to a change of sign of the observations, but their asymptotic properties are unchanged.

The *signed rank* tests form a vast and frequently used group of tests of symmetry. We consider such tests based on the statistics

$$Z_n = n^{-1} \sum_{i=1}^{n} a_n\big(R_i^+/(n+1)\big)\, \mathbf{1}_{\{X_i>0\}}, \qquad (4.1.10)$$

where $\big(R_1^+,\ldots,R_n^+\big)$ is the vector of ranks for the sample $|X_1|, |X_2|, \ldots,$ $|X_n|$ and the score function a_n, as in Section 3.1, is assumed to be constant on all intervals of the form $\big((i-1)/n,\, i/n\big]$, $1 \leq i \leq n$. We assume that

$$a_n(u) \xrightarrow{L^2} J(u),$$

where $J(u)$ is the limiting score function on $[0,\, 1]$. The simplest examples are $a_n(u) \equiv 1$, corresponding to sign statistic (4.1.2), and

$$a_n(u) = \sum_{i=1}^{n} \frac{i}{n}\, \mathbf{1}_{((i-1)/n,\, i/n]},$$

which corresponds to the Wilcoxon statistic.

The statistic (4.1.10) may be viewed as an integral functional of the empirical d.f. F_n due to the representation

$$Z_n = \int\limits_{0}^{\infty} a_n\Big(n\big(F_n(x+0) - F_n(-x)\big)/(n+1) \Big)\, dF_n(x). \qquad (4.1.11)$$

In the present chapter we shall find the values of various types of ARE for the statistics under consideration. We proceed now to study their large deviations under the null hypothesis.

4.2 Large Deviations of Kolmogorov-Type and Signed Rank Statistics

We start out with sign statistic (4.1.2). Since E_n is a sum of independent identically distributed random variables, it follows from Theorem 1.6.2 that under the hypothesis H_2 we have for $0 < a < 1/2$

$$\lim_{n \to \infty} n^{-1} \ln P(E_n \geq a) = -g_4(a), \tag{4.2.1}$$

where

$$g_4(a) := (\tfrac{1}{2} + a) \ln(1 + 2a) + (\tfrac{1}{2} - a) \ln(1 - 2a).$$

It is evident that

$$g_4(a) = 2a^2 (1 + o(1)) \qquad \text{as } a \to 0. \tag{4.2.2}$$

Now we proceed to study the statistic I_n, defined in (4.1.3), and its one-sided variants I_n^+ and I_n^-.

Let us consider for any $0 < a < 1$ the function

$$g_5(a) := -\inf \left\{ -t(a+1) + \ln \left[\tfrac{1}{2} (e^{2t} + 1) \right] : \ 0 \leq t \leq 1 \right\}.$$

The infimum on the right-hand side is easily found and we obtain

$$g_5(a) = \tfrac{1}{2}(1+a) \ln(1+a) + \tfrac{1}{2}(1-a) \ln(1-a) = \tfrac{1}{2} a^2 (1+o(1)), \quad a \to 0. \tag{4.2.3}$$

Theorem 4.2.1 *Let T_n be any of statistics I_n, I_n^+, or I_n^-. Under the hypothesis H_2 we have*

$$\lim_{n \to \infty} n^{-1} \ln P(T_n \geq a) = -g_5(a) \tag{4.2.4}$$

for $0 < a < 1$.

Proof As in the proof of Theorem 2.2.1 it is sufficient to consider the statistic I_n^-. After the Smirnov transform it takes the form

$$I_n^- = \sup_{0 \leq x \leq 1/2} \left[F_n(x) + F_n(1-x) - 1 \right] = \sup_{0 \leq x \leq 1/2} \left| \Delta F_n(x) \right|,$$

where the empirical d.f. F_n is constructed on the basis of observations uniformly distributed on $[0, 1]$. Arguing as in the proof of Theorem 2.2.3

and applying the Chernoff theorem we obtain the inequality

$$\lim_{n\to\infty} n^{-1}\ln P\big(I_n^- \ge a\big)$$

$$\le \sup_{0\le x\le 1/2}\ \inf_{0\le t\le 1}\left\{-t\,(a+1)+\ln\left(x\,e^{2t}+(1-2\,x)\,e^t+x\right)\right\}.$$

$$(4.2.5)$$

The function in braces is monotone in x for any fixed a and t, therefore the monotonicity is preserved also for the infimum of this function. Hence the supremum on the right-hand side is attained at the point $x=\frac{1}{2}$ and we get there the function $-g_5(a)$.

An upper bound may be obtained as in Theorem 2.2.3. Let k be again a large positive integer such that $\varepsilon_k=(\ln\ln k)^{-1}<a$. Then

$$P\left(I_n^- \ge a\right) \le P\left(\max_{1\le i\le k}\left[F_n\left(\tfrac{i-1}{k}\right)+F_n\left(1-\tfrac{i-1}{k}\right)-1\right]\ge a-\varepsilon_k\right)$$

$$+P\left(\max_{1\le i\le k}\left[F_n\left(\tfrac{i}{k}\right)-F_n\left(\tfrac{i-1}{k}\right)\right]\ge \varepsilon_k\right)=P_1+P_2.$$

The term P_1 may be estimated as in Theorem 2.2.3:

$$P_1 \le k\,\exp\left\{-n\,g_5(a-\varepsilon_k)\right\}.$$

The term P_2 has been bounded already in (2.2.15):

$$P_2 \le k\,\exp\left\{-n\big((1-\varepsilon_k)\ln(1-\varepsilon_k)+\varepsilon_k\,\ln\varepsilon_k+\varepsilon_k\,\ln(k-1)\big)\right\}.$$

Therefore

$$\overline{\lim_{n\to\infty}}\ n^{-1}\ln P\big(I_n^- \ge a\big) \le \max\left\{-g_5(a-\varepsilon_k),\,\ln 2-\varepsilon_k\,\ln(k-1)\right\}.$$

It remains to pass to the limit as $k\to\infty$. The continuity of the function $g_5(a)$ implies

$$\overline{\lim_{n\to\infty}}\ n^{-1}\ln P\big(I_n^- \ge a\big) \le -g_5(a).$$

$$(4.2.6)$$

Combining (4.2.5) and (4.2.6) we establish (4.2.4). $\qquad\square$

Initially Theorem 4.2.1 was proved by Chatterjee and Sen (1973) by applying another method based on the random walk theory.

Large deviations of the statistics H_n and $H_n^{\pm}$ may be analyzed in the same way as in Theorem 4.2.1. The principal difference, however, lies in the fact that the functional

$$\Delta F_n(x) - \int_{-\infty}^{\infty} \Delta F_n(y)\,dF_n(y)$$

cannot be represented as a sum of i.i.d. random variables but it is a von Mises functional of degree 2 with a bounded kernel. The large deviations of such functionals may be easily reduced with the aid of (1.6.12) to the large deviations of corresponding U-statistics and then one should use relation (1.6.11). We prove along these lines the following result.

Theorem 4.2.2 (Abbakumov and Nikitin 1993) *Let T_n be any of statistics H_n, H_n^+, or H_n^-. Then under the hypothesis H_2 we have*

$$\lim_{n \to \infty} n^{-1} \ln P\big(T_n \geq a\big) = g_6(a) = -\tfrac{3}{2} a^2 \left(1 + o(1)\right), \qquad a \to 0,$$

$$(4.2.7)$$

where $g_6(a)$ is continuous for small $a > 0$.

Proof Arguing as in previous chapters by analysis of the Kolmogorov–Smirnov-type statistics we see that it is sufficient to consider only the statistic H_n^+. Without loss of generality we may assume that the distribution of initial observations X_i, $i = 1, \ldots, n$, is uniform on $[-1, 1]$. Denote for any $u \in [-1, 1]$

$$H_n^+(u) := F_n(u) + F_n(-u) - \int_{-1}^{1} \big(F_n(t) + F_n(-t)\big)\, dF_n(t)$$

and observe that

$$H_n^+(u) = n^{-2} \sum_{i,j=1}^{n} \varphi_u(X_i, X_j) + n^{-1},$$

where the symmetric kernel φ_u is

$$\varphi_u(X_i, X_j)$$

$$:= \tfrac{1}{2}\big\{\mathbf{1}_{\{X_i < u\}} + \mathbf{1}_{\{X_i < -u\}} + \mathbf{1}_{\{X_j < u\}} + \mathbf{1}_{\{X_j < -u\}} - 2\,\mathbf{1}_{\{X_i + X_j < 0\}} - 1\big\}.$$

In order to apply Theorem 1.6.4 we introduce

$$\widetilde{H}_n^+(u) := \binom{n}{2}^{-1} \sum_{1 \leq i < j \leq n} \varphi_u(X_i, X_j)$$

and notice that for any $u \in [-1, 1]$

$$\big| H_n^+(u) - \widetilde{H}_n^+(u) \big| \leq 7\, n^{-1}. \qquad (4.2.8)$$

It is easy to verify that

$$\psi_u(t) := E\big[\, \varphi_u(X_1, X_2) \,|\, X_1 = t\,\big] = \tfrac{1}{2}\big\{\mathbf{1}_{\{t < u\}} + \mathbf{1}_{\{t < -u\}} + t - 1\big\}.$$

Calculating the moment generating function of the random variable $Y :=$ $\psi_u(X_1)$ we obtain for $t \geq 0$

$$\lambda_u(t) := E \exp\{tY\} = \frac{4}{t} \sinh \frac{t}{4} \cosh \left(\frac{t}{4} - \frac{t|u|}{2} \right).$$

Since the function $\lambda_u(t)$ is monotone in u on each of the sets $\{u : u > 0\}$ and $\{u : u < 0\}$ it follows that

$$\sup_{-1 \leq u \leq 1} \inf_{t \geq 0} \left[-at + \ln \lambda_u(t) \right] = \inf_{t \geq 0} \left[-at + \ln \lambda_1(t) \right] = g_6(2a), \quad (4.2.9)$$

where

$$g_6(a) := \inf_{t \geq 0} \left\{ -at + \ln(\sinh t/t) \right\}, \qquad a \geq 0.$$

It follows immediately from the implicit function theorem that the function $g_6(a)$ is continuous for small a and, moreover,

$$g_6(a) = -\frac{3}{2} a^2 \left(1 + o(1)\right), \qquad a \to 0. \quad (4.2.10)$$

Now for any $u \in [-1, 1]$ we have, due to (4.2.8),

$$P\left(H_n^+ \geq a \right) \geq P\left(H_n^+(u) \geq a \right) \geq P\left(\tilde{H}_n^+(u) \geq a + \delta \right) \quad (4.2.11)$$

for arbitrary $\delta > 0$ and sufficiently large n. Applying Theorem 1.6.4 for $\Phi = \varphi_u$ to inequality (4.2.11) we obtain

$$\lim_{n \to \infty} n^{-1} \ln P\left(H_n^+ \geq a \right) \geq \lim_{n \to \infty} n^{-1} \ln P\left(n^{-1} \sum_{i=1}^{n} \psi_u(X_i) \geq \tfrac{1}{2}(a+\delta) \right).$$

The limit on the right-hand side may be again evaluated by the Chernoff theorem (Theorem 1.6.2), which gives

$$\lim_{n \to \infty} n^{-1} \ln P\left(n^{-1} \sum_{i=1}^{n} \psi_u(X_i) \geq \tfrac{1}{2}(a + \delta) \right)$$
$$= \inf_{t \geq 0} \left[-\tfrac{1}{2}(a + \delta) t + \ln \lambda_u(t) \right].$$

Since $u \in [-1, 1]$ is arbitrary, we have, taking $u = 1$, the relation

$$\lim_{n \to \infty} n^{-1} \ln P\left(H_n^+ \geq a \right) \geq \inf_{t \geq 0} \left[-\tfrac{1}{2}(a + \delta) t + \ln \lambda_1(t) \right] \equiv g_6(a + \delta).$$

Passing to the limit as $\delta \to 0$ one obtains, owing to the continuity of g_6, that

$$\lim_{n \to \infty} n^{-1} \ln P\left(H_n^+ \geq a \right) \geq g_6(a). \quad (4.2.12)$$

To prove the opposite inequality we fix some large positive integer N and consider an integer m, $-N \leq m \leq N$. We may write

$$P\left(H_n^+ \geq a\right) = P\left(\sup_{-1 \leq u \leq 1} n^{-2} \sum_{i,j=1}^{n} \varphi_u(X_i, X_j) \geq a - n^{-1}\right)$$

$$\leq P\left(\max_{-N \leq m \leq N} \left[n^{-2} \sum_{i,j=1}^{n} \varphi_{m/N}(X_i, X_j)\right.\right.$$

$$\left.\left. + (2n)^{-1} \sum_{i=1}^{n} \mathbf{1}_{\{-m/N < X_i < -(m-1)/N\}}\right] > a - n^{-1}\right)$$

$$\leq \sum_{m=-N}^{N} P\left(H_n^+(m/N)\right.$$

$$\left. + (2n)^{-1} \sum_{i=1}^{n} \mathbf{1}_{\{-m/N < X_i < -(m-1)/N\}} > a\right)$$

$$\leq \sum_{m=-N}^{N} P\left(H_n^+(m/N) > a - \tau_N\right)$$

$$+ \sum_{m=-N}^{N} P\left((2n)^{-1} \sum_{i=1}^{n} \mathbf{1}_{\{-m/N < X_i < -(m-1)/N\}} > \tau_N\right)$$

$$= \Sigma_1 + \Sigma_2 \,,$$

say, where $\tau_N > 0$ will be prescribed subsequently.

To estimate Σ_1 we will use Theorem 1.6.1. Due to (4.2.9) and (1.6.3) we have for any $\delta_1 > \tau_N + 7n^{-1}$

$$\Sigma_1 \leq 2N \sup_{-1 \leq u \leq 1} P\left(H_n^+(u) \geq a - \tau_N\right)$$

$$\leq 2N \sup_{-1 \leq u \leq 1} P\left(\tilde{H}_n^+(u) \geq a - \delta_1\right)$$

$$= 2N \sup_{-1 \leq u \leq 1} P\left(n^{-1} \sum_{i=1}^{n} \psi_u(X_i) \geq \tfrac{1}{2}(a - \delta_1)\right)(1 + o(1))$$

$$\leq 2N \exp\left(n\left[\sup_{-1 \leq u \leq 1} \inf_{t \geq 0}\left(-\tfrac{1}{2}(a - \delta_1)t + \ln \lambda_u(t)\right)\right]\right)(1 + o(1))$$

$$= 2N \exp\left(n\, g_6(a - \delta_1)\right)(1 + o(1))\,.$$

As to the sum Σ_2 we have with the aid of (1.6.3) (cf. (2.2.15))

$$\Sigma_2 \leq 2N \sup_{-1 \leq u \leq 1} P\left(n^{-1} \sum_{i=1}^{n} \mathbf{1}_{\{-u < X_i < -u+N^{-1}\}} > 2\tau_N\right)$$

$$\leq 2N \exp\left\{-n\left[(1-2\tau_N)\ln(1-2\tau_N)\right.\right.$$

$$\left.\left. +2\tau_N \ln(2\tau_N) + 2\tau_N \ln(N-1)\right]\right\}$$

$$\leq 2N \exp\left\{n \ln 2 - 2n\tau_N \ln(N-1)\right\}.$$

If we take now

$$2\tau_N = \left(\ln(N-1)\right)^{-1/2},$$

then we see that Σ_2 is exponentially decreasing much faster than the first sum Σ_1. It follows that

$$\overline{\lim_{n\to\infty}} \, n^{-1} \ln P\left(H_n^+ \geq a\right) \leq g_6(a - \delta_1)$$

and consequently, owing to the continuity of g_6,

$$\overline{\lim_{n\to\infty}} \, n^{-1} \ln P\left(H_n^+ \geq a\right) \leq g_6(a). \tag{4.2.13}$$

Combining the two inequalities (4.2.12) and (4.2.13) with (4.2.10) we establish the conclusion of Theorem 4.2.2. □

The problem of studying large deviations of signed rank statistics (4.1.10) under the hypothesis of symmetry has been well studied. First results were obtained by Klotz (1965), and more general ones belong to Ho (1974), who used the same methods as Woodworth (1970), and to Groeneboom (1980), who exploited the Sanov theorem and variational arguments. To formulate the main result of Groeneboom (1980) let us assume that the sequence of statistics (4.1.10) satisfies all conditions indicated in Section 4.1, specifically, that

$$a_n(u) \xrightarrow{L^2} J(u), \qquad n \to \infty,$$

for a square integrable score function J.

It will be convenient to impose the normalization condition on the score function J in the form

$$\tfrac{1}{4} \int_0^1 J^2(u)\,du = 1. \tag{4.2.14}$$

Denote for brevity

$$J_0 = \tfrac{1}{2} \int\limits_0^1 J(u)\, du\,. \tag{4.2.15}$$

Theorem 4.2.3 (Groeneboom 1980) *Let the sequence of statistics* Z_n *given by (4.1.10) satisfy all conditions formulated earlier and*

$$\int\limits_0^1 \exp\left\{r\, J(u)\right\} du < \infty \qquad \text{for some } r > 0\,. \tag{4.2.16}$$

Under the hypothesis H_2 *we have for any real sequence* $\{\gamma_n\}$, $\gamma_n \to 0$, *the following relation:*

$$\lim_{n\to\infty} n^{-1} \ln P\big(Z_n - J_0 \geq a + \gamma_n\big) = g_7(a) = -\tfrac{1}{2}\, a^2 \left(1 + o(1)\right)\,, \qquad a \to 0\,, \tag{4.2.17}$$

where $g_7(a)$ *is continuous for small* $a > 0$.

4.3 Large Deviations of Integral Statistics for Testing Symmetry

To find large-deviation asymptotics of the integral statistic $R_{n,q}^k$ we will use the "variational" method elaborated in previous chapters. Consequently the proofs will be given in less detail. It is clear that under the hypothesis H_2 the distribution of the statistic $R_{n,q}^k$ for continuous F does not depend on the distribution of observations. Therefore we can write

$$R_{n,q}^k = \int\limits_0^1 \big(\Delta\, F_n(x)\big)^k q\big(F_n(x)\big)\, dF_n(x)\,, \tag{4.3.1}$$

assuming that $F_n(x)$ is the empirical d.f. based on a sample from the uniform distribution on $[0, 1]$.

Consider now the set $\mathbf{R}_a$, $a > 0$, consisting of those absolutely continuous on $[0, 1]$ d.f.s g for which the following condition holds:

$$\int\limits_0^1 \big(1 - g(t) - g(1-t)\big)^k q\big(g(t)\big)\, dg(t) \geq a\,. \tag{4.3.2}$$

We put again, in accordance with (1.2.10),

$$K(\mathbf{R}_a) := \inf\left\{\int\limits_0^1 g'(t) \ln g'(t)\, dt : g \in \mathbf{R}_a\right\}.$$

The following result will be proved in the sequel.

Theorem 4.3.1 (Nikitin 1982) *Let the weight q be positive, symmetric with respect to the point $\frac{1}{2}$, and expandable in a convergent power series on $(0, 1)$. Then for sufficiently small $a > 0$ the quantity $K(\mathbf{R}_a)$ may be represented as a convergent series with numerical coefficients*

$$K(\mathbf{R}_a) = -\sum_{j=2}^{\infty} c_j\, a^{j/k}, \qquad (4.3.3)$$

where $c_2 = \frac{1}{8}\lambda_0(q;k)$ and $\lambda_0(q;k)$ is the "principal" eigenvalue of problem (2.4.5) (see also (4.3.17)).

For sufficiently small a it follows from (4.3.3) that $K(\mathbf{R}_a)$ is a continuous function of a. We prove now that the statistic $R_{n,q}^k$ is a uniformly continuous functional of the empirical d.f. F_n in the ρ-topology. Denote for brevity

$$M := \sup_x q(x), \qquad \varphi_k(F) := \bigl(1 - F(x) - F(1 - x)\bigr)^k,$$

$$|F - G| := \sup_x |F(x) - G(x)|,$$

where F and G are arbitrary d.f.s. It is evident that

$$|\varphi_k(F) - \varphi_k(G)| \le 2\,k\,|F - G|,$$

$$\int_0^1 |d\varphi_k(F)| \le 2\,k.$$

Clearly the following chain of inequalities holds:

$$\left| \int_0^1 \varphi_k(F)\, q(F)\, dF - \int_0^1 \varphi_k(G)\, q(G)\, dG \right|$$

$$\le \left| \int_0^1 \varphi_k(F)\, q(F)\, dF - \int_0^1 \varphi_k(F)\, q(G)\, dG \right|$$

$$+ \left| \int_0^1 \varphi_k(F)\, q(G)\, dG - \int_0^1 \varphi_k(G)\, q(G)\, dG \right|$$

$$\le \left| \int_0^1 \left(\int_0^F q(u)\, du - \int_0^G q(u)\, du \right) d\varphi_k(F) \right| + 2\,k\,M\,|F - G|$$

$$\le 4\,k\,M\,|F - G|.$$

This implies the continuity of the functional $R_{n,q}^k$ of F_n in the ρ-topology and hence in the τ-topology. Using Theorem 1.6.7 we obtain the following result.

Theorem 4.3.2 *For sufficiently small $a > 0$*

$$\lim_{n \to \infty} n^{-1} \ln P\left(R_{n,q}^k \geq a\right) = -K(\mathbf{R}_a), \qquad (4.3.4)$$

where $K(\mathbf{R}_a)$ may be calculated by (4.3.3).

Proof of Theorem 4.3.1 The computation of $K(\mathbf{R}_a)$ is connected with the solution of the extremal problem of the minimization of entropy on the set $K(\mathbf{R}_a)$. Lemma 1.7.5 shows that this solution exists. By analogy with the proof of Theorem 2.3.1 it is natural to restrict the class of admissible elements to the subset of $\overset{\circ}{\mathbf{W}}_{\infty,1}$ where the variation exists and then to use the Lagrange principle for conditional extrema.

Reasoning as in Lemma 2.3.1 we state the following inequality: For sufficiently small $\delta > 0$

$$K\left(\mathbf{R}_{a+\alpha(\delta)} \cap V_{\delta/2}\right) - \delta \leq K(\mathbf{R}_a) \leq K\left(\mathbf{R}_a \cap V_{\delta/2}\right), \qquad (4.3.5)$$

where $\alpha(\delta) \to 0$ as $\delta \to 0$ and V_δ has been defined in (2.3.10). To compute $K\left(\mathbf{R}_{a_1} \cap V_{\delta_1}\right)$, where $a_1 = a + \alpha(\delta)$ and $\delta_1 = \delta/2$, we enlarge the set V_{δ_1} up to $V'_{\delta_1/2}$ (see the definition in (2.3.11)); we shall see later that this will not change the value of K.

Now we pass from absolutely continuous d.f.s g to the class of absolutely continuous functions

$$h(t) = g(t) - t, \qquad 0 \leq t \leq 1,$$

and look for the infimum of the functional

$$f_0(h) := \int_0^1 (h' + 1) \ln (h' + 1) \, dt, \qquad (4.3.6)$$

defined on an open set $\mathcal{U}$ of the Banach space $\overset{\circ}{\mathbf{W}}_{\infty,1}$ containing those functions h for which the inequality

$$h'(t) + 1 > \tfrac{1}{4} \delta \qquad (4.3.7)$$

holds for almost all $t \in [0, 1]$ under the additional constraint

$$f_1(h) := (-1)^k \int_0^1 \bigl(h(t) + h(1 - t)\bigr)^k q\bigl(t + h(t)\bigr) \bigl(1 + h'(t)\bigr) \, dt - a_1 \geq 0.$$

$$(4.3.8)$$

Under such a formalization of the problem Theorem 1.8.1 is applicable and it guarantees the existence of Lagrange multipliers $\widehat{\lambda}_0$ and $\widehat{\lambda}_1$,

$\widehat{\lambda}_0^2 + \widehat{\lambda}_1^2 > 0$, such that the extremal $\widehat{h}(t)$ of problem (4.3.6)–(4.3.8) satisfies the equation

$$\mathcal{L}_h\left(\widehat{h},\widehat{\lambda}_0,\widehat{\lambda}_1\right) = 0, \qquad (4.3.9)$$

where

$$\mathcal{L}\left(h,\widehat{\lambda}_0,\widehat{\lambda}_1\right) = \widehat{\lambda}_0\, f_0(h) + \widehat{\lambda}_1\, f_1(h)\,,$$

as well as the complementary condition

$$\widehat{\lambda}_1\, f_1\left(\widehat{h}\right) = 0\,.$$

Calculating (4.3.9) explicitly we obtain after standard manipulations that $\widehat{h}$ is twice differentiable and satisfies the following Euler–Lagrange equation

$$\widehat{\lambda}_0\, \widehat{h}''(t) = k\,\widehat{\lambda}_1\left(\widehat{h}(t) + \widehat{h}(1-t)\right)^{k-1}\left(1 + \widehat{h}'(t)\right)$$

$$\times \left(1 + \widehat{h}'(1-t)\right)\left[q(t + \widehat{h}(t)) + q(1 - t + \widehat{h}(1-t))\right]$$

(4.3.10)

together with the boundary conditions

$$\widehat{h}(0) = \widehat{h}(1) = 0\,. \qquad (4.3.11)$$

The assumption $\widehat{\lambda}_0 = 0$ contradicts normalization condition (4.3.8). Therefore we can take $\widehat{\lambda}_0 = 1$. It follows from (4.3.10) that

$$\widehat{h}''(t) = \widehat{h}''(1-t)\,.$$

This implies because of the boundary conditions that

$$\widehat{h}(t) = \widehat{h}(1-t)\,,$$

and this relation enables one to simplify nonlinear equation (4.3.10), which takes now the form

$$\widehat{h}''(t) = k\, 2^{k-1}\,\widehat{\lambda}_1\, \widehat{h}^{k-1}(t)\left(1 - \widehat{h}'^{\,2}(t)\right)\left[\, q(t + \widehat{h}(t)) + q\left(1 - t + \widehat{h}(t)\right)\,\right].$$

(4.3.12)

As to the normalization condition it may be written as

$$(-1)^k \int_0^1 \widehat{h}^k(t)\, q\left(\widehat{h}(t) + t\right)\left(\widehat{h}'(t) + 1\right) = a_1\, 2^{-k}\,.$$

Using the new notation

$$\varepsilon = \tfrac{1}{2}\, a_1^{1/k}\,, \qquad \widehat{h} = -f\,\varepsilon\,,$$

we can rewrite it in the form

$$\int_0^1 f^k(t)\, q\big(t - f(t)\varepsilon\big)\left(1 - f'(t)\,\varepsilon\right) dt = 1\,. \qquad (4.3.13)$$

Then equation (4.3.12) becomes

$$f'' = \tfrac{1}{2}\, v\, f^{k-1}\left(1 - f'^2\,\varepsilon^2\right)\left[q(t - f\varepsilon) + q(1 - t - f\varepsilon)\right] \qquad (4.3.14)$$

with $v = (-1)^k\,\widehat{\lambda}_1\, k\, 2^k\, \varepsilon^{k-2}$.

Now we use the smoothness properties of the weight q and obtain

$$\frac{1}{2}\left[q(t - f\varepsilon) + q(1 - t - f\varepsilon)\right] = q(t) + \sum_{j=1}^{\infty} \frac{(f\varepsilon)^{2j}}{(2j)!}\, q^{(2j)}(t)\,.$$

Hence equation (4.3.12) takes the form

$$f'' = v\, f^{k-1}\left(1 - f'^2\varepsilon^2\right)\left(q + \sum_{j=1}^{\infty} \frac{(f\varepsilon)^{2j}}{(2j)!}\, q^{(2j)}\right), \qquad (4.3.15)$$

and normalization condition (4.3.13) turns into

$$\int_0^1 f^k\left(q + \sum_{j=1}^{\infty} \frac{(-f\varepsilon)^{j}}{j!}\, q^{(j)}\right)(1 - f'\varepsilon)\, dt = 1\,. \qquad (4.3.16)$$

Let us prove that system (4.3.15)–(4.3.16) admits an analytic solution in ε in a neighborhood of zero.

Consider the auxiliary system resulting from (4.3.15)–(4.3.16) for $\varepsilon = 0$:

$$f'' = v\, f^{k-1}\, q\,, \qquad \int_0^1 f^k\, q\, dt = 1\,, \qquad f(0) = f(1) = 0\,. \qquad (4.3.17)$$

We have already met the same system in Section 2.4. Denote by x_0 and $\lambda_0 = \lambda_0(q;k)$ the "principal" solution of (4.3.17), that is, assume that λ_0 is the maximal eigenvalue and x_0 is a corresponding eigenfunction (possibly not unique).

Now put

$$x = f - x_0\,, \qquad \lambda = v - \lambda_0\,.$$

Then equation (4.3.15) takes the form

$$x'' - (k - 1)\,\lambda_0\, x_0^{k-2}\, q\, x = \lambda\, x_0^{k-1}\, q + R_1(x, \varepsilon, \lambda), \qquad (4.3.18)$$

where $R_1(x, \varepsilon, \lambda)$ contains the terms with x, ε, and λ in powers not less than 2 and ε only in even powers.

Equation (4.3.16) may be rewritten in the same way:

$$k \int_0^1 x\, x_0^{k-1}\, q\, dt - \varepsilon \int_0^1 x_0^k\, x_0'\, q\, dt - \varepsilon \int_0^1 x_0^{k+1}\, q'\, dt + R_2(x, \varepsilon, \lambda) = 0, \quad (4.3.19)$$

where $R_2(x, \varepsilon, \lambda)$ again contains terms with powers not less than 2.

We have already studied in Section 2.4 a system similar to (4.3.18)–(4.3.19). The difference consists only in the terms of "high" power and in the smoothness of the weight q. It is natural to consider equation (4.3.18) as an implicit analytic operator from the space $\overset{\circ}{C}{}^2[0,1]$ of twice continuously differentiable functions with zero boundary conditions into the space $C\,[0,1]$. Using the same technique as in Section 2.4 (including the Lyapunov–Schmidt branching equation), we obtain that the solution x of problem (4.3.18)–(4.3.19) may be represented in the form

$$x(t) = \sum_{j=1}^{\infty} d_j(t)\, \varepsilon^j, \quad (4.3.20)$$

and the series in the right-hand side of (4.3.20) is absolutely converging for sufficiently small $\varepsilon > 0$. This yields the representation

$$\widehat{h}(t) = -x_0(t)\, \varepsilon - \sum_{j=1}^{\infty} d_j(t)\, \varepsilon^{j+1}, \quad (4.3.21)$$

where $\varepsilon = \frac{1}{2}\, a_1^{1/k}$. We see that for sufficiently small $\varepsilon > 0$ inequality (4.3.7) is valid and, moreover, $\widehat{h}(t) + t$ is a d.f. Substituting (4.3.21) in the functional being minimized we get

$$K\big(\mathbf{R}_{a_1} \cap V_{\delta_1}\big) = -\tfrac{1}{8}\, \lambda_0\, a_1^{2/k} - \sum_{j=3}^{\infty} c_j\, a_1^{j/k}.$$

Passing to the limit as $\delta \to 0$ in this equality and using (4.3.5), we establish the conclusion of Theorem 4.3.1. $\qquad\square$

Taking the weight q to be infinitely differentiable, we have striven in the first place for simplicity of formulation and for similarity with the proof of Theorem 2.3.1. If the weight q has only a finite smoothness, one should use in equations (4.3.15) and (4.3.16) only a truncated series expansion of q. It gives the approximate branching equation (cf. Vainberg and Trenogin (1974), Chapter 10) and as a result we can obtain only a finite number of terms in the expressions on the right-hand sides of (4.3.21) and (4.3.3).

In the case when the weight q is nonsymmetric with respect to the point $\frac{1}{2}$, one should make an obvious change in the proof due to the more complicated and clumsy form of some basic equations.

To prove Theorems 4.3.1–4.3.2 we used the previous method, which may be applied to analyze large deviations of the Hill–Rao statistic N_n^2 given by (4.1.9). Abbakumov (1986) formulated the following result and it was proved by Abbakumov (1987).

Theorem 4.3.3 *Under the hypothesis H_2 for sufficiently small $a > 0$ we have*

$$\lim_{n\to\infty} n^{-1}\ln P\left(N_n^2 \geq a\right) = -\tfrac{1}{2}\pi^2 a + \sum_{j=3}^{\infty} b_j\, a^{j/2}, \qquad (4.3.22)$$

where the series on the right-hand side is convergent and the coefficients b_j may be computed explicitly.

The proof is connected again with the minimization of the Kullback–Leibler information on the appropriate set of d.f.s. The Euler–Lagrange equation for an extremal f has the form

$$f''(t) - \lambda\left(f(t) - \int_0^1 f(s)\,ds - \varepsilon\, f'^2(t)\left(f(t) - \int_0^1 f(s)\,ds\right)\right) = 0\,, \quad (4.3.23)$$

which should be considered together with the boundary conditions

$$f(0) = f(1) = 0 \qquad (4.3.24)$$

and the normalization condition

$$\int_0^1 f^2(t)\,dt - \left(\int_0^1 f(s)\,ds\right)^2 = 1\,.$$

The principal and very essential difference between equation (4.3.23) and equations (4.3.15) and (2.4.2) with $k = 2$ consists in that its linear part

$$f''(t) - \lambda\left(f(t) - \int_0^1 f(s)\,ds\right) = 0 \qquad (4.3.25)$$

considered together with (4.3.24) has two eigenfunctions corresponding to the principal eigenvalue $\lambda_0 = -4\pi^2$, namely,

$$f_1(t) = C\,\sin 2\pi t\,, \qquad\qquad f_2(t) = C\,(1 - \cos 2\pi t)\,.$$

It means that we are faced with the two-dimensional case of branching, as in Section 2.5 in connection with U_n^2. It is known (see Vainberg and

Trenogin (1974)) that the construction of such solutions requires more cumbersome considerations than in the one-dimensional case. Abbakumov (1986, 1987) has overcome these difficulties but the detailed proof of Theorem 4.3.3 is too lengthy to fit within the scope of this book.

4.4 Bahadur Efficiency of Symmetry Tests

The results of Sections 4.2–4.3 enable one to compare the symmetry tests we are considering on the basis of the Bahadur ARE. As an alternative to the symmetry hypothesis H_2 we consider the hypothesis A_2 introduced at the beginning of Section 4.1. We possess now the necessary information on large deviations of our test statistics under the null hypothesis. Therefore, in conformity with Theorem 1.2.2, it remains to determine their limiting behavior in P_θ-probability, corresponding to the alternative d.f. $G(x; \theta)$. In order to distinguish the resulting functions $b(\theta)$ from each other we shall equip them with the subscripts that correspond to their respective statistics.

In the simplest case of the sign statistic E_n it is sufficient to use the classical law of large numbers. It implies

$$E_n \xrightarrow{P_\theta} b_E(\theta) = \tfrac{1}{2} - G(0, \theta), \qquad n \to \infty. \qquad (4.4.1)$$

To treat the other statistics we use their representations as functionals of empirical d.f.s F_n. If one proves the continuity of these functionals in the uniform topology, then the corresponding functions $b(\theta)$ can be obtained via the Glivenko–Cantelli theorem (see, e.g., Borovkov (1984), Ch. 1). This continuity is obvious for the statistics I_n and $I_n^\pm$ and has been proved in Section 4.3 for the statistic $R_{n,q}^k$. The continuity of functionals corresponding to the statistics H_n, $H_n^\pm$, and N_n^2 may be proved in a similar way. Consider as an example the statistic H_n. Let x and y be arbitrary d.f.s on the real line, and put

$$R(x) := \int_{-\infty}^{\infty} \Delta x(t)\, dx(t), \qquad H(x) := \sup_t \big| \Delta x(t) - R(x) \big|.$$

We already proved in Section 4.3

$$\big| R(x) - R(y) \big| \le 4 \sup_t \big| x(t) - y(t) \big|.$$

Therefore

$$\left| H(x) - H(y) \right| \leq \sup_t \left| \Delta x(t) - R(x) - \Delta y(t) + R(y) \right|$$

$$\leq \sup_t \left| \Delta x(t) - \Delta y(t) \right| + \left| R(x) - R(y) \right|$$

$$\leq 6 \sup_t \left| x(t) - y(t) \right|.$$

Finally the law of large numbers for linear rank statistics has been well explored (see, e.g., Ho (1974), Groeneboom (1980), and Müller-Funk (1983)). Applying it to statistic (4.1.10) under the conditions of Theorem 4.2.3 we obtain

$$Z_n \xrightarrow{P_\theta} \int\limits_0^\infty J\big(G(x;\theta) - G(-x;\theta)\big) \, dG(x;\theta) \equiv b_Z(\theta).$$

Now we are able, as in Sections 2.6 and 3.3, to find the local exact Bahadur slopes of the statistics under consideration. Assuming that in all cases the alternative d.f. $G(x;\theta)$ is such that $b(\theta) > 0$ for $\theta > 0$ and $b(\theta) \to 0$ as $\theta \to 0$, we obtain

$$c_E(\theta) \sim 4\left[G(0;\theta) - \tfrac{1}{2} \right]^2,$$

$$c_I(\theta) \sim \sup_x \left[\Delta G(x;\theta) \right]^2,$$

$$c_H(\theta) \sim 3\left[\sup_x \left| \Delta G(x;\theta) - \int\limits_{-\infty}^\infty \Delta G(y;\theta) \, dG(y;\theta) \right| \right]^2,$$

$$c_{R_q^k}(\theta) \sim -\frac{1}{4} \lambda_0(q;k) \left[\int\limits_{-\infty}^\infty \big(\Delta G(x;\theta)\big)^k q\big(G(x;\theta)\big) \, dG(x;\theta) \right]^{2/k},$$

$$c_{N^2}(\theta) \sim \pi^2 \left[\int\limits_{-\infty}^\infty \Big(\Delta G(x;\theta) - \int\limits_{-\infty}^\infty \Delta G(y;\theta) \, dG(y;\theta) \Big)^2 dG(x;\theta) \right],$$

$$c_Z(\theta) \sim \left[\int\limits_0^\infty J\big(G(x;\theta) - G(-x;\theta)\big) \, dG(x;\theta) - J_0 \right]^2.$$

For brevity we have omitted in this list the local exact slopes of the statistics $I_n^\pm$ and $H_n^\pm$ which differ insignificantly from the slopes of the statistics I_n and H_n.

The subsequent simplification is possible under supplementary assumptions on the structure of the alternative. As an example let us consider the case of the simple location alternative when $G(x;\theta) = G(x-\theta)$, $\theta \geq 0$, and G is some symmetric with respect to zero absolutely continuous d.f. with a differentiable and bounded density g. Under this assumption it is easy to derive the principal part of all the local exact slopes we have found. As in Sections 2.6 and 3.3 we shall compare the local indices, that is, the coefficients of θ^2 as $\theta \to 0$. These indices now have the forms:

$$i_E = 4\,g^2(0)\,,$$

$$i_I = 4 \sup_x g^2(x) = i_{I^+}\,,$$

$$i_H = 12\left[\sup_x \left(g(x) - \int\limits_{-\infty}^{\infty} g^2(y)\,dy\right)^2\right],$$

$$i_{H^+} = 12\left[\sup_x \left(g(x) - \int\limits_{-\infty}^{\infty} g^2(y)\,dy\right)\right]^2,$$

$$i_{H^-} = 12\left[\int\limits_{-\infty}^{\infty} g^2(y)\,dy\right]^2,$$

$$i_{R_q^k} = -\lambda_0(q;k)\left[\int\limits_{-\infty}^{\infty} g^k(y)\,q\big(G(y)\big)\,dG(y)\right]^{2/k},$$

$$i_{N^2} = 4\,\pi^2\left[\int\limits_{-\infty}^{\infty} g^3(y)\,dy - \left(\int\limits_{-\infty}^{\infty} g^2(y)\,dy\right)^2\right].$$

To find the local index of signed rank statistic (4.1.10) we need to impose supplementary conditions on the score function J. We suppose that it satisfies (4.2.14) and is smooth enough so that

$$c_Z(\theta) \sim i_Z \cdot \theta^2\,, \qquad \theta \to 0\,,$$

where

$$i_Z = \left[\int\limits_0^{\infty} J\big(2\,G(t) - 1\big)\,g'(t)\,dt\right]^2. \tag{4.4.2}$$

Under typical conditions it really happens to be so. The expressions for indices of statistics of such a kind in the general case have been given by Kremer (1979a,b, 1982).

Before finding the mentioned indices let us discuss the question of an upper bound for all indices of considered statistics. In conformity with the general inequality (1.2.15) the exact slope of any sequence of statistics $\{T_n\}$ for testing H_2 against A_2 should satisfy the inequality

$$c_T(\theta) \leq 2 \inf\left\{ \int\limits_{-\infty}^{\infty} \ln \frac{g(x;\theta)}{h(x)}\, g(x;\theta)\,dx : h \in \mathfrak{H}\right\}, \tag{4.4.3}$$

where $\mathfrak{H}$ is the class of all symmetric densities. The infimum on the right-hand side may be easily found. It is attained for just the same density $h \in \mathfrak{H}$ for which is attained

$$\sup\left\{\int\limits_{-\infty}^{\infty} g(x;\theta)\,\ln h(x)\,dx : h \in \mathfrak{H}\right\}.$$

The symmetry of h implies that this supremum coincides with

$$\sup\left\{ \int_{-\infty}^{\infty} \tfrac{1}{2}\left(g(x;\theta) + g(-x;\theta)\right) \ln h(x)\, dx : \; h \in \mathfrak{H} \right\},$$

which is obviously attained for

$$h(x) = \tfrac{1}{2}\left(g(x;\theta) + g(-x;\theta)\right) \in \mathfrak{H}.$$

Therefore one has always (see also Ho (1974))

$$c_T(\theta) \le 2 \int_{-\infty}^{\infty} g(x;\theta) \ln \frac{2\,g(x;\theta)}{g(x;\theta) + g(-x;\theta)}\, dx. \tag{4.4.4}$$

In typical cases

$$\int_{-\infty}^{\infty} g(x;\theta) \ln \frac{2\,g(x;\theta)}{g(x;\theta) + g(-x;\theta)}\, dx \sim \frac{1}{8}\, I_1(g) \cdot \theta^2, \qquad \theta \to 0, \tag{4.4.5}$$

where

$$I_1(g) := \int_{-\infty}^{\infty} \frac{\left(g'_\theta(x;0) - g'_\theta(-x;0)\right)^2}{g(x;0)}\, dx.$$

Therefore in the class of densities $g(x;\theta)$ for which (4.4.5) holds the local index i_T of any sequence of statistics satisfies the inequality

$$i_T \le \tfrac{1}{4}\, I_1(g).$$

Now we present Table 5, a table of the local indices calculated for the three "model" densities, Gaussian, logistic, and Cauchy, that we already used in previous chapters. For all these densities in the case of a location parameter relation (4.4.5) is true, which enables one to consider $\frac{1}{4} I_1(g)$ as an upper bound for indices. As a representative of the class of signed rank statistics we choose the Wilcoxon rank statistic W_n with $J(u) = \sqrt{12}\, u$.

The coincidence of some rows with Table 3 of Section 2.6 may be explained by the identity of the local indices of several statistics for goodness-of-fit and symmetry testing in the case of location alternative.

For any pair of the statistics under consideration we can find the local Bahadur ARE by calculating the ratio of the corresponding indices. For instance, in the Gaussian case

$$e^{B}_{I, R_1^2} \approx 0.702, \qquad e^{B}_{N^2, W} \approx 0.509.$$

Table 5. *Local indices of symmetry tests under the location alternative*

Statistics	Gaussian	Logistic	Cauchy
E_n	0.637	0.250	0.407
I_n	0.637	0.250	0.407
H_n	0.955	0.333	0.304
H_n^+	0.164	0.083	0.304
H_n^-	0.955	0.333	0.304
$R_{n,1}^2$	0.907	0.329	0.375
$R_{n,1}^1$	0.955	0.333	0.304
N_n^2	0.486	0.219	0.500
W_n	0.955	0.333	0.304
$\frac{1}{4} I_1(g)$	1.000	0.333	0.500

4.5 Hodges–Lehmann Efficiency of Symmetry Tests

Consider again the problem of testing H_2 that we have studied in Sections 4.1–4.4. Let $X_1, X_2, \ldots, X_n$ be a sample of size n with absolutely continuous d.f. F. We will assume that under H_2 (4.1.1) is true whereas under the alternative A_2' d.f. F coincides with some nonsymmetric d.f. R with a positive density r.

Let Θ_0 be the set of all absolutely continuous symmetric d.f.s and Θ_1 be the set of nonsymmetric d.f.s with corresponding positive densities. We may assume that we are testing the hypothesis

$$H_2 \colon \ \theta \in \Theta_0$$

against the simple alternative

$$A_2' \colon \ \theta = R \in \Theta_1 \,.$$

Let $\{T_n\}$ be a sequence of nonparametric statistics assigned to solve this problem and $\{1 - \beta_{2n}(\alpha; R)\}$ be the corresponding sequence of probabilities of errors of the second kind where $\alpha \in (0, 1)$ is the value of the level.

Theorem 4.5.1 (Nikitin 1986c) *The inequality*

$$\varliminf_{n \to \infty} n^{-1} \ln \left[1 - \beta_{2n}(\alpha; R) \right] \geq \ln \int_{-\infty}^{\infty} \sqrt{r(x) \, r(-x)} \, dx \qquad (4.5.1)$$

holds.

Proof Arguing as in the proof of Theorem 1.3.1 we obtain

$$\varliminf_{n \to \infty} n^{-1} \ln \left[1 - \beta_{2n}(\alpha; R) \right] \geq - \inf_h \left\{ \int_{-\infty}^{\infty} \ln \frac{h(x)}{r(x)} \, h(x) \, dx \right\},$$

where the infimum is taken over the class $\mathfrak{H}$ of symmetric densities h. For any $h \in \mathfrak{H}$ we have

$$\int_{-\infty}^{\infty} \ln \frac{h(x)}{r(x)} \, h(x) \, dx = \int_{-\infty}^{\infty} \ln \frac{h(x)}{\sqrt{r(x) \, r(-x)}} \, h(x) \, dx \,. \qquad (4.5.2)$$

It only remains to notice that the lower bound of the functional on the right-hand side of (4.5.2) is attained for the symmetric density

$$h(x) = \frac{\sqrt{r(x) \, r(-x)}}{\int\limits_{-\infty}^{\infty} \sqrt{r(x) \, r(-x)} \, dx} \,. \qquad \square$$

It follows from Theorem 4.5.1 that if the Hodges–Lehmann index $d_T(R)$ of the sequence $\{T_n\}$ exists, it satisfies the inequality

$$d_T(R) \leq -2 \ln \int_{-\infty}^{\infty} \sqrt{r(x) \, r(-x)} \, dx \,. \qquad (4.5.3)$$

As in previous chapters the sequence $\{T_n\}$ is said to be Hodges–Lehmann asymptotically optimal if equality holds in (4.5.3).

Now we will show that the statistics I_n, H_n, $R_{n,q}^k$ (for even k), and N_n^2, introduced in Section 4.1, are Hodges–Lehmann asymptotically optimal. This is in agreement with the Hodges–Lehmann asymptotic optimality of similar goodness-of-fit and two-sample statistics stated in Chapters 2 and 3.

Extending the class of integral statistics we consider the statistics of the form

$$\bar{R}_{n,q}^{\lambda} = \int_{-\infty}^{\infty} |\Delta F_n(x)|^{\lambda} \, q\big(F_n(x)\big) \, dF_n(x),$$

where q is a positive weight function with a bounded derivative and $\lambda \geq 1$. It differs from the statistics $R_{n,q}^k$ introduced in Section 4.1 by the presence of the absolute value under the integral sign and the arbitrariness of the exponent λ. One may generalize in a similar way the statistic N_n^2, too.

Theorem 4.5.2 *The sequences of statistics* $\{I_n\}$, $\{H_n\}$, $\{N_n^2\}$, *and* $\{R_{n,q}^\lambda\}$ *are Hodges–Lehmann asymptotically optimal.*

Proof As all sequences of statistics under consideration correspond to continuous in the ρ-topology functionals of empirical distributions, we may apply Theorem 1.6.8. Let $\{T_n\}$ be any of these sequences of statistics, $T_n := T(F_n)$ and $1 - \beta_{2n}(\alpha; R)$ be the corresponding sequence of probabilities of errors of second kind. It follows from (1.6.20) that

$$\varlimsup_{n \to \infty} \; n^{-1} \ln \left[1 - \beta_{2n}(\alpha; R) \right]$$

$$\leq - \inf \left\{ \int\limits_{-\infty}^{\infty} \ln \frac{f(x)}{r(x)} \; f(x) \, dx : \; T(F) \leq 0 \right\}, \qquad (4.5.4)$$

where the infimum is taken over all absolutely continuous d.f.s F with corresponding densities f. But the condition $T(F) \leq 0$ is equivalent to the condition of symmetry of F; hence the right-hand side of (4.5.4) coincides with the right-hand side of (4.5.1), which implies the conclusion of Theorem 4.5.2. $\qquad \square$

At the end of this section we shall establish an upper bound for the Chernoff indices of test statistics in the symmetry problem. Let $X_1, X_2, \ldots$ be again a sequence of independent identically distributed observations having under the hypothesis H_2' some unknown symmetric and positive density f whereas under the alternative A_2'' being distributed according to some positive density $g(x; \theta)$ that is symmetric only for $\theta = 0$. Let $\{T_n\}$ be a sequence of test statistics used for testing H_2' against A_2'' with the Chernoff index $\rho_T(\theta)$. Denote for brevity

$$\Re(g; \theta) := \sup_{0 < t < 1} \left\{ -t \ln \int\limits_{-\infty}^{\infty} \left[\tfrac{1}{2} \left(g^t(x; \theta) + g^t(-x; \theta) \right) \right]^{1/t} dx \right\}$$

Theorem 4.5.3 (Kopylev and Nikitin 1992) *For any $\theta \neq 0$ the inequality*

$$\rho_T(\theta) \leq \mathfrak{K}(g; \theta) \qquad (4.5.5)$$

holds.

Proof Denote by $\mathfrak{F}_0$ the set of all positive symmetric densities f on the real line. From (1.5.4) it follows that

$$\rho_T(\theta) \leq -\sup_f \inf_{0 \leq t \leq 1} \left[\ln \int_{-\infty}^{\infty} g^t(x; \theta) f^{1-t}(x) \, dx \right]$$

for any $\theta \neq 0$, with the supremum being taken over the set $\mathfrak{F}_0$. To simplify the right-hand side of this inequality we apply the asymmetric minimax theorem (see, e.g., Aubin and Ekeland (1984)), which permits one to exchange the operations of taking the sup and the inf. It is obvious that we can take the lower bound only over $t \in (0, 1)$. Consider now the even function

$$b(x; \theta) := \tfrac{1}{2} \left(g^t(x; \theta) + g^t(-x; \theta) \right)$$

and the corresponding symmetric positive density

$$b_0(x; \theta) := c_t \, b^{1/t}(x; \theta) \quad \text{with } c_t := \left(\int_{-\infty}^{\infty} b^{1/t}(x; \theta) \, dx \right)^{-1}.$$

The symmetry of f implies

$$\ln \int_{-\infty}^{\infty} g^t(x; \theta) f^{1-t}(x) \, dx = \ln \int_{-\infty}^{\infty} b(x; \theta) f^{1-t}(x) \, dx . \qquad (4.5.6)$$

Simple variational arguments show now that the supremum over $\mathfrak{F}_0$ of the right-hand side of (4.5.6) is attained at b_0. Hence inequality (4.5.5) is valid. $\qquad \square$

Likewise, the sequence $\{T_n\}$ is said to be *Chernoff locally asymptotically optimal* (in some class of densities $g(x; \theta)$ satisfying appropriate regularity conditions) in the case of testing H_2' against A_2'' if

$$\rho_T(\theta) \sim \mathfrak{K}(g; \theta) \qquad \text{as } \theta \to 0 . \qquad (4.5.7)$$

The verification of this condition may sometimes be facilitated by the asymptotics

$$\mathfrak{K}(g; \theta) \sim \tfrac{1}{32} I_1(g) \cdot \theta^2 \qquad \text{as } \theta \to 0 , \qquad (4.5.8)$$

which holds in typical cases, the corresponding Fisher information $I_1(g)$ being given by (4.4.5).

4.6 Hodges–Lehmann and Chernoff Efficiencies of the Sign and the Wilcoxon Statistics

In this section we shall find the Hodges–Lehmann and Chernoff local indices of the two simplest statistics for testing symmetry: the sign statistic and the Wilcoxon statistic. The simplicity of their structure enables one to do it relatively easily under weak regularity conditions imposed on the distribution of observations. In Section 4.7, we compute the local indices of general signed rank statistics (4.1.10) including the Wilcoxon statistic as a particular case. But the results of Section 4.7 are obtained under substantially more stringent conditions. Therefore the consideration of the sign and Wilcoxon tests in a separate section seems to be quite justified. The results have been obtained by Nikitin (1989).

We are again testing the hypothesis H_2 against the alternative A_2 under which the initial sample $X_1, X_2, \ldots, X_n$ has for some $\theta > 0$ a known d.f. $G(x; \theta)$, $x \in \mathbf{R}^1$, being symmetric only for $\theta = 0$. It is natural to assume that the considered tests are consistent in this testing problem. To that end we introduce two additional conditions

$$G(0; \theta) < \tfrac{1}{2}, \qquad \theta > 0, \qquad (4.6.1)$$

$$\int\limits_{-\infty}^{\infty} G(-x; \theta) \, dG(x; \theta) < \tfrac{1}{2}, \qquad \theta > 0, \qquad (4.6.2)$$

which seem to be quite natural and are fulfilled for most frequently encountered families of distributions.

In the sequel we shall need also some conditions on the smoothness of $G(x; \theta)$ in θ that will be specified later.

The large-deviation asymptotics of sign statistic (4.1.2) under the null hypothesis are described by relations (4.2.1) and (4.2.2). The corresponding result under the alternative depends certainly on d.f. $G(x; \theta)$. Denote for brevity

$$\lambda = \tfrac{1}{2} - G(0; \theta).$$

The inequality (4.6.1) implies that $\lambda > 0$ for all $\theta > 0$. Then Theorem 1.6.2 enables one to obtain the following result.

Theorem 4.6.1 *Suppose that A_2 is true and condition (4.6.1) holds.*

Then for any sequence $\{\gamma_n\} \to 0$ and for $\delta \uparrow 0$ we have the relation

$$\lim_{n\to\infty} n^{-1} \ln P(E_n - \lambda \leq \delta + \gamma_n) = -\tfrac{1}{2} \delta^2 \left(\tfrac{1}{4} - \lambda^2\right)^{-1} (1 + o(1)) . \quad (4.6.3)$$

If the function $\theta \mapsto G(0; \theta)$ is differentiable in θ at zero then

$$\lambda = -G'_\theta(0; 0) \cdot \theta + o(\theta), \qquad \theta \to 0.$$

Putting $\delta = -\lambda$ in (4.6.3) we find

$$\lim_{n\to\infty} n^{-1} \ln P_\theta \left(E_n \leq \gamma_n\right) = -2 \left[G'_\theta(0; 0)\right]^2 \cdot \theta^2 \left(1 + o(1)\right), \quad \theta \to 0.$$
$$(4.6.4)$$

Let $c_{n,\alpha}$ be the critical value at a level $\alpha \in (0, 1)$ for the sign test based on the statistic E_n and $1 - \beta_n^E(\alpha; \theta)$ be the corresponding probability of error of the second kind. Then substituting $\gamma_n = c_{n,\alpha}$ in (4.6.4) we have

$$\lim_{n\to\infty} n^{-1} \ln \left[1 - \beta_n^E(\alpha; \theta)\right] = -2 \left[G'_\theta(0; 0)\right]^2 \cdot \theta^2 \left(1 + o(1)\right), \quad \theta \to 0.$$

Therefore the Hodges–Lehmann index $d_E(\theta)$ of the sequence of statistics $\{E_n\}$ satisfies the relation

$$d_E(\theta) \sim 4 \left[G'_\theta(0; 0)\right]^2 \cdot \theta^2 \left(1 + o(1)\right), \qquad \theta \to 0.$$

This result seems to have been obtained first by Hodges and Lehmann (1956).

Local representation for the Chernoff index $\rho_E(\theta)$ of the sequence of sign statistics may be found using Theorem 1.5.1. After substituting $a = \tfrac{1}{2}\lambda$ in (4.2.1) and $\delta = -\tfrac{1}{2}\lambda$, $\gamma_n = 0$ in (4.6.3) one obtains

$$\rho_E(\theta) \sim \tfrac{1}{2} \left[G'_\theta(0; 0)\right]^2 \cdot \theta^2 \left(1 + o(1)\right), \qquad \theta \to 0.$$

It is convenient to write the Wilcoxon signed rank statistic in the equivalent form (see Hettmansperger (1984), Chapter 2), namely,

$$W_n = \frac{2}{n(n+1)} \sum_{1 \leq i \leq j \leq n} \left[\mathbf{1}_{\{X_i + X_j > 0\}} - \tfrac{1}{2}\right].$$

The statistic W_n is asymptotically equivalent to the following statistic

$$W'_n = \binom{n}{2}^{-1} \sum_{1 \leq i < j \leq n} \left[\mathbf{1}_{\{X_i + X_j > 0\}} - \tfrac{1}{2}\right],$$

due to the elementary inequality

$$\left| W_n - \frac{n-1}{n+1} W'_n \right| \leq \frac{1}{n+1} .$$

Clearly both sequences of statistics have the same asymptotics of large-deviation probabilities. The statistic W'_n is a U-statistic with the kernel

$$\Phi_1(X_1, X_2) := 1_{\{X_1 + X_2 > 0\}} - \tfrac{1}{2} .$$

Obviously,

$$\varphi_1(t) := E\{\Phi_1(X_1, X_2) \mid X_1 = t\} = F(t) - \tfrac{1}{2},$$

where F is the d.f. of X_1 under H_2 and

$$E\varphi_1^2(X_1) = \tfrac{1}{12} .$$

Moreover, we have

$$E\Phi_1(X_1, X_2) = 0, \qquad E\exp\{t\,\Phi_1(X_1, X_2)\} < \infty, \quad \forall t \in \mathbf{R}^1 .$$

Applying (1.6.11) we obtain the following result.

Theorem 4.6.2 *Under the hypothesis H_2 we have*

$$\lim_{n \to \infty} n^{-1}\ln P\big(W'_n \geq a\big) = -\tfrac{3}{2}\,a^2\,\big(1 + o(1)\big), \qquad a \to 0.$$

Similar arguments enable us to treat the case when the observations have, accordingly to the alternative A_2, d.f. $G(x; \theta)$. Denote by E_θ the expectation under A_2. Put

$$m(\theta) := E_\theta\,\Phi_1(X_1, X_2) = \tfrac{1}{2} - \int\limits_{-\infty}^{\infty} G(-x; \theta)\,dG(x; \theta),$$

$$\Phi_2(X_1, X_2) := m(\theta) - \Phi_1(X_1, X_2).$$

After simple calculations we obtain

$$E_\theta\,\Phi_2(X_1, X_2) = 0,$$

$$\varphi_2(t) := E_\theta\,\{\Phi_2(X_1, X_2) \mid X_1 = t\} = m(\theta) - \tfrac{1}{2} + G(-t; \theta),$$

$$\sigma^2(\theta) := E_\theta\,\varphi_2^2(X_1) = \int\limits_{-\infty}^{\infty} G^2(-x; \theta)\,dG(x; \theta) - \big(m(\theta) - \tfrac{1}{2}\big)^2.$$

Moreover, it is clear that for all t and θ

$$E_\theta\,\exp\{t\,\Phi_2(X_1, X_2)\} < e^t < \infty.$$

Using again (1.6.11) we establish the following assertion.

Theorem 4.6.3 *Under A_2 and condition (4.6.2) we have*

$$\lim_{n\to\infty} n^{-1}\ln P\big(W'_n - m(\theta) \le \tau + \gamma_n\big) = -\frac{\tau^2}{8\,\sigma^2(\theta)}\,\big(1+o(1)\big)$$

for any sequence $\gamma_n \to 0$ and $\tau \uparrow 0$.

Suppose now that $\theta \to 0$. If the family of d.f.s $\{G(x;\theta)\}$ is regular enough, for example, if

$$\sup_x \sup_\theta \big|G''_{\theta\theta}(x;\theta)\big| \le M\,, \qquad G'_\theta(-\infty;0) = 0\,,$$

then the asymptotic relations

$$
\begin{aligned}
m(\theta) &\sim -2\int_{-\infty}^{\infty} G'_\theta(x;0)\,dG(x;0)\cdot\theta\,, \\
\sigma^2(\theta) &\sim \tfrac{1}{12}
\end{aligned}
\tag{4.6.5}
$$

hold as $\theta \to 0$. In the sequel let us denote for brevity

$$\nu(G) := -\int_{-\infty}^{\infty} G'_\theta(x;0)\,dG(x;0)\,.$$

It follows from (4.6.2) that $m(\theta) > 0$. Therefore if we put $\tau = -m(\theta)$ in Theorem 4.6.3, we obtain, due to (4.6.5), under regularity conditions stated earlier the representation

$$\lim_{n\to\infty} n^{-1}\ln P\big(W'_n \le \gamma_n\big) = -6\,\nu^2(G)\cdot\theta^2\,\big(1+o(1)\big)\,, \qquad \theta \to 0\,.$$
$$\tag{4.6.6}$$

As for the sign statistic, the critical value $l_{n,\alpha}$ at a level α is of order $O(n^{-1/2})$ for the statistic W'_n, too. Therefore substituting $l_{n,\alpha}$ in (4.6.6) instead of γ_n we obtain that the Hodges–Lehmann index $d_W(\theta)$ of the sequence of statistics $\{W'_n\}$ admits the representation

$$d_W(\theta) \sim 12\,\nu^2(G)\cdot\theta^2\,, \qquad \theta \to 0\,.$$

The local Chernoff index $\rho_W(\theta)$ of this sequence may be found if we put

$$a \sim \tfrac{1}{2}\,m(\theta)$$

in Theorem 4.6.2 and

$$\tau \sim -\tfrac{1}{2}\,m(\theta)\,, \qquad \gamma_n = 0$$

in Theorem 4.6.3. Applying Theorem 1.5.1 we easily obtain

$$\rho_W(\theta) \sim \tfrac{3}{2}\,\nu^2(G)\cdot\theta^2\,, \qquad \theta \to 0\,.$$

The results just obtained show that the local Hodges–Lehmann and Chernoff AREs of the sign test with respect to the Wilcoxon test are given under natural regularity conditions by means of the formula

$$\left[\, G_\theta'(0;0)\,\right]^2 \big/ \left(3\,\nu^2(G)\right).$$

An analogous expression for the local Bahadur ARE of these tests follows from the results of Section 4.4. In the case of the Pitman efficiency it follows from Hodges and Lehmann (1956) and Hettmansperger (1984).

4.7 Hodges–Lehmann and Chernoff Efficiencies of Linear Signed Rank Tests

In this section we consider again testing the hypothesis H_2 against the alternative A_2, but as a test statistic we take the linear signed rank statistic

$$Z_n = n^{-1} \sum_{j=1}^{n} J\big(R_j^+/(n+1)\big)\, \mathbf{1}_{\{X_j>0\}} - J_0, \qquad (4.7.1)$$

where J is a nonconstant smooth score function on $[0, 1]$,

$$J_0 = \tfrac{1}{2} \int_0^1 J(u)\, du,$$

and R_j^+ is the rank of $|X_j|$ among the observations $|X_1|, |X_2|, \ldots, |X_n|$. The sequence $\{Z_n\}$ is asymptotically equivalent to sequence (4.1.10) and we will use the same notation for them.

To obtain large-deviation asymptotics of $\{Z_n\}$ under the alternative we need again some strengthening of the regularity conditions imposed on a score function J and d.f. $G(x;\theta)$ of observations. These conditions are entirely analogous to the conditions formulated in Section 3.5 by analysis of the two-sample linear rank statistics and consist in the following.

As is usually assumed in the theory of signed rank tests (see Pratt and Gibbons (1981) and Hettmansperger (1984)) we suppose that the function J is nonnegative and $J(0) = 0$. Moreover, let J be twice continuously differentiable on $[0, 1]$ and normalization condition (4.2.9) be fulfilled.

As to the family $\{G(x;\theta)\}$ defined for any $x \in \mathbf{R}^1$ and $\theta \in [0, \theta_0]$, $\theta_0 > 0$, we suppose that d.f.s $G(x;\theta)$ are absolutely continuous in x for all considered θ and that corresponding densities $g(x;\theta)$ are strictly positive.

As in Section 3.5 we introduce the auxiliary d.f.

$$r(x; \theta) := G\big(G^{-1}(x; 0); \theta\big)$$

and assume that for all x and θ the function $r(x; \theta)$ possesses all mixed derivatives up to second order in x and up to first order in θ that are jointly continuous in x and θ. The last condition imposed on the family $\{G(x; \theta)\}$ is

$$\lim_{x \to \infty} G'_\theta(x; 0) = 0.$$

When formulating the results we will use the constant

$$\varsigma := - \int\limits_{1/2}^{1} J'(2t - 1)\big(r_\theta(t; 0) + r_\theta(1 - t; 0)\big)\, dt,$$

supposing $\varsigma > 0$ (this is true in typical cases, e.g., for the location family $G(x; \theta) = G(x - \theta)$, $\theta \geq 0$). This condition ensures the consistency of the test based on large values of Z_n for small θ.

To obtain large-deviation asymptotics of the sequence $\{Z_n\}$ under the alternative we use the well-known representation (cf. (4.1.11))

$$Z_n = \int\limits_{0}^{\infty} J\left(\frac{n}{n+1}\big(F_n(x + 0) - F_n(-x)\big)\right) dF_n(x) - J_0. \qquad (4.7.2)$$

(See, e.g., Groeneboom (1980) and Hettmansperger (1984).) In order to apply results on large deviations of empirical measures we introduce for any absolutely continuous d.f. f on $[0, 1]$ the entropy functional

$$I(f; r) := \int\limits_{0}^{\infty} f' \ln\big(f'/r'\big)\, dt,$$

and for any set Ω of d.f.s on $[0, 1]$ put

$$I(\Omega; r) := \inf\Big\{I(f; r)\colon f \in \Omega\Big\}.$$

For any $a \in \mathbf{R}^1$ consider the following set of absolutely continuous d.f.s on $[0, 1]$ connected with the statistic Z_n:

$$\Omega_a^Z := \Big\{f\colon \int\limits_{1/2}^{1} J\big(f(x) - f(1 - x)\big) f'\, dx - J_0 \leq a\Big\}.$$

Theorem 4.7.1 (Nikitin 1990b) *Under the regularity conditions imposed above on the score function J and the parametric family $\{G(x; \theta)\}$*

for sufficiently small $|a|$ *and* $\theta > 0$ *such that* $\varsigma\,\theta > a$, *the function*

$$a \mapsto I\left(\Omega_a^Z; r\right)$$

is continuous. Moreover, if $a \to 0$ *and* $\theta \to 0$ *so that* $\varsigma\,\theta > a$, *we have*

$$I\left(\Omega_a^Z; r\right) = \tfrac{1}{2}\left(\varsigma\,\theta - a\right)^2 + o\left(\theta^2 + a^2\right). \tag{4.7.3}$$

This theorem enables one to obtain the following result, directly connected with large deviations of statistics (4.7.1) under the alternative.

Theorem 4.7.2 (Nikitin 1990b) *Suppose the alternative* A_2 *holds and the regularity conditions imposed earlier are fulfilled. Then for any sequence* $\{\gamma_n\}$, $\gamma_n \to 0$, *and sufficiently small* $|a|$ *and* $\theta > 0$ *such that* $\varsigma\,\theta > a$, *we have*

$$\lim_{n\to\infty} n^{-1}\ln P\big(Z_n \le a + \gamma_n\big) = -I\left(\Omega_a^Z; r\right).$$

Proof of Theorem 4.7.2 For statistic (4.7.1) written in the form (4.7.2) apply the Smirnov transform $x = G^{-1}(t;0)$. As a result this statistic will take the form

$$Z_n' = \int\limits_{1/2}^{1} J\left(\frac{n}{n+1}\Big(\widetilde{F}_n(x+0) - \widetilde{F}_n(1-x)\Big)\right)d\widetilde{F}_n(x) - J_0,$$

where the empirical d.f. $\widetilde{F}_n$ is based already on a sample with d.f. $r(t;\theta)$. Because of the smoothness of the score function there exists a numerical sequence $\{\gamma_n\}$ such that with probability 1

$$\left|Z_n' - Z_n''\right| \le \gamma_n \longrightarrow 0, \qquad n \to \infty,$$

where

$$Z_n'' := \int\limits_{1/2}^{1} J\big(\widetilde{F}_n(x) - \widetilde{F}_n(1-x)\big)\,d\widetilde{F}_n(x) - J_0.$$

Under the conditions of Theorem 4.7.2 the statistic Z_n'' is continuous in the uniform topology functional of the empirical d.f. $\widetilde{F}_n$. Taking into account the statement on the continuity of $I\left(\Omega_a^Z; r\right)$ in a given by Theorem 4.7.1, we can apply Theorem 1.6.7. Hence the conclusion of Theorem 4.7.2 is obtained. $\qquad\square$

Proof of Theorem 4.7.1 As in previous chapters let us restrict the set of admissible elements to the set $\mathcal{N}_\delta$ of absolutely continuous d.f.s on $[0, 1]$ of the form

$$\mathcal{N}_\delta := \{f: \ f' \geq \delta \quad \text{a.e. on } [0, 1]\},$$

where δ is an arbitrary small positive number. Arguing as in the proofs of Theorems 2.3.1 and 3.5.1 it is possible to show the existence of a function $\beta(\delta) \to 0$ as $\delta \to 0$ such that for sufficiently small δ and θ

$$I\left(\Omega_{a_1}^Z \bigcap \mathcal{N}_{\delta/2};\, r\right) - \delta \leq I\left(\Omega_a^Z;\, r\right) \leq I\left(\Omega_a^Z \bigcap \mathcal{N}_{\delta/2};\, r\right), \qquad (4.7.4)$$

where $a_1 = a + \beta(\delta) \to a$ as $\delta \to 0$. Therefore we must calculate $I\left(\Omega_a^Z \cap \mathcal{N}_{\delta/2};\, r\right)$ for small a and then pass in (4.7.4) to the limit as $\delta \to 0$.

In order to apply the Lagrange principle let us make the substitution $h = f - r$. It is necessary to find a lower bound of the functional

$$\int_0^1 (h' + r') \ln\left(\frac{h'}{r'} + 1\right) dt, \qquad (4.7.5)$$

defined on an open subset $\mathcal{U}$ of the Banach space $\overset{\circ}{\mathbf{W}}_{\infty,1}[0, 1]$ that contains the functions satisfying for a.e. $t \in [0, 1]$ the inequality

$$h' + r' > \tfrac{1}{4}\delta \qquad (4.7.6)$$

under the complementary condition

$$\int_{1/2}^1 J\big(h(t) + r(t) - h(1 - t) - r(1 - t)\big)\big(h'(t) + r'(t)\big)\, dt - J_0 \leq a. \quad (4.7.7)$$

Since the Gâteaux derivatives of functionals (4.7.5) and (4.7.7) are continuous on $\mathcal{U}$ in the uniform operator topology (it follows from inequality (4.7.6)), both functionals are strictly differentiable on $\mathcal{U}$ due to Theorem 1.8.2.

Under such a formalization of the initial problem we can apply the Lagrange principle (Theorem 1.8.1). Calculating the variation we obtain after standard transforms the Euler–Lagrange equation for extremal h with undetermined multiplier $\lambda \geq 0$

$$h''(t)\, r'(t) - h'(t)\, r''(t) + \lambda\left(h'(t) + r'(t)\right)\left(h'(1 - t) + r'(1 - t)\right)$$

$$\times\, r'(t)\left[J'\big(v(1 - t)\big)\mathbf{1}_{[0,\,1/2]}(t) + J'\big(v(t)\big)\mathbf{1}_{[1/2,\,1]}(t)\right] = 0$$

$$(4.7.8)$$

where

$$v(t) = h(t) + r(t) - h(1 - t) - r(1 - t).$$

The equation (4.7.8) should be considered simultaneously with the boundary conditions

$$h(0) = h(1) = 0 \qquad (4.7.9)$$

and the normalization condition

$$\int\limits_{1/2}^{1} J\big(v(t)\big)\,\big(h'(t) + r'(t)\big)\,dt - J_0 = a. \qquad (4.7.10)$$

There are two small parameters θ and a in the problem (4.7.8)–(4.7.10) under consideration. We begin by finding the solutions when $\theta = a = 0$.

Lemma 4.7.1 *The system (4.7.8)–(4.7.10) has the unique solution $h \equiv 0$, $\lambda = 0$ when $\theta = a = 0$.*

Proof For $\theta = a = 0$ equation (4.7.8) takes the form

$$h''(t) + \lambda\,\big(h'(t) + 1\big)\,\big(h'(1 - t) + 1\big)$$
$$\times \left[J'\big(v(1 - t)\big)\,\mathbf{1}_{[0,\,1/2]}(t) + J'\big(v(t)\big)\,\mathbf{1}_{[1/2,\,1]}(t) \right] = 0. \qquad (4.7.11)$$

It follows that $h''(t) = h''(1 - t)$; therefore by (4.7.9) $h(t) = h(1 - t)$ and $v(t) \equiv 2t - 1$ for $t \in [0, 1]$.

Then normalization condition (4.7.10) will be taken to be

$$\int\limits_{1/2}^{1} J\big(2t - 1\big)\,\big(h'(t) + 1\big)\,dt - J_0 = a$$

and equation (4.7.11) may be written in the form

$$h''(t) = \begin{cases} -\lambda\,J'(1 - 2t)\,\big(1 - h'^2(t)\big), & 0 \le t \le \tfrac{1}{2}, \\[2ex] -\lambda\,J'(2t - 1)\,\big(1 - h'^2(t)\big), & \tfrac{1}{2} \le t \le 1. \end{cases}$$

Integrating this equation we obtain

$$
h'(t) + 1 = \begin{cases} \dfrac{2\,C\,\exp\left\{\lambda\,J(1-2t)\right\}}{1 + C\,\exp\left\{\lambda\,J(1-2t)\right\}}\,, & 0 \le t \le \tfrac{1}{2}\,, \\[4mm] \dfrac{2\,C\,\exp\left\{\lambda\,J(2t-1)\right\}}{1 + C\,\exp\left\{\lambda\,J(2t-1)\right\}}\,, & \tfrac{1}{2} \le t \le 1\,, \end{cases}
$$

where C is some positive constant.

The equality $h'(t) = -h'(1-t)$ implies that $h'(\tfrac{1}{2}) = 0$, and hence $C = 1$. The normalization condition transforms into

$$
\int\limits_{1/2}^{1} J(2t-1)\,\frac{2\exp\left\{\lambda\,J(2t-1)\right\}}{1 + \exp\left\{\lambda\,J(2t-1)\right\}}\,dt - J_0 = 0\,. \tag{4.7.12}
$$

It is fulfilled for $\lambda = 0$ but cannot be true for $\lambda \ne 0$ because the left-hand side of (4.7.12) is monotone in λ. Due to (4.7.11) $\lambda = 0$ and $h \equiv 0$. $\quad\square$

Now we return to basic equation (4.7.8) and study the "perturbation" of the zero solution for $\theta \ne 0$ and $a \ne 0$.

Equation (4.7.8) together with boundary conditions (4.7.9) may be considered as an implicit operator

$$
\mathcal{B}(h;\theta,\lambda) = 0\,, \tag{4.7.13}
$$

from a neighborhood of zero $\mathcal{U}$ in the space $\overset{\circ}{C}{}^{2}[0,1] \times \mathbf{R}^2$ into the space $C[0,1]$. Clearly $\mathcal{B}(0;0,0) = 0$.

Let us apply the implicit operator theorem (Theorem 1.8.5) to obtain the solutions of (4.7.13). To check its conditions note first of all that the Fréchet derivative of the operator $\mathcal{B}$

$$
\mathcal{B}'_h(0;0,0)\,(u) = u''
$$

is a continuously invertible operator. Note, in the second place, that $\mathcal{B}$ belongs to the class $C^1(\mathcal{U})$. For this it is sufficient to prove that all partial derivatives in h, θ, and λ of $\mathcal{B}$ are continuous. Their existence follows from the regularity conditions listed in the formulation of Theorem 4.7.1. We write them out, denoting for brevity

$$
w_1(t) := J'\big(v(1-t)\big)\,\mathbf{1}_{[0,\,1/2]}(t) + J'\big(v(t)\big)\,\mathbf{1}_{[1/2,\,1]}(t)\,,
$$

$$
w_2(t) := J''\big(v(1-t)\big)\,\mathbf{1}_{[0,\,1/2]}(t) - J''\big(v(t)\big)\,\mathbf{1}_{[1/2,\,1]}(t)\,.
$$

We have

$$\mathcal{B}_h(h;\theta,\lambda)(u) = u''(t)\,r'(t) - u'(t)\,r''(t) + \lambda\,r'(t)\,w_1(t)$$

$$\times \Big(u'(t)\,\big(h'(1-t)+r'(1-t)\big) + u'(1-t)$$

$$\times \big(h'(t)+r'(t)\big)\Big) + \lambda\,r'(t)\,w_2(t)\,\big(h'(t)+r'(t)\big)$$

$$\times \big(h'(1-t)+r'(1-t)\big)\,\big(u(1-t)-u(t)\big),$$

$$\mathcal{B}_\theta(h;\theta,\lambda) = h''(t)\,r'_\theta(t) - h'(t)\,r''_\theta(t) + \lambda\,r'_\theta(t)\,r'(t)\,w_1(t)$$

$$\times \big(h'(1-t)+r'(1-t)\big) + \lambda\,r'_\theta(1-t)\,r'(t)\,w_1(t)$$

$$\times \big(h'(t)+r'(t)\big) + \lambda\,r'_\theta(t)\,w_1(t)\,\big(h'(t)+r'(t)\big)$$

$$\times \big(h'(1-t)+r'(1-t)\big) + \lambda\,r'(t)\,w_2(t)\,\big(h'(t)+r'(t)\big)$$

$$\times \big(h'(1-t)+r'(1-t)\big)\,\big(r'_\theta(1-t)-r'_\theta(t)\big),$$

$$\mathcal{B}_\lambda(h;\theta,\lambda) = \big(h'(t)+r'(t)\big)\big(h'(1-t)+r'(1-t)\big)r'(t)\,w_1(t).$$

It is easily verified that under the conditions of Theorem 4.7.1 the partial derivatives $\mathcal{B}_h$, $\mathcal{B}_\theta$, and $\mathcal{B}_\lambda$ are really continuous. Therefore by Theorem 1.8.5 there exists a ball $\mathcal{D}_\varepsilon(0,\mathbf{R}^2)$ with sufficiently small radius $\varepsilon > 0$ and a locally unique mapping

$$h\colon \mathcal{D}_\varepsilon(0,\mathbf{R}^2) \mapsto \overset{\circ}{C}{}^2[0,1]$$

of the class $C^1\big(\mathcal{D}_\varepsilon(0,\mathbf{R}^2)\big)$ such that $h(0,0)=0$ and $\mathcal{B}\big(h(\theta,\lambda);\theta,\lambda\big)=0$. Expanding this mapping in a neighborhood of zero we obtain

$$h(t;\theta,\lambda) = h_{01}(t)\,\theta + h_{10}(t)\,\lambda + o\Big(\sqrt{\theta^2+\lambda^2}\Big), \qquad (4.7.14)$$

where

$$\big\| o\Big(\sqrt{\theta^2+\lambda^2}\Big)\big\| \longrightarrow 0 \qquad \text{as } \theta \to 0,\quad \lambda \to 0$$

with respect to the norm in the space $\overset{\circ}{C}{}^2[0,1]$. Substituting (4.7.14) in (4.7.8) we find the undetermined coefficients h_{01} and h_{10} using the expansions of r, r', and r'' for small θ. We have

$$h_{01}(t) \equiv 0, \qquad h''_{10}(t) = \begin{cases} -J'(1-2t), & 0 \le t \le \frac{1}{2}, \\ -J'(2t-1), & \frac{1}{2} \le t \le 1, \end{cases}$$

so that because of the boundary conditions and condition $J(0) = 0$ we obtain

$$
h_{10}(t) = \begin{cases} \frac{1}{2} \int\limits_0^t J(1 - 2s)\, ds\,, & 0 \le t \le \frac{1}{2}\,, \\[3mm] \frac{1}{2} \int\limits_t^1 J(2s - 1)\, ds\,, & \frac{1}{2} \le t \le 1\,. \end{cases} \tag{4.7.15}
$$

Condition (4.2.9) implies

$$
\int\limits_0^1 h_{10}^{\prime 2}(t)\, dt = 1\,. \tag{4.7.16}
$$

Now put (4.7.14) in normalization condition (4.7.10) in order to eliminate parameter λ. Consider the equation

$$
\mathcal{F}(\theta, \lambda, a) := \int\limits_{1/2}^1 J(v)\left(h' + r'\right) dt - J_0 - a = 0 \tag{4.7.17}
$$

from the point of view of the theory of implicit functions. Clearly $\mathcal{F}(0, 0, 0) = 0$ and $\mathcal{F}$ is continuously differentiable in θ and λ. Moreover,

$$
\begin{aligned}
\mathcal{F}_\lambda'(0, 0, 0) &= \int\limits_{1/2}^1 J(2t - 1)\, h_{10}'(t)\, dt \\
&\quad + \int\limits_{1/2}^1 J'(2t - 1)\left(h_{10}(t) - h_{10}(1 - t)\right) dt \\
&= -\frac{1}{2} \int\limits_{1/2}^1 J^2(2t - 1)\, dt = -1 \ne 0\,.
\end{aligned}
$$

The implicit function theorem guarantees that for sufficiently small θ and a there exists a function $\lambda(\theta, a)$, differentiable in each variable, such that $\lambda(0, 0) = 0$ and

$$
\mathcal{F}\big(\lambda(\theta, a), \theta, a\big) = 0\,.
$$

We can write

$$
\lambda(\theta, a) = \lambda_{01}\theta + \lambda_{10}\, a + o\left(\sqrt{\theta^2 + \lambda^2}\right)\,, \qquad \theta^2 + a^2 \to 0\,, \tag{4.7.18}
$$

for the undetermined constants λ_{01} and λ_{10} that may be found by substituting (4.7.18) into (4.7.17):

$$\lambda_{10} = -1\,,$$

$$\lambda_{01} = \int\limits_{1/2}^{1} J'(2t-1)\,\bigl(r_\theta(t;0)-r_\theta(1-t;0)\bigr)\,dt + \int\limits_{1/2}^{1} J(2t-1)\,r'_\theta(t;0)\,dt = \varsigma\,.$$

Therefore we have

$$\lambda(\theta,a) = \varsigma\,\theta - a + o\!\left(\sqrt{\theta^2+\lambda^2}\right),\qquad \theta^2+a^2 \to 0\,,$$

and if $\varsigma\,\theta > a$ then $\lambda = \lambda(\theta,a) > 0$ for sufficiently small θ and a. It follows that the unique solution of the considered extremal problem satisfies

$$h(t;\theta,a) = h_{10}(t)\,(\varsigma\,\theta - a) + o\!\left(\sqrt{\theta^2+a^2}\right),\qquad \theta^2+a^2 \to 0\,.$$

By construction the solution h belongs to $\overset{\circ}{C}{}^2[0,1]$ and we can write

$$h'(t;\theta,a) = h'_{10}(t)\,(\varsigma\,\theta - a) + o\!\left(\sqrt{\theta^2+a^2}\right),$$

where the remainder term tends to 0 uniformly for $t \in [0,1]$.

Clearly our extremal h satisfies (4.7.6) and at the same time $h + r = f \in \mathcal{N}_{\delta/2}$. Substituting f in the minimized functional and passing in (4.7.4) to the limit as $\delta \to 0$ we obtain, due to (4.7.16), the conclusion of Theorem 4.7.1. □

Using Theorems 4.7.1 and 4.7.2 we may compute the local Hodges–Lehmann and Chernoff indices of the sequence of statistics (4.7.1). Since the critical value $\gamma_n(\alpha)$ for any $\alpha \in (0,1)$ is of order $O\!\left(n^{-1/2}\right)$, we can apply Theorem 4.7.2 with $a = 0$ in combination with Theorem 4.7.1 to obtain that

$$d_Z(\theta) \sim \varsigma^2 \cdot \theta^2\,,\qquad \theta \to 0\,. \qquad (4.7.19)$$

In order to calculate the Chernoff index $\rho_Z(\theta)$ we must, in accordance with Theorem 1.5.1, compare (4.2.12) with Theorem 4.7.2 (in the case $\gamma_n = 0$) taking into account (4.7.3). It becomes

$$\rho_Z(\theta) \sim \tfrac{1}{8}\varsigma^2 \cdot \theta^2\,,\qquad \theta \to 0\,.$$

Analyzing the local representation of exact slopes of the statistics Z_n given in Section 4.4, we get under some regularity conditions, less restrictive than in the present section,

$$c_Z(\theta) \sim \varsigma^2 \cdot \theta^2\,,\qquad \theta \to 0\,.$$

Note also that under appropriate regularity conditions the Pitman efficacy of the sequence $\{Z_n\}$ is proportional to the quantity ς^2 (see Pratt and Gibbons (1981) and Hettmansperger (1984)). Therefore, as in the case of homogeneity testing, the local ordering in a broad class of signed rank tests does not depend on the type of efficiency.

4.8 Approximate Bahadur and Pitman Efficiencies of Symmetry Tests

As nonparametric symmetry tests considered in the present chapter have more complicated structure than corresponding goodness-of-fit tests, we shall calculate in this section their local approximate Bahadur efficiency and then, using Theorem 1.4.3, their limiting Pitman efficiency. We underline that for asymptotically normal statistics (4.1.2) and (4.1.10) these two types of efficiency coincide as proved by Bahadur (1960b).

We begin by discussing the rate of decreasing of the tails for limiting distributions of the statistics under consideration if the hypothesis of symmetry holds. We must find for any sequence of statistics $\{T_n\}$ the coefficient $a = a_T$ from (1.2.18).

Let $D\,[0,1]$ be the usual space of left-continuous functions on $[0,1]$ (right-continuous at zero) having a finite number of jumps endowed with the uniform metric ρ. We recall that the functional L, defined on $D\,[0,1]$, is continuously differentiable of order $k \geq 1$ at the point F_0 if there exists a functional $l(F_0, v)$ for which the following statement is true:

> *For any sequence* $\{v_h\}$, $h \in \mathbf{R}^1$, $v_h \in D[0,1]$, *such that* $\rho(v_h, v) \to 0$ *as* $h \to 0$

$$\frac{L\big(F_0 + hv_h\big) - L\big(F_0\big)}{h^k} \longrightarrow l\big(F_0, v\big),$$

$$l\big(F_0, v_h\big) \longrightarrow l\big(F_0, v\big),$$

> *with some* $v \in C[0,1]$.

The functional $l(F_0, v)$ is called the *derivative of L in the direction v of order k*.

Theorem 4.8.1 (Borovkov 1984, Sec. 8) *Let F_n be the empirical distribution function corresponding to a sample with continuous* *distribution function F_0 and let L be a continuously differentiable functional of order k on $D[0,1]$. Then*

$$n^{k/2}\left[L(F_n) - L(F_0)\right] \longrightarrow l(F_0, w^0), \qquad n \to \infty,$$

in distribution where w^0 is the Brownian bridge.

Since all the statistics considered in this chapter are "distribution-free" under H_2 we may suppose that $F_0(t) = t$, $0 \leq t \leq 1$.

It is easy to verify that these statistics considered as functionals of F_n are continuously differentiable of order 1, except statistics (4.1.7) and (4.1.9) where the order is equal to 2 and statistic (4.1.8) where the order is equal to k. Their derivatives in the direction v are given by the formulas

$$l_I(t,v) = \sup_{0 \leq t \leq 1} |\Delta v(t) - 1|,$$

$$l_H(t,v) = \sup_{0 \leq t \leq 1} \left|\Delta v(t) - \int_0^1 \Delta v(t)\, dt\right|,$$

$$l_{R^k}(t,v) = \int_0^1 (\Delta v(t) - 1)^k\, dt, \qquad k = 1, 2, \ldots,$$

$$l_{N^2}(t,v) = \int_0^1 \left(\Delta v(t) - \int_0^1 \Delta v(t)\, dt\right)^2 dt.$$

Formulas for the one-sided statistics $I_n^{\pm}$ and $H_n^{\pm}$ are connected with obvious changes. Therefore Theorem 4.8.1 enables us to find the limiting distributions for the considered statistics. In order to calculate constants a from (1.2.18) for the statistics I_n, H_n, and their one-sided counterparts we will use the following result.

Lemma 4.8.1 (Fernique 1971; Marcus and Shepp 1972) *Let $X(t)$ be a Gaussian process on $[0,1]^d$, $d \geq 1$, with $EX(t) = 0$ and*

$$\mathcal{R}(t,s) := EX(t)X(s), \qquad t, s \in [0,1]^d,$$

such that

$$P\left(\sup_{t \in [0,1]^d} |X(t)| < \infty\right) = 1.$$

Denote

$$\kappa^2 := \sup_{t \in [0,1]^d} \mathcal{R}(t,t). \tag{4.8.1}$$

Then

$$\lim_{z \to \infty} z^{-2} \ln P\Big(\sup_{t \in [0,1]^d} |X(t)| \geq z \Big) = -\frac{1}{2\kappa^2}\,. \qquad (4.8.2)$$

The same is also true for $\sup_{t \in [0,1]^d} X(t)$.

This lemma has been proved by Marcus and Shepp (1972) in the case $d = 1$. For $d > 1$ it follows from the results of Fernique (1971) who proved that for $\alpha \in \big(0,\, (2\kappa^2)^{-1}\big)$

$$E \exp \Big\{ \alpha \sup_{t \in [0,1]^d} |X(t)|^2 \Big\} < \infty\,.$$

Lemma 4.8.1 may also be deduced from Theorem 5.2 of Borell (1975).

Consider now the Gaussian random processes

$$X_I(t) := \Delta w^0(t) = w^0(t) + w^0(1-t)\,,$$

$$X_H(t) := \Delta w^0(t) - \int_0^1 \Delta w^0(s)\, ds\,.$$

Clearly

$$EX_I(t) = EX_H(t) = 0 \qquad \text{for all } t \in [0,1]\,.$$

Simple calculations show that

$$\mathcal{R}_I(t,s) := EX_I(t)X_I(s) = 2 \min(t,\, s,\, 1-t,\, 1-s)\,;$$

consequently

$$\kappa_I^2 := \sup_{0 \leq t \leq 1} \mathcal{R}_I(t,t) = 1\,.$$

In the same way, one has

$$\mathcal{R}_H(t,s) := EX_H(t)X_H(s)$$

$$= 2\big[\min(t,\, s,\, 1-t,\, 1-s) - t(1-t) - s(1-s) \big] + \tfrac{1}{3}\,,$$

so

$$\kappa_H^2 := \sup_{0 \leq t \leq 1} \mathcal{R}_H(t,t) = \tfrac{1}{3}\,.$$

Using Lemma 4.8.1 we obtain now

$$a_I = 1\,, \qquad\qquad a_H = 3\,.$$

The same values of constants hold for the statistics $I_n^\pm$ and $H_n^\pm$.

We apply again, as in Section 2.6, the results of Kallianpur and Oodaira (1978) to obtain the corresponding constants for integral-type statistics. In this case the constants a_{R^k} and a_{N^2} are the values of the corresponding variational problems:

$$a_{R^k} = \inf \left\{ \int\limits_0^1 g'^2 \, dt: \ g \in \Omega^R \right\},$$

$$a_{N^2} = \inf \left\{ \int\limits_0^1 g'^2 \, dt: \ g \in \Omega^N \right\},$$

where Ω^R and Ω^N are sets of absolutely continuous functions g on $[0, 1]$ such that $g(0) = g(1) = 0$ and, correspondingly,

$$\int\limits_0^1 \big(g(t) + g(1 - t)\big)^k \, dt \geq 1,$$

$$\int\limits_0^1 \left(g(t) + g(1 - t) - \int\limits_0^1 (g(s) + g(1 - s))\, ds \right)^2 dt \geq 1.$$

These variational problems may be solved using the Lagrange principle again, but in a much simpler way than in Section 4.3. The Euler–Lagrange equation with the boundary conditions and the normalization condition has for the statistic $R_{n,1}^k$ the form

$$g'' = \lambda g^{k-1}, \qquad g(0) = g(1) = 0, \qquad \int\limits_0^1 g^k \, dt = 2^{-k}.$$

Comparing with problem (2.4.5) we see that

$$a_{R^k} = \tfrac{1}{4}\, \lambda_0(1; k),$$

where $\lambda_0(1; k)$ is the principal eigenvalue in this problem.

Similarly we obtain equation (4.3.25) for the statistic N_n^2, which implies

$$a_{N^2} = \pi^2.$$

Since the functions $b(\theta)$ have already been found in Section 4.4, we can find the local approximate slopes of the statistics under consideration. Clearly they coincide with the local exact slopes.

In order to prove that the limiting Pitman ARE coincides with the local approximate Bahadur ARE for any pair of our statistics we must verify the Wieand condition (1.4.3) and then use Theorem 1.4.3. The verification of this condition is very similar in all cases and we take as an example only the statistic I_n.

Note first of all that

$$\left| I_n - b_I(\theta) \right| = \left| \sup_{-\infty < x < \infty} \left| \Delta F_n(x) \right| - \sup_{-\infty < x < \infty} \left| \Delta G(x; \theta) \right| \right|$$

$$\leq \sup_{-\infty < x < \infty} \left| \left| \Delta F_n(x) \right| - \left| \Delta G(x; \theta) \right| \right|$$

$$\leq 2 \sup_{-\infty < x < \infty} \left| F_n(x) - G(x; \theta) \right|.$$

Now let the hypothesis H_2 be true. It is clear that the sequence of statistics

$$U_{n,\theta} = n^{1/2} \sup_{-\infty < x < \infty} \left| F_n(x) - G(x; \theta) \right|$$

satisfies the conditions of Theorem 1.4.1, so for any $\varepsilon > 0$ and $\delta \in (0, 1)$ there exists a constant C such that for all $\theta \in (0, \theta^*)$ and $0 < d(\theta) < 1$ the inequality

$$P_\theta \left(U_{n,\theta} \, n^{-1/2} < \varepsilon \, d(\theta) \right) > 1 - \delta$$

is valid for $n > C/d^2(\theta)$. Choose

$$d(\theta) = \tfrac{1}{2} \, b_I(\theta) < 1$$

for sufficiently small θ^*. Obviously

$$P_\theta \left(\left| I_n - b_I(\theta) \right| < \varepsilon \, b_I(\theta) \right) \geq P_\theta \left(2 \, U_{n,\theta} \, n^{-1/2} < \varepsilon \, b_I(\theta) \right) > 1 - \delta,$$

which means that the Wieand condition is fulfilled.

5
Asymptotic Efficiency of Nonparametric Independence Tests

5.1 Problem Statement and Tests under Consideration

Let $(X_1, Y_1), (X_2, Y_2), \ldots, (X_n, Y_n)$ be n independent two-dimensional observations with continuous d.f. $F(x, y)$. Denote by $G(x)$ and $H(y)$ marginal d.f.s of $F(x, y)$, that is, d.f.s of X_1 and Y_1. In this chapter we are interested in testing the independence hypothesis

$$H_3: \ F(x, y) = G(x) H(y)$$

for all x and y against the general alternative

$$A_3: \ F(x, y) \neq G(x) H(y)$$

for at least one point (x, y) as well as against particular alternatives described in the sequel.

It is important to note that G and H are supposed to be unknown. At the end of the chapter we shall consider changes connected with the case when they are known.

Denote by $F_n(x, y)$ the empirical d.f. based on the initial sample and by $G_n(x)$ and $H_n(y)$ empirical d.f.s based separately on the samples $X_1, X_2, \ldots, X_n$ and $Y_1, Y_2, \ldots, Y_n$. The statistic

$$\Gamma_n = \sup_{-\infty < x, y < \infty} \left| F_n(x, y) - G_n(x) H_n(y) \right|, \qquad (5.1.1)$$

first proposed apparently by Blum, Kiefer, and Rosenblatt (1961), is a natural analog of the Kolmogorov statistic (2.1.1) for testing H_3 against A_3. The one-sided variants of this statistic, namely,

$$\Gamma_n^+ = \sup_{-\infty < x, y < \infty} \left[F_n(x, y) - G_n(x) H_n(y) \right], \qquad (5.1.2)$$

$$\Gamma_n^- = \sup_{-\infty < x, y < \infty} \left[G_n(x) H_n(y) - F_n(x, y) \right], \qquad (5.1.3)$$

169

may be useful, for example, for alternatives of the form $F(x,y) \geq G(x)\,H(y)$ or $F(x,y) \leq G(x)\,H(y)$ with the strict inequality for at least one point (x,y).

Unfortunately, the limiting distributions of all three statistics (after normalizing by $\sqrt{n}$) are still unknown. However it is quite possible to obtain their critical values by simulation and therefore to use these statistics for testing independence. This argument justifies the study of their asymptotic efficiency.

The well-known statistics for testing independence are

$$B^2_{n,q_1,q_2} = \int\limits_{-\infty}^{\infty} \int\limits_{-\infty}^{\infty} \Big[\, F_n(x,y) - G_n(x)\,H_n(y)\,\Big]^2$$

$$\times q_1\big(G_n(x)\big)\, q_2\big(H_n(y)\big)\, dG_n(x)\, dH_n(y), \qquad (5.1.4)$$

where q_1 and q_2 are nonnegative weight functions on $(0, 1)$. A statistic equivalent to (5.1.4) was proposed by Hoeffding (1948) for $q_1 = q_2 = 1$, with its properties later being studied by Blum et al. (1961), and by de Wet (1980) for weights q_1 and q_2 not equal to 1. Tables of the limiting distribution of the statistic $n\,B^2_{n,1,1}$ may be found in the paper of Blum et al. (1961) and corresponding critical values were given by Cotterill and Csörgö (1985).

It is possible to consider the statistics B^k_{n,q_1,q_2} generalizing statistics (5.1.4) for positive integer k. But in the case $k > 2$ there appear substantial difficulties in the study of large deviations connected with the lack of information concerning the spectrum and eigenfunctions of corresponding nonlinear boundary-value problems (for $k = 2$ they are linear and this circumstance simplifies the matter).

As the results obtained for $k > 2$ have to a certain extent a conditional character we restrict ourselves to a formulation of them and a short discussion. For $k = 1$ the statistics $n\,B^1_{n,q_1,q_2}$ are close in structure to rank statistics; for example, the statistic $n\,B^1_{n,1,1}$ is asymptotically equivalent to the Spearman rank correlation coefficient (see Kendall (1970)). One should distinguish also the special case $k = 1$, $q_1(x) = q_2(x) = \sin \pi x$ corresponding to the "first component" of the integral statistic $n\,B^2_{n,1,1}$ and considered by Koziol and Nemec (1979).

As to linear rank statistics for testing H_3, they usually are written in the form

$$T_n = n^{-1} \sum_{i=1}^{n} a_{n1}\big(R_i/(n+1)\big)\, a_{n2}\big(S_i/(n+1)\big), \qquad (5.1.5)$$

where R_i is the rank of X_i among $X_1, X_2, \ldots, X_n$, S_i is the rank of Y_i among $Y_1, Y_2, \ldots, Y_n$, and a_{n1}, a_{n2} are some real functions on $[0, 1]$, satisfying the same conditions as the function a_n in (3.1.9). Statistic (5.1.5) may be considered as a special case of statistic (1.6.26). By concrete choice of a_{n1} and a_{n2} we obtain particular linear rank statistics, for example, the Spearman rank correlation coefficient

$$O_n = \sum_{i=1}^{n} R_i \, S_i \,. \qquad (5.1.6)$$

Large deviations of statistic (5.1.5) under the independence hypothesis were described by Theorem 1.6.11. The analysis of large deviations of statistics (5.1.1)–(5.1.4) is substantially more difficult and we proceed to it in the next sections.

5.2 Large Deviations of Kolmogorov-Type Statistics

It is convenient now to introduce the empirical field

$$\alpha_n(x, y) := G_n(x) \, H_n(y) - F_n(x, y)\,, \qquad x, y \in \mathbf{R}^1\,.$$

We can write then

$$\Gamma_n = \sup_{x,y} |\alpha_n(x, y)|\,,$$

$$\Gamma_n^+ = -\inf_{x,y} \alpha_n(x, y)\,,$$

$$\Gamma_n^- = \sup_{x,y} \alpha_n(x, y)\,.$$

Asymptotic properties of the statistic Γ_n and, particulary, corresponding large deviations were deeply investigated in a series of papers by Deheuvels (1980, 1981, 1982). But the methods he used were not effective enough. Deheuvels (1982) proved only that for all $a > 0$

$$\varlimsup_{n \to \infty} n^{-1} \ln \mathbf{P} \left(\Gamma_n \geq a \right) \leq -4 \ln 4 \; a^2\,.$$

A significantly more exact result is given (in the most interesting case of small a) by the following proposition.

Theorem 5.2.1 (Nikitin and Pankrashova 1988) *Under the independence hypothesis H_3 one has*

$$\lim_{n \to \infty} n^{-1} \ln P \left(\Gamma_n \geq a \right) = g_8(a) = -8 \, a^2 \left(1 + o(1) \right), \qquad a \to 0\,,$$

$$(5.2.1)$$

where the function g_8 is continuous for small $a > 0$.

Proof First of all we shall make some comments. Arguing as in previous chapters, it is not difficult to show that the rough asymptotics of large-deviation probabilities are the same for all three sequences of statistics $\{\Gamma_n\}$, $\{\Gamma_n^+\}$, and $\{\Gamma_n^-\}$ and we consider only the last of them. All of these statistics are obviously distribution-free, and we can suppose without loss of generality that $G(x) = x$, $0 \le x \le 1$, and $H(y) = y$, $0 \le y \le 1$.

An essential obstacle to the proof consists in the fact that it is impossible to reduce the problem to investigating the sums of independent random variables, but it turns out that it is sufficient to single out and study only typical summands.

Denote for brevity for $a > 0$

$$\beta_n := n \, \alpha_n(\tfrac{1}{2}, \tfrac{1}{2}) - n \, a \, .$$

It is obvious that for any $a > 0$

$$\varlimsup_{n \to \infty} n^{-1} \ln \mathbf{P}\left(\Gamma_n^- \ge a\right) \ge \varlimsup_{n \to \infty} n^{-1} \ln \mathbf{P}\left(\beta_n \ge 0\right)$$

and that

$$P(\beta_n \ge 0) = \sum_{s=0}^{n} \binom{n}{s} \, 2^{-n} P\left(\beta_n \ge 0 \,|\, G_n(\tfrac{1}{2}) = \frac{s}{n}\right)$$

$$= \sum_{s=0}^{n} \binom{n}{s} \, 2^{-n} P\left(Z_{s,n}(\tfrac{1}{2}) \ge a\right),$$

where we have put for $y \in (0, 1)$

$$Z_{s,n}(y) := n^{-1} \sum_{i=1}^{s} \frac{s-n}{n} \mathbf{1}_{\{Y_i < y\}} + n^{-1} \sum_{i=s+1}^{n} \frac{s}{n} \mathbf{1}_{\{Y_i < y\}} \, . \quad (5.2.2)$$

One of the principal means of proving Theorem 5.2.1 is the following theorem of Bahadur and Ranga Rao (1960) (see also Bahadur (1971)).

Let ξ be a random variable with the moment generating function $\chi(t)$ and

$$t^* := \sup\left\{t\colon \chi(t) < \infty\right\}.$$

It is said that ξ satisfies *standard conditions* (Bahadur 1971, Section 2) if $0 < t^* \le \infty$, $\chi'(0+) < 0$, and $\chi'(b) > 0$ for some $b \in (0, t^*)$. If standard conditions are fulfilled, the lower bound of χ is attained at a unique point $\tau \in (0, t^*)$ such that $\chi'(\tau) = 0$.

Theorem 5.2.2 *Let* $\{Z_n\}$ *be a sequence of random variables satisfying standard conditions. Denote by* $\widetilde{F}_n$ *the distribution function of* Z_n *and by* $\widetilde{\varphi}_n$ *the moment generating function of* Z_n *and let*

$$\tilde{\rho}_n := \inf\big\{\, \widetilde{\varphi}_n(u)\colon\ u \geq 0 \,\big\} = \widetilde{\varphi}_n(\widetilde{\tau}_n)\,.$$

Let $\widetilde{G}_n$ *be the distribution function conjugate to* $\widetilde{F}_n$*, that is,*

$$\widetilde{G}_n(x) := \tilde{\rho}_n^{-1} \int\limits_{-\infty}^{x} \exp\{\widetilde{\tau}_n z\}\, d\widetilde{F}_n(z)\,,$$

and let $P_n := P\,(Z_n \geq 0)$*,* $\widetilde{H}_n(x) := \widetilde{G}_n(x\,\sigma_n)$*, and*

$$\sigma_n^2 := \int\limits_{-\infty}^{\infty} x^2\, d\widetilde{G}_n(x) > 0\,.$$

If the following two conditions:

1. $\widetilde{\tau}_n\, \widetilde{\sigma}_n = O(n)\,,$

2. $\ln\big[\, \widetilde{H}_n(\varepsilon) - \widetilde{H}_n(0)\,\big] = o(n)$ *for any* $\varepsilon > 0\,,$

hold as $n \to \infty$*, then we have the relation*

$$\ln P_n - \ln \tilde{\rho}_n = o(n)\,, \qquad\qquad n \to \infty\,.$$

We shall apply this result to the analysis of $\ln P\,(\beta_n \geq 0)$. To verify its conditions we shall need Lemmas 5.2.1–5.2.9, formulated and proved subsequently, in which we investigate in more detail properties and interdependence of moment generating functions of β_n and $n\, Z_{s,\,n}(\tfrac{1}{2}) - na$. The greatest difficulty is caused by verifying the second condition for β_n; however, the lemmas allow us to reduce this condition to an analogous condition for $n\, Z_{s,\,n}(\tfrac{1}{2}) - na$, which may be verified in a significantly simpler manner.

We omit or shorten the proof of those lemmas that have a technical character and can be easily supplied by the reader.

First we compute the moment generating functions of β_n and $n\, Z_{s,\,n}(\tfrac{1}{2}) - na$, denoting them by φ_n and $\omega_{s,\,n}$ respectively. For any $u \geq 0$ and $0 \leq x \leq 1$ we introduce the notation

$$\psi(x,u) := \tfrac{1}{2}\, e^{-au} \left(e^{(x-1)\,u} + 1 \right)^{x} \left(e^{xu} + 1 \right)^{1-x}\,. \tag{5.2.3}$$

Lemma 5.2.1 *For any $u \geq 0$ the relations*

$$\omega_{s,n}(u) := E \exp\left\{n\, u\left(Z_{s,n}\left(\tfrac{1}{2}\right) - a\right)\right\} = \psi^n\left(\tfrac{s}{n}, u\right),$$

$$\varphi_n(u) := E \exp\left\{u\, \beta_n\right\} = \sum_{s=0}^{n} \binom{n}{s} 2^{-n}\, \psi^n\left(\tfrac{s}{n}, u\right)$$

are valid.

The following lemma gives information about the infimum ρ_x and the infimum point τ_x of the function $\psi(x, u)$ with respect to u (for small a).

Lemma 5.2.2 *For $a \to 0$ and $0 < x < 1$*

$$\ln \rho_x = \inf\left\{\ln \psi(x, u)\colon u \geq 0\right\} = -\frac{2\,a^2}{x\,(1-x)}\left(1 + o(1)\right),$$

$$\tau_x = \arg\inf\left\{\ln \psi(x, u)\colon u \geq 0\right\} = \frac{4\,a}{x\,(1-x)}\left(1 + o(1)\right).$$

This lemma is proved by direct calculation using implicit function theory and the smoothness of the function ψ.

It is important to notice that the function $\psi(x, y)$ simplifies for $x = \tfrac{1}{2}$ to $\exp(-au)\cosh u/4$. Therefore it is easy to prove that a new function

$$g_8(a) := \ln \rho_{1/2}$$

is continuous in a for sufficiently small $a > 0$ and satisfies the asymptotic relation:

$$g_8(a) = -8\,a^2\left(1 + o(1)\right), \qquad\qquad a \to 0.$$

Additional and relatively easily verifiable information on the function ψ is contained in the following lemma.

Lemma 5.2.3 *For all sufficiently small $u > 0$ the function $\psi(x, u)$ possesses the properties:*

1. *$\psi(x, u)$ is increasing in x for $0 \leq x \leq 1/2$ and is decreasing in x for $1/2 \leq x \leq 1$;*

2. *$\psi_u'(x, \tau_{x_0}) > 0$ for $0 < x_0 < x \leq 1/2$;*

3. *$\psi_u'(x, \tau_{x_0}) < 0$ for $0 < x < x_0 \leq 1/2$;*

4. *$\psi_{uu}''(x, u) \geq \tfrac{1}{4}\, x\,(1-x)\,\psi(x, u)\,(\cosh \tfrac{u}{2})^{-2}$ $0 < x < x_0 \leq 1/2$.*

$$(5.2.4)$$

Lemma 5.2.4 *Let τ_n and τ_x be the points of infimum for the functions $\varphi_n(u)$ and $\psi(x, u)$, respectively; set $m = \lceil n/2 \rceil$ and $\lambda = m/n$. Then for sufficiently small $a > 0$ we have*

1. $\tau_{1/2} \leq \tau_\lambda \leq \tau_n$ for all $n \geq 1$;

2. $\tau_n \to \tau_{1/2}$ as $n \to \infty$.

Proof First of all we note that from Lemma 5.2.2, for small $a > 0$, it follows that

$$\tau_{1/2} \leq \tau_x, \qquad 0 \leq x \leq 1,$$

and, in addition, if for the sake of definiteness we put $x \leq 1/2$, then $\tau_{x_0} \leq \tau_x$ for $0 \leq x_0 \leq x$ or $1 - x \leq x_0 < 1$. Thus, we have $\tau_{1/2} \leq \tau_\lambda$. From Lemma 5.2.3 it follows that, for a fixed x, the function $\psi(x, u)$ decreases with respect to u for $0 < u < \tau_x$. If we had the inequality $\tau_n < \tau_\lambda$, then, by the monotonicity of ψ, we would obtain

$$\varphi_n(\tau_n) > \varphi_n(\tau_\lambda),$$

which contradicts the definition of τ_n. Thus, $\tau_\lambda \leq \tau_n$. Now we select an arbitrary $x_0 \in [0, 1]$, $x_0 \neq 1/2$, and show that $\tau_n < \tau_{x_0}$ for sufficiently large n. Since τ_x depends continuously on x and

$$\tau_{x_0} \to \tau_{1/2} \qquad \text{as } x_0 \to 1/2,$$

assertion 2 of the lemma will follow from assertion 1.

For the proof of the inequality $\tau_n < \tau_{x_0}$ it is sufficient to show that $\varphi_n'(\tau_{x_0}) > 0$ for sufficiently large n. For the sake of definiteness we assume that $x_0 < 1/2$. Now we have

$$\varphi_n'(\tau_{x_0}) = \sum_{s=0}^{n} \binom{n}{s} n \, 2^{-n} \, \psi^{n-1}\left(\tfrac{s}{n}, \tau_{x_0}\right) \psi_u'\left(\tfrac{s}{n}, \tau_{x_0}\right).$$

By Lemma 5.2.3 some of the terms for $\frac{s}{n} < x_0$, $\frac{s}{n} > 1 - x_0$ are negative and in them we replace $\psi(\frac{s}{n}, \tau_{x_0})$ by the larger quantity $\psi(x_0, \tau_{x_0})$; in the remaining terms the derivative ψ_u' is positive and there we estimate $\psi(\frac{s}{n}, \tau_{x_0})$ from below by means of $\psi(x_0, \tau_{x_0})$. As a result we obtain

$$\varphi_n'(\tau_{x_0}) \geq n \, \psi^{n-1}(x_0, \tau_{x_0}) \sum_{s=0}^{n} \binom{n}{s} 2^{-n} \, \psi_u'\left(\tfrac{s}{n}, \tau_{x_0}\right).$$

But by the Weierstrass–Bernstein theorem (see Feller (1971), Chapter 7), the sum on the right-hand side converges as $n \to \infty$ to the quantity

$\psi'_u(\frac{1}{2}, \tau_{x_0})$, which is positive by Lemma 5.2.3. Therefore, $\varphi'_n(\tau_{x_0}) > 0$ for sufficiently large n. Hence the lemma is proved.   □

Combining Lemmas 5.2.2 and 5.2.4, we establish the following statement.

Lemma 5.2.5 *There exists a positive constant A such that for sufficiently large n and sufficiently small a we have*

$$\tau_n < A.$$

Lemma 5.2.6 *There exists a constant C such that for sufficiently large n and sufficiently small a we have*

$$g_n(\tau_n) := \sum_{s=0}^{n} \binom{n}{s} \frac{s(n-s)}{n^2} \psi^n(\tfrac{s}{n}, \tau_n) \Big/ \sum_{s=0}^{n} \binom{n}{s} \psi^n(\tfrac{s}{n}, \tau_n) > C > 0.$$

Proof The quantity τ_n is a root of the equation $\varphi'_n(\tau_n) = 0$, which can be written in the form

$$f_n(\tau_n) = \tau_n \, a^{-1},$$

where

$$f_n(u) := \frac{\displaystyle\sum_{s=0}^{n} \binom{n}{s} \psi^n(\tfrac{s}{n}, u)}{\displaystyle\sum_{s=0}^{n} \binom{n}{s} \frac{s(n-s)}{n^2} \psi^n(\tfrac{s}{n}, u) h_{n,s}(u)},$$

$$h_{n,s}(u) := \frac{1 - e^{-u}}{u} \cdot \frac{\exp\{\tfrac{s}{n} u\}}{\left(1 + \exp\{\tfrac{s}{n} u\}\right)\left(1 + \exp\{\tfrac{s-n}{n} u\}\right)}.$$

Clearly $h_{n,s}(u) < 1$. Taking into account the inequality $\tau_n < \tau_{x_0}$, $x_0 \in (0, 1)$, $x_0 \neq 1/2$, we obtain the chain of inequalities

$$\tau_{x_0} \, a^{-1} > \tau_n \, a^{-1} = f_n(\tau_n) > \frac{1}{g_n(\tau_n)}.$$

Now the assertion of Lemma 5.2.6 follows from Lemma 5.2.2.   □

Lemma 5.2.7

$$\frac{\varphi''_n(\tau_n)}{\varphi_n(\tau_n)} = O(n^2), \qquad n \to \infty.$$

Proof Performing the differentiation we have

$$\frac{\varphi_n''(\tau_n)}{\varphi_n(\tau_n)} = \left[\sum_{s=0}^{n} \binom{n}{s} \psi^n(\tfrac{s}{n}, \tau_n) \right]^{-1} \sum_{s=0}^{n} \binom{n}{s} \psi^{n-2}(\tfrac{s}{n}, \tau_n)$$

$$\times \left[n(n-1)\, \psi'^2(\tfrac{s}{n}, \tau_n) + n\, \psi''(\tfrac{s}{n}, \tau_n)\, \psi(\tfrac{s}{n}, \tau_n) \right].$$

From here it is clear that for the proof of lemma it is sufficient to verify that for any $x \in [0, 1]$ and large n we have

$$\frac{|\psi_u'(x, \tau_n)|}{\psi(x, \tau_n)} < C_1, \qquad \frac{|\psi_{uu}''(x, \tau_n)|}{\psi(x, \tau_n)} < C_1,$$

where C_1 is some constant. But by Lemma 5.2.5

$$\frac{\psi_u'(x, \tau_n)}{\psi(x, \tau_n)} \equiv \frac{x(1-x)(e^{\tau_n} - 1)}{(1 + e^{x\tau_n})(1 + e^{(1-x)\tau_n})} - a < \frac{e^A}{4}.$$

The second inequality is proved in a similar manner. □

Lemma 5.2.8 *Let σ_n^2 and Δ_n^2 be the variances of the conjugate distributions corresponding to the distributions of the random variables β_n and $n\, Z_{m,n}(\tfrac{1}{2}) - na$, where $m = \lceil n/2 \rceil$. Then for sufficiently large n we have*

$$\sigma_n^2 > K^2 \Delta_n^2,$$

where the constant K^2 may depend on a but not on n.

Proof We denote again $\lambda := \lambda_n = m/n$ and note that, as $n \to \infty$, one has $\lambda_n \to 1/2$, $\tau_\lambda \to \tau_{1/2}$, and

$$\psi(\lambda, \tau_\lambda) \longrightarrow \psi(\tfrac{1}{2}, \tau_{1/2}),$$

$$\psi''(\lambda, \tau_\lambda) \longrightarrow \psi''(\tfrac{1}{2}, \tau_{1/2}).$$

From the theory of conjugate distributions it is known (see, for example, Bahadur (1971), Section 2) that

$$\sigma_n^2 = \frac{\varphi_n''(\tau_n)}{\varphi_n(\tau_n)}, \qquad \Delta_n^2 = \frac{\omega_{m,n}''(\tau_\lambda)}{\omega_{m,n}(\tau_\lambda)}. \tag{5.2.5}$$

Taking into account that $\psi'(\lambda, \tau_\lambda) = 0$, and using the relations between φ_n and $\omega_{m,n}$, stated in Lemma 5.2.1, and inequality (5.2.4), we obtain the estimate

$$\frac{\sigma_n^2}{\Delta_n^2} > \frac{\rho_\lambda^2\, g_n(\tau_n)}{4\, \psi''(\lambda, \tau_\lambda)}.$$

Then we estimate $g_n(\tau_n)$ from below by Lemma 5.2.6 and make use of the fact that, for sufficiently large n, ρ_λ^2 and $\psi''(\lambda, \tau_\lambda)$ can be approximated in an arbitrarily accurate manner by means of $\rho_{1/2}^2 > 0$ and $\psi''(\frac{1}{2}, \tau_{1/2}) < \infty$. $\qquad\square$

Lemma 5.2.9

$$\lim_{n\to\infty} \varphi_n^{1/n}(\tau_n) \geq \psi(\tfrac{1}{2}, \tau_{1/2}).$$

Proof Let ζ_n be an auxiliary random variable having the binomial distribution with parameters n and $\frac{1}{2}$. Then by Lemma 5.2.1 one obtains

$$\varphi_n(u) = E\psi^n(n^{-1}\zeta_n, u).$$

From Jensen's inequality and Lemma 5.2.3 it follows that

$$\varphi_n^{1/n}(u) \geq E\psi(n^{-1}\zeta_n, u).$$

Setting $u = \tau_n$ and taking the limit as $n \to \infty$, by Lemma 5.2.4 and the continuity of ψ we establish the conclusion of Lemma 5.2.9. $\qquad\square$

We proceed now directly to the proof of Theorem 5.2.1 on the basis of Theorem 5.2.2. We verify that the random variable β_n satisfies the conditions of this theorem. Clearly, the standard conditions are satisfied since $\varphi_n'(0+) = -na < 0$, while the fact that $\varphi_n'(\tau_{x_0}) > 0$ for $0 < x < 1/2$ has been verified in the proof of Lemma 5.2.4.

The first condition of Theorem 5.2.2 is satisfied by virtue of Lemma 5.2.5 in combination with Lemma 5.2.7 and the first of formulas (5.2.5).

We pass to the verification of the second condition. We denote by $\widetilde{F}_n$ and $F_{m,n}$ d.f.s of random variables β_n and $nZ_{m,n}(\frac{1}{2}) - na$, respectively, and by $\widetilde{G}_n$ and $G_{m,n}$ the corresponding conjugate d.f.s. Assume, in addition, that $\widetilde{H}_n(x) = \widetilde{G}_n(x\sigma_n)$. Clearly, for any $\varepsilon > 0$, we have

$$\overline{\lim_{n\to\infty}} \; n^{-1}\ln\left[\widetilde{H}_n(\varepsilon) - \widetilde{H}_n(0)\right] \leq 0,$$

and, therefore, it is sufficient to verify the opposite inequality.

We introduce more notation:

$$\rho_n := \varphi_n(\tau_n), \qquad d_n := \psi(\lambda, \tau_\lambda) / \psi(\tfrac{1}{2}, \tau_n).$$

Clearly, $d_n \to 1$ as $n \to \infty$. In addition, by Lemma 5.2.3

$$\frac{\rho_\lambda^n}{\rho_n} = \frac{\psi^n(\lambda, \tau_\lambda)}{\sum_{s=0}^{n}\binom{n}{s} 2^{-n}\psi^n(\frac{s}{n}, \tau_n)} > \frac{\psi^n(\lambda, \tau_\lambda)}{\psi^n(\frac{1}{2}, \tau_n)} = d_n^n. \qquad (5.2.6)$$

With the aid of (5.2.6) and Lemmas 5.2.4 and 5.2.8, we obtain

$$\widetilde{H}_n(\varepsilon) - \widetilde{H}_n(0) = \rho_n^{-1} \int_0^{\varepsilon \sigma_n} \exp\{z\,\tau_n\}\, d\widetilde{F}_n(z)$$

$$> d_n^n\, \rho_\lambda^{-n} \int_0^{\varepsilon K \Delta_n} \exp\{z\,\tau_\lambda\}\, d\widetilde{F}_n(z)$$

$$= d_n^n\, \rho_\lambda^{-n} \sum_{s=0}^{n} \binom{n}{s} 2^{-n} \int_0^{\varepsilon K \Delta_n} \exp\{z\,\tau_\lambda\}\, dF_{s,\,n}(z)$$

$$> d_n^n \binom{n}{m} 2^{-n} \left[G_{m,\,n}(\varepsilon\, K\Delta_n) - G_{m,\,n}(0) \right].$$

It is easy to see that $\binom{n}{m} 2^{-n} = O(n^{-1/2})$ and, by virtue of the central limit theorem, we derive

$$G_{m,\,n}(\varepsilon K \Delta_n) - G_{m,n}(0) = \Phi(\varepsilon K) - \Phi(0) + o(1), \qquad n \to \infty.$$

Taking the logarithm of the obtained inequality and then passing to the limit, we have

$$\varliminf_{n \to \infty} n^{-1} \ln \left[\widetilde{H}_n(\varepsilon) - \widetilde{H}_n(0) \right] \geq 0,$$

which concludes the verification of the second condition of Theorem 5.2.2.

By the conclusion of this theorem and by Lemma 5.2.9 we easily obtain

$$\varliminf_{n \to \infty} n^{-1} \ln P\big(\beta_n \geq 0\big) \geq \varliminf_{n \to \infty} n^{-1} \ln \rho_n$$

$$\geq \ln \psi(\tfrac{1}{2}, \tau_{1/2}) \equiv \ln \rho_{1/2}.$$

It remains to make use of Lemma 5.2.2 in order to have

$$\varliminf_{n \to \infty} n^{-1} \ln P\big(\Gamma_n^- \geq a\big) \geq \ln \rho_{1/2} = g_8(a). \qquad (5.2.7)$$

The estimate from above requires less effort and is proved analogously to corresponding estimates in Theorems 2.2.1 and 4.2.1. We introduce for sufficiently small $\delta > 0$ the random process

$$Z_{s,n}(y,\delta) := n^{-1} \sum_{i=1}^{s} \frac{s-n}{n}\, \mathbf{1}_{\{Y_i < y\}} + n^{-1} \sum_{i=s+1}^{n} \frac{s}{n}\, \mathbf{1}_{\{Y_i < y+\delta\}}, \qquad (5.2.8)$$

turning into $Z_{s,n}(\frac{1}{2})$ for $y = 1/2$ and $\delta = 0$. It is easy to see that the moment generating function of $n\,Z_{s,\,n}(y,\,\delta) - na$ has the form

$$E \exp\left\{ u\left(n\,Z_{s,n}(y,\delta) - n\,a \right) \right\} = \psi_\delta^n\left(\tfrac{s}{n}, y, u \right),$$

where

$$\psi_\delta(x,y,u) := e^{-au}\left(y\,e^{(x-1)u} + 1 - y \right)^x \left((y+\delta)\,e^{xu} + 1 - y - \delta \right)^{1-x}.$$

Let

$$\rho_\delta(x,y) := \inf\left\{ \psi_\delta(x,y,u)\colon\ u \geq 0 \right\}$$

for any fixed x and y. The following lemma plays the key role in the estimation from above.

Lemma 5.2.10 *For any $u_0 > 0$ the following estimate holds:*

$$\sup_{0 \leq x \leq 1}\ \sup_{0 \leq y \leq 1}\ \rho_\delta(x,y) \leq \psi(\tfrac{1}{2}, u_0)\left(1 + \delta\,e^{u_0} \right),$$

where the function $\psi(x,y)$ has been defined in (5.2.4).

Proof It is elementary to verify that for all admissible values of arguments

$$\psi_\delta(x,y,u_0) \leq \psi_0(x,y,u_0)\left(1 + \delta\,e^{u_0} \right). \tag{5.2.9}$$

It is also comparatively easy to show that the function $\psi_0(x,\,y,\,u_0)$ for a fixed u_0 attains its maximal value for $x = y = 1/2$. Indeed, the critical points may be found from the system of equations

$$\frac{\partial\,\ln\,\psi_0(x,y,u_0)}{\partial x} = 0, \qquad \frac{\partial\,\ln\,\psi_0(x,y,u_0)}{\partial y} = 0,$$

which has the unique solution $x = y = 1/2$. Clearly the function $\ln\,\psi_0(x,\,y,\,u_0)$ is positive for $x = y = 1/2$ and equals 0 on the boundary of a unit square.
Therefore

$$\psi_0(x,y,u_0) \leq \psi_0(\tfrac{1}{2}, \tfrac{1}{2}, u_0) \equiv \psi(\tfrac{1}{2}, u_0). \tag{5.2.10}$$

Estimating $\rho_\delta(x,\,y)$ from above by $\psi_\delta(x,\,y,\,u_0)$ and then consequently using (5.2.9) and (5.2.10), we establish the conclusion of lemma. $\square$

Now let k be a sufficiently large positive integer. Using the definition of the processes $Z_{s,n}(y)$ from (5.2.2) and $Z_{s,n}(y, \delta)$ from (5.2.8), we obtain the following chain of inequalities:

$$P\big(\Gamma_n^- \geq a\big) \leq P\Big(\sup_{0 \leq y \leq 1} \max_{0 \leq s \leq n} Z_{s,n}(y) \geq a \Big)$$

$$\leq P\Big(\max_{0 \leq j \leq k-1} \sup_{(j/k) \leq y \leq (j+1)/k} \max_{0 \leq s \leq n} Z_{s,n}(y) \geq a \Big)$$

$$\leq P\Big(\max_{0 \leq j \leq k-1} \max_{0 \leq s \leq n} Z_{s,n}\big(\tfrac{j}{k}, \tfrac{1}{k}\big) \geq a \Big)$$

$$\leq \sum_{j=0}^{k-1} \sum_{s=0}^{n} P\Big(Z_{s,n}\big(\tfrac{j}{k}, \tfrac{1}{k}\big) \geq a \Big).$$

Taking into account estimate (1.6.3) and Lemma 5.2.10 for $u_0 = \tau_{1/2}$ we have further

$$P\big(\Gamma_n^- \geq a\big) \leq \sum_{s=0}^{n-1} \sum_{j=0}^{k-1} \rho_{1/k}^n\big(\tfrac{s}{n}, \tfrac{j}{k}\big)$$

$$\leq 2nk \sup_{0 \leq x \leq 1} \sup_{0 \leq y \leq 1} \rho_{1/k}^n(x,y)$$

$$\leq 2nk \, \psi^n\big(\tfrac{1}{2}, \tau_{1/2}\big) \Big(1 + k^{-1} \exp\{\tau_{1/2}\} \Big)^n.$$

Taking logarithms, multiplying by n^{-1}, and then passing to the limit first as $n \to \infty$ and then as $k \to \infty$, we obtain

$$\varlimsup_{n \to \infty} n^{-1} \ln \mathbf{P}\big(\Gamma_n^- \geq a\big) \leq \ln \psi\big(\tfrac{1}{2}, \tau_{1/2}\big)$$

$$= \ln \rho_{1/2} = g_8(a)$$

$$= -8a^2 \big(1 + o(1)\big), \qquad a \to 0.$$

Combining this inequality with (5.2.7) we establish the conclusion of Theorem 5.2.1. $\qquad \Box$

5.3 Large Deviations of Integral Statistics for Independence Testing

Investigating the distribution and large deviations of statistics (5.1.4) under the independence hypothesis, we can again assume that the initial sample has been extracted from the uniform distribution on the unit square $I^2 = [0, 1] \times [0, 1]$. We also require in this section that weight functions q_1 and q_2, which are positive on $(0, 1)$, are differentiable and their derivatives are bounded in absolute value:

$$\sup_{0 \le t \le 1} |q_i'(t)| \le L, \qquad i = 1, 2. \qquad (5.3.1)$$

For any d.f. F on I^2 with marginal d.f.s G and H we consider the functional

$$B(F) := \int_0^1 \int_0^1 \Big[F(s,t) - G(s)\,H(t) \Big]^2 q_1\big(G(s)\big)\, q_2\big(H(t)\big)\, dG(s)\, dH(t).$$

Elementary estimates using the triangle inequality and (5.3.1) show that this functional is continuous in the topology of uniform convergence of d.f.s on I^2. Consequently, by Lemma 2.1 of Groeneboom et al. (1979) the functional B is also τ-continuous.

We denote by $\mathcal{F}_2$ the set of all absolutely continuous d.f.s on I^2 and consider for any $a > 0$ the set

$$\Omega_a^B := \Big\{ F\colon F \in \mathcal{F}_2, B(F) \ge a \Big\}.$$

If $f(s,t)$ is the density corresponding to d.f. $F \in \mathcal{F}_2$, we define the Kullback–Leibler information for each set $A \subset \mathcal{F}_2$ by the formula

$$K(A) := \inf \Big\{ \int_0^1 \int_0^1 f \ln f \, ds\, dt \colon F \in A \Big\}.$$

Following Nikitin (1984c, 1986a) we shall prove two theorems describing large-deviation behavior of the statistics $B_{n,\,q_1,\,q_2}^2$ under the hypothesis of independence.

Theorem 5.3.1 *Let the conditions imposed on weight functions q_1 and q_2 be fulfilled. Then for sufficiently small $a > 0$*

$$K(\Omega_a^B) = \tfrac{1}{2}\lambda_0 a + \sum_{j=3}^{\infty} c_j\, a^{j/2}, \qquad (5.3.2)$$

where the series is convergent and $\lambda_0 = \lambda_0(q_1, q_2)$ is the principal eigenvalue of the linear integral equation

$$u(s,t) = \lambda \int_0^1 \int_0^1 K_1(s,t;\,v,w)\, u(v,w)\, q_1(v)\, q_2(w)\, dv\, dw \qquad (5.3.3)$$

with the kernel

$$K_1(s,t;\,v,w) = \big(\min(s,v) - s\,v\big)\big(\min(t,w) - t\,w\big). \qquad (5.3.4)$$

If Theorem 5.3.1 holds then the function $a \mapsto K(\Omega_a^B)$ is continuous, consequently Theorem 1.6.7 implies the following result.

Theorem 5.3.2 *Under the conditions of Theorem 5.3.1 for sufficiently small $a > 0$ one has*

$$\lim_{n \to \infty} n^{-1} \ln P\left(B_{n,q_1,q_2}^2 \geq a\right) = -K(\Omega_a^B). \qquad (5.3.5)$$

Proof of Theorem 5.3.1 The problem of the computation of $K(\Omega_a^B)$ is solved essentially by the same methods as have been used in previous chapters, but there exist also some distinctions caused by the complicated structure of statistics (5.1.4). Subsequently we shall first dwell upon these distinctions.

One of them consists in that it is insufficient now to restrict the class of admissible elements to the class of densities uniformly separated from zero. We are forced to introduce the complementary condition of their uniform boundedness from above.

For any real $M > 0$ we consider the set

$$W_M := \left\{ F\colon F \in \mathcal{F}_2, \ f \leq M \quad \text{a.e. on } I^2 \right\}.$$

As in (2.3.10) we introduce for any $\delta > 0$ the set

$$V_\delta := \left\{ F\colon F \in \mathcal{F}_2, \ f \geq \delta \right\}.$$

The proof of Lemma 2.3.1 is valid also for the set Ω_a^B. Therefore for sufficiently small $\delta > 0$ we have

$$K\left(\Omega_{a+\alpha(\delta)}^B \cap V_{\delta/2}\right) - \delta \leq K(\Omega_a^B) \leq K\left(\Omega_a^B \cap V_{\delta/2}\right), \qquad (5.3.6)$$

where $\alpha(\delta) \to 0$ as $\delta \to 0$.

The following lemma enables us to still more restrict the class of admissible elements.

Lemma 5.3.1 *For any $a_1 > 0$, $\delta_1 > 0$, and sufficiently large $M > 0$ we have*

$$K\left(\Omega_{a_1}^B \cap V_{\delta_1}\right) \geq K\left(\Omega_{a_1+\beta_1(M)}^B \cap V_{\delta_1} \cap W_{2M}\right) - \beta_2(M), \qquad (5.3.7)$$

where $\beta_i(M) \to 0$ as $M \to \infty$ for $i = 1, 2$.

Proof Assume that the lower bound of the entropy functional on the set $\Omega_a^B \cap V_{\delta_1}$ is attained for d.f. F with density f. We consider the truncation

$$f^{(M)} := \begin{cases} f & \text{if } f \le M, \\ M & \text{if } f > M, \end{cases}$$

and construct the density

$$f_M := f^{(M)} \Big/ \int_0^1 \int_0^1 f^{(M)} \, dx \, dy,$$

and then from it d.f. F_M. For sufficiently large M we have

$$F_M \in \Omega_{a_1+\beta_1(M)}^B \bigcap V_{\delta_1} \bigcap W_{2M}.$$

It is easy to show that

$$\int_0^1 \int_0^1 f_M \ln f_M \, ds \, dt - \int_0^1 \int_0^1 f \ln f \, ds \, dt \le \beta_2(M) \longrightarrow 0, \qquad M \to \infty.$$

Consequently

$$K\left(\Omega_{a_1}^B \bigcap V_{\delta_1}\right) = \int_0^1 \int_0^1 f \ln f \, ds \, dt$$
$$\ge K\left(\Omega_{a_1+\beta_1(M)} \bigcap V_{\delta_1} \bigcap W_{2M}\right) - \beta_2(M). \qquad \square$$

It follows from (5.3.6) and (5.3.7) that

$$K\left(\Omega_{a_2}^B \bigcap V_{\delta_1} \bigcap W_{2M}\right) - \gamma(M, \delta) \le K\left(\Omega_a^B\right) \le K\left(\Omega_a^B \bigcap V_\delta \bigcap W_M\right),$$

(5.3.8)

where $a_2 \to a$, $\delta_1 \to \delta$, $\gamma(M, \delta) \to 0$ as $\delta \to 0$ and $M \to \infty$.

Now we proceed to the computation of $K\left(\Omega_{a_2}^B \cap V_{\delta_1} \cap W_{M_1}\right)$, intending to take then the limit as $\delta \to 0$ and $M \to \infty$. The existence of a solution follows again from the τ-closedness of considered sets. Let F be the extremal d.f., G and H be marginal d.f.s, and f, g, and h be the corresponding densities. Since $F \in V_{\delta_1}$, there exist strictly monotone and absolutely continuous inverse d.f.s G^{-1} and H^{-1}. We perform a change of the unknown function

$$\psi(s,t) = F\left(G^{-1}(s), H^{-1}(t)\right) - s\,t. \qquad (5.3.9)$$

Straightforward computations show that for almost all s and t we have

$$\left(\psi_{st} + 1\right) g\left(G^{-1}(s)\right) h\left(H^{-1}(t)\right) = f\left(G^{-1}(s), H^{-1}(t)\right). \qquad (5.3.10)$$

Consequently,

$$\int_0^1 \int_0^1 f \ln f \, ds \, dt = \int_0^1 \int_0^1 (\psi_{st} + 1) \ln (\psi_{st} + 1) \, ds \, dt$$

$$+ \int_0^1 g(s) \ln g(s) \, ds + \int_0^1 h(t) \ln h(t) \, dt \quad (5.3.11)$$

and this functional has to be minimized on the set of those $F \in \mathcal{F}_2$ such that

$$\int_0^1 \int_0^1 \psi^2(s,t) \, q_1(s) \, q_2(t) \, ds \, dt \geq a_2 \,, \qquad (5.3.12)$$

$$\delta_1 \leq f(s,t) \leq M_1 \quad \text{a.e. on } I^2 . \qquad (5.3.13)$$

We extend this set somewhat, replacing (5.3.13) with the aid of (5.3.10) by its consequence:

$$\delta_1 / M_1^2 < \psi_{st} + 1 < M_1 / \delta_1^2 \quad \text{a.e. on } I^2 . \qquad (5.3.14)$$

Subsequently we see that for sufficiently small a_2 the extremal automatically obeys both (5.3.14) and (5.3.13); consequently, the value of the extremal problem does not change. We also note that by virtue of (5.3.9)

$$\psi \big|_{\partial I^2} = 0 . \qquad (5.3.15)$$

We proceed to the determination of the infimum of the functional

$$\int_0^1 \int_0^1 (v + 1) \ln (v + 1) \, ds \, dt$$

on the set $\mathcal{U}$ of functions v from $\mathcal{S} = L^\infty(I^2)$ for which the function

$$\psi(x,y) = \int_0^x \int_0^y v(s,t) \, ds \, dt$$

satisfies (5.3.12), (5.3.14), and (5.3.15). The extremal d.f. F can be recovered from the extremal ψ within the accuracy of marginal d.f.s G and H. For G and H we select a uniform d.f. and then for such F all three functionals on the right-hand side of (5.3.11) attain their greatest lower bounds.

Now we apply the Lagrange principle (Theorem 1.8.1). The condition of strict differentiability is fulfilled due to inequalities (5.3.14). It follows from Theorem 1.8.1 that the extremal v should satisfy equation (1.8.2), which reduces after simplifications to the form

$$
\int_0^1\int_0^1 \tau(x,y)\left[\ln(v+1)+1+\mu_1 \int_x^1\int_y^1 q_1(s)\,q_2(t) \right.
$$

$$
\left. \times \left(\int_0^s\int_0^t v(\xi,\eta)\,d\xi\,d\eta\right)ds\,dt \right] dx\,dy = 0,
$$

where μ_1 is a Lagrange multiplier and τ is a smooth "trial" function from S. Taking as τ the elements of the trigonometric system of functions, by the uniqueness theorem for multiple Fourier series, we obtain (see Stein and Weiss (1971), Chapter 7) that for almost all s and t

$$
\ln(\psi_{st}+1)+\mu_1 \int_s^1\int_t^1 \psi(\xi,\eta)\,q_1(\xi)\,q_2(\eta)\,d\xi\,d\eta = C\,. \tag{5.3.16}
$$

One can consider (5.3.16) as being valid for all s and t since this assumption does not change the values of the minimized functional. From (5.3.16) we obtain the Euler–Lagrange equation (with μ_1 replaced by $-\mu$)

$$
\psi_{sstt}\left(\psi_{st}+1\right) - \psi_{stt}\,\psi_{sst} = \mu\,\psi\left(\psi_{st}+1\right)^2 q_1\,q_2\,, \tag{5.3.17}
$$

which has to be considered together with the condition

$$
\int_0^1\int_0^1 \psi^2(s,t)\,q_1(s)\,q_2(t)\,ds\,dt = a_2\,. \tag{5.3.18}
$$

We introduce a new small parameter $\varepsilon = \sqrt{a_2}$ and a new unknown function $u = \varepsilon^{-1}\psi$. Then (5.3.17) takes the form

$$
u_{sstt} - \mu\,u\,q_1\,q_2 - \mu u\left(u_{st}^2\,\varepsilon^2 + 2\,u_{st}\,\varepsilon\right)q_1\,q_2 + \left(u_{sstt}\,u_{st} - u_{stt}\,u_{sst}\right)\varepsilon = 0\,,
$$
$$
\tag{5.3.19}
$$

while the additional conditions (5.3.15) and (5.3.18) become

$$
\int_0^1\int_0^1 u^2\,q_1\,q_2\,ds\,dt = 1\,, \tag{5.3.20}
$$

$$
u\,\big|_{\partial I^2} = 0\,. \tag{5.3.21}
$$

Consider the auxiliary equation, corresponding to $\varepsilon = 0$ in (5.3.19):

$$u_{sstt} = \mu\, u\, q_1\, q_2, \qquad (5.3.22)$$

together with condition (5.3.21). It can be rewritten in the form of the integral equation (5.3.3) with kernel (5.3.4). Denote by λ_0 the smallest of the values of λ for which the solution of (5.3.3) exists and let x_0 be the corresponding eigenfunction. For example, for $q_1 = q_2 = 1$ we have

$$\lambda_0 = \pi^4, \qquad x_0(s,t) = C\, \sin \pi s\, \sin \pi t$$

by Blum et al. (1961).

We emphasize that λ_0 has necessarily multiplicity 1 since, under the conditions imposed on q_1 and q_2, the linear integral operator with the kernel

$$\big(\min\,(s,v) - s\,v\big)\big(\min\,(t,w) - t\,w\big)\, q_1(v)\, q_2(w)$$

is u_0-positive (see Krasnoselskii (1964), Theorem 2.10). This circumstance excludes the difficult multidimensional branching case.

We now set $x = u - x_0$ and $\lambda = \mu - \lambda_0$ and rewrite equation (5.3.19) in the form

$$x_{sstt} - \lambda_0\, x\, q_1\, q_2 = R_3(x, \varepsilon, \lambda), \qquad (5.3.23)$$

where $R_3(x, \varepsilon, \lambda)$ contains terms with x, ε, and λ in powers higher than 1, under the additional conditions

$$\int_0^1 \int_0^1 (x + x_0)^2\, q_1\, q_2\, ds\, dt = 1, \qquad (5.3.24)$$

$$x\big|_{\partial I^2} = 0. \qquad (5.3.25)$$

We consider on I^2 the space of all functions of two variables, twice continuously differentiable with respect to all variables and subject to condition (5.3.25) with the natural norm

$$\|\,x\,\| := \sup_{s,\,t}\,|\,x_{sstt}\,| + \sum_{|\alpha|\leq 3} \sup\,|D^\alpha x|,$$

and denote it by $\mathbf{E}_1$. Let $\mathbf{E}_2$ be the usual space of all continuous functions on I^2. We denote by $\mathcal{B}$ the operator on the left-hand side of (5.3.23) mapping the space $\mathbf{E}_1$ into the space $\mathbf{E}_2$ according to the formula

$$\mathcal{B}\, x = x_{sstt} - \lambda_0\, x\, q_1\, q_2\,.$$

If the operator $\mathcal{B}$ were invertible, then, by virtue of the theorem on implicit analytic operators (see Section 1.8), one could assert that the solution of equation (5.3.23) is unique and analytic as a function of the pair (ε, λ) of parameters.

However, the operator $\mathcal{B}$ is noninvertible, so we have the so-called branching case and for the determination of the solutions it is necessary to form the Lyapunov–Schmidt branching equation. The derivation of this equation in the considered case is similar to arguments presented in a detailed manner in Section 2.3 by the analysis of goodness-of-fit tests. It is essential that the operator $\mathcal{B}$ is a Fredholm operator with the number of zeros equal to 1, but its concrete form and the form of superior terms in $R_3(x, \varepsilon, \lambda)$ (see (5.3.23)) does not affect the result much.

Using, as in previous chapters, Schmidt's lemma, we obtain that the solution can be represented in the form

$$x = \sum_{i,j,k=1}^{\infty} x_{ijk}\, \xi^i\, \varepsilon^j\, \lambda^k, \qquad (5.3.26)$$

where ξ is an auxiliary real parameter and the series converges with respect to the norm of $\mathbf{E}_1$ for sufficiently small $|\xi|$, ε, and $|\lambda|$. The parameters ξ, ε, and λ are connected by the branching equation and by normalization condition (5.3.24). Excluding ξ and λ with the aid of implicit function theory, and passing from x to u and then to ψ, we obtain

$$\psi = x_0\, \varepsilon + \sum_{j=1}^{\infty} b_j\, \varepsilon^{j+1}, \qquad (5.3.27)$$

where b_j are elements of $\mathbf{E}_1$ and the series converges with respect to the norm of $\mathbf{E}_1$ for sufficiently small ε. Differentiating (5.3.27) we find a similar series also for ψ_{st}, converging uniformly on I^2. From here, for sufficiently small $\varepsilon = \sqrt{a_2}$ there follows condition (5.3.14) and also (5.3.13). Thus, for d.f. F, obtained from (5.3.27) with the aid of (5.3.9) for $G(x) = H(x) = x$, $0 \leq x \leq 1$, the value of $K\left(\Omega_{a_2}^B \cap V_{\delta_1} \cap W_{M_1}\right)$ is attained.

Computing the series for $(\psi_{st} + 1)\ln(\psi_{st} + 1)$ and integrating it term by term, for sufficiently small $\delta_1 > 0$ and $a_2 > 0$ and sufficiently large $M_1 > 0$, we obtain

$$K\left(\Omega_{a_2}^B \cap V_{\delta_1} \cap W_{M_1}\right) = \tfrac{1}{2}\lambda_0\, a_2 + \sum_{j \geq 3}^{\infty} c_j\, a_2^{j/2}, \qquad (5.3.28)$$

where c_j are numerical coefficients depending only on q_1 and q_2, and the series converges for sufficiently small a_2.

Now we return to inequality (5.3.8) and take the limit as $\delta \to 0$ and $M \to \infty$, obtaining the conclusion of Theorem 5.3.1. $\square$

At the same time as the statistic B^2_{n,q_1,q_2} it is quite natural to consider its generalization

$$B^k_{n,q_1,q_2} = \int\limits_{-\infty}^{\infty} \int\limits_{-\infty}^{\infty} \Big[F_n(x,y) - G_n(x)\,H_n(y)\Big]^k q_1\big(G_n(x)\big)$$

$$\times\, q_2\big(H_n(y)\big)\, dG_n(x)\, dH_n(y),$$

where k is a positive integer, $k \neq 2$.

An analog of Theorems 5.3.1 and 5.3.2 for such statistics is the relation

$$\lim_{n\to\infty} n^{-1} \ln P\left(B^k_{n,q_1,q_2} \geq a\right) = -\tfrac{1}{2}\lambda_{0k}\, a^{2/k} + \sum_{j\geq 3}^{\infty} c_{jk}\, a^{j/k}, \quad (5.3.29)$$

valid for sufficiently small $a > 0$ (the constant λ_{0k} is defined subsequently).

The proof of the validity of (5.3.29), up to the construction of the solution of the Euler–Lagrange equations, is almost the same as for the case $k = 2$. An essential difference arises in the consideration of the properties of the principal solutions λ_{0k} and x_{0k} of the equation

$$u_{sstt} = \lambda\, u^{k-1}\, q_1\, q_2, \quad (5.3.30)$$

considered together with normalization condition (5.3.20) and boundary condition (5.3.21). Another difference is connected with the properties of operator $\mathcal{B}_k$ given by the formula

$$\mathcal{B}_k\, x = x_{sstt} - \lambda_{0k}\,(k-1)\, x\, x_{0k}^{k-2}\, q_1\, q_2 \quad (5.3.31)$$

and considered together with condition (5.3.25).

In the case $k = 1$ the values λ_{01} and x_{01} are explicitly computed with the aid of Green function (5.3.4), while the operator $\mathcal{B}_1$ is invertible. We note that for $q_1 = q_2 = 1$ we have

$$\lambda_{01} = 144 \quad \text{and} \quad x_{01}(s,t) = 36\, s\,(1-s)\, t\,(1-t).$$

The case $k > 2$ is considerably more complicated. The author is unaware of any results regarding the multiplicity of λ_{0k} and the properties of the eigenfunctions x_{0k} in equation (5.3.30). In connection with this, we do not have a clear answer to the key question of the present method:

What is the dimension d of the subspace of the zeros of the operator $\mathcal{B}_k$? In the regular $(d = 0)$ and the one-dimensional $(d = 1)$ branching cases, the methods of the present work lead to relation (5.3.29).

In order to exclude the multidimensional branching case, we assume that the exponent k and the weights q_1 and q_2 are such that $d \leq 1$ (the corresponding triple (k, q_1, q_2) will be said to be *admissible*). For this it is sufficient, for example, that the principal eigenfunction x_{0k} should vanish on I^2 on a set of zero measure (see Krasnoselskii (1964)). For $k = 1$ and $k = 2$ all the triples satisfying (5.3.1) are admissible (and even under the weaker restrictions). Now we formulate a generalization of Theorems 5.3.1 and 5.3.2.

Theorem 5.3.3 *Assume that the triple (k, q_1, q_2) is admissible and (5.3.1) holds. Then for sufficiently small $a > 0$ relation (5.3.29) is valid.*

5.4 Bahadur Efficiency of Independence Tests

As in previous chapters we shall compute the Bahadur efficiency of non-parametric independence tests for simple parametric alternatives. We assume now that under the alternative A_3' the d.f. of the initial sample has the form $F(x, y; \theta)$, $x, y \in \mathbf{R}^1$, $\theta \in [0, \theta_0]$, for some $\theta_0 > 0$. It is supposed that the relation

$$F(x, y; \theta) = G(x)\,H(y)\,, \tag{5.4.1}$$

where G and H are marginal d.f.s, holds only for $\theta = 0$ and that the derivative with respect to parameter $F_\theta'(x, y; \theta)$ exists and is uniformly bounded for all x, y, and θ. We assume also that d.f. F is absolutely continuous in (x, y) for every θ and denote the corresponding density by $f(x, y; \theta)$.

The continuity of the functional $B(F)$ proved in Section 5.3 together with the Glivenko–Cantelli theorem imply that with probability 1 (and consequently in probability) we have under A_3' the relation

$$B_{n,\,q_1,\,q_2}^k \longrightarrow b_k(\theta)\,, \qquad n \to \infty\,,$$

where, as $\theta \to 0$,

$$b_k(\theta) \sim \int\limits_{-\infty}^{\infty} \int\limits_{-\infty}^{\infty} \left[\, F_\theta'(x, y; 0)\,\right]^k q_1\big(G(x)\big)\, q_2\big(H(y)\big)\, dG(x)\, dH(y) \cdot \theta^k\,.$$

It follows that under the conditions of Theorem 5.3.3 the local Bahadur exact slope of the sequence $\{B^k_{n,q_1,q_2}\}$ satisfies the relation

$$c_{B^k}(\theta) \sim \lambda_{0k} \left[b_k(\theta) \right]^{2/k}, \qquad \theta \to 0. \qquad (5.4.2)$$

The local exact slopes of the sequences $\{\Gamma_n\}$ and $\{\Gamma_n^{\pm}\}$ may be computed analogously. Using Theorem 5.2.1 and the Glivenko–Cantelli theorem we obtain

$$c_{\Gamma}(\theta) \sim 16 \sup_{-\infty < x, y < \infty} \left| F'_\theta(x, y; 0) \right|^2 \cdot \theta^2, \qquad \theta \to 0. \qquad (5.4.3)$$

Formulas for $c_{\Gamma+}(\theta)$ and $c_{\Gamma-}(\theta)$ are connected with obvious changes.

The intermediate position between the statistics of Kolmogorov–Smirnov type and the integral statistics belongs to the statistics of Durbin type:

$$M_n = \sup_{-\infty < x < \infty} \left| \int_{-\infty}^{\infty} \Big(F_n(x, y) - G_n(x) H_n(y) \Big) \, d H_n(y) \right|$$

and their one-sided counterparts

$$M_n^+ = \sup_{-\infty < x < \infty} \left[\int_{-\infty}^{\infty} \Big(F_n(x, y) - G_n(x) H_n(y) \Big) \, d H_n(y) \right],$$

$$M_n^- = \sup_{-\infty < x < \infty} \left[\int_{-\infty}^{\infty} \Big(G_n(x) H_n(y) - F_n(x, y) \Big) \, d H_n(y) \right].$$

The statistics of such a kind were first proposed by Durbin (1970) for goodness-of-fit testing in the two-dimensional case. Dmitrieva (1988) investigated large deviations of these statistics under the independence hypothesis. She proved that as $a \to 0$

$$\lim_{n \to \infty} n^{-1} \ln P \left(M_n \geq a \right) = g_9(a) = -24 \, a^2 \left(1 + o(1) \right), \qquad (5.4.4)$$

where the function $g_9(a)$ is continuous for small $a > 0$. Analogous results are valid for $M_n^{\pm}$.

An indisputable merit of these statistics consists in that they have, unlike statistics Γ_n, $\Gamma_n^{\pm}$, and $B^2_{n,1,1}$, a remarkably simple limiting distribution. It has been shown in Dmitrieva (1988) that the statistics $\sqrt{12 \, n} \; M_n^{\pm}$ and $\sqrt{12 \, n} \; M_n$ have the same limiting distributions as the Kolmogorov–Smirnov statistics $\sqrt{n} \, D_n^{\pm}$ and $\sqrt{n} \, D_n$ for goodness-of-fit testing.

Using relation (5.4.4), the continuity of M_n as a functional of the empirical d.f. F_n, and the Glivenko–Cantelli theorem we can write out the local exact slope of the sequence $\{M_n\}$:

$$c_M(\theta) \sim 48 \sup_{-\infty < x < \infty} \left| \int_{-\infty}^{\infty} F'_\theta(x, y; 0) \, dH(y) \right|^2 \cdot \theta^2, \qquad \theta \to 0.$$

$$(5.4.5)$$

The expressions for local exact slopes of the statistics $M_n^{\pm}$ are again connected with obvious changes.

We can compute the local exact slope of the sequence of linear rank statistics (5.1.5) on the basis of Theorems 1.6.12 and 1.6.11. Consider in the notation of Section 5.1 the sequence

$$T_n = n^{-1} \sum_{i=1}^{n} a_{n1}\big(R_i/(n+1)\big) \, a_{n2}\big(S_i/(n+1)\big),$$

assuming that

$$a_{ni}(u) \xrightarrow{L^2} J_i(u), \qquad i = 1, 2,$$

where the score functions $J_i(u)$ satisfy the relations:

$$\int_0^1 J_i(u) \, du = 0, \qquad \int_0^1 J_i^2(u) \, du = 1, \qquad i = 1, 2;$$

$$\int_0^1 \int_0^1 \exp\big\{r J_1(u) J_2(v)\big\} \, du \, dv < \infty \qquad \text{for some } r > 0.$$

Then we have under the independence hypothesis

$$\lim_{n \to \infty} n^{-1} \ln P\big(T_n \geq a\big) = -\tfrac{1}{2} a^2 \big(1 + o(1)\big), \qquad a \to 0. \qquad (5.4.6)$$

The statistic T_n admits the representation

$$T_n = \int_{-\infty}^{\infty} \int_{-\infty}^{\infty} a_{n1}\big((nG_n(x) + 1)/(n+1)\big)$$

$$\times a_{n2}\big((nH_n(y) + 1)/(n+1)\big) \, dF_n(x, y),$$

which enables one to find the limit in probability of T_n under the hypothesis A'_3. Using the law of large numbers for linear rank statistics (see Woodworth (1970) and Müller-Funk (1983)), we obtain under A'_3 that in probability

$$T_n \longrightarrow b_T(\theta) := \int_{-\infty}^{\infty} \int_{-\infty}^{\infty} J_1\big(F(x, \infty; \theta)\big) J_2\big(F(\infty, y; \theta)\big) \, dF(x, y; \theta).$$

Assuming that $b_T(\theta) > 0$ for $\theta > 0$ and $b_T(\theta) \to 0$ as $\theta \to 0$, we deduce from (5.4.6) the following formula for the local exact slope of the sequence $\{T_n\}$

$$c_T(\theta) = b_T^2(\theta)\left(1 + o(1)\right), \qquad \theta \to 0, \qquad (5.4.7)$$

which can be simplified if $F(x, y; \theta)$ is smooth enough.

For the comparison we mention here also the Kendall rank correlation coefficient τ_n (see Kendall (1970) and Hettmansperger (1984)) that is often used for independence testing but by its structure is a nonlinear rank statistic. One of the possible forms in which to introduce τ_n is

$$\tau_n = \frac{1}{n(n-1)} \sum_{i \neq j} \mathrm{sign}\,(R_i - R_j)\,\mathrm{sign}\,(S_i - S_j)\,.$$

Large deviations and exact slopes of the sequence $\{\tau_n\}$ have been investigated by Sievers (1969) and Woodworth (1970). It has been proved that for sufficiently regular d.f. $F(x, y; \theta)$, as $\theta \to 0$, the following relation holds:

$$c_\tau(\theta) \sim 36 \left\{ \int\limits_{-\infty}^{\infty} \int\limits_{-\infty}^{\infty} F_\theta'(x, y; 0)\, dG(x)\, dH(y) \right.$$

$$\left. + \int\limits_{-\infty}^{\infty} \int\limits_{-\infty}^{\infty} G(x)\, H(y)\, dF_\theta'(x, y; 0) \right\}^2 \cdot \theta^2\,.$$

We proceed to find the maximal values of exact slopes. It follows from Theorem 1.2.3 that for any sequence of statistics $\{T_n\}$ used for testing H_3 against A_3' the following inequality holds:

$$c_T(\theta) \leq 2 \inf_{l,m} \int\limits_{-\infty}^{\infty} \int\limits_{-\infty}^{\infty} \ln \frac{f(x, y; \theta)}{l(x)\,m(y)}\, f(x, y; \theta)\, dx\, dy,$$

where the infimum is taken over all one-dimensional densities l, m on the real line. Groeneboom, Lepage, and Ruymgaart (1976) proved that this infimum is attained for the case $l = g$, $m = h$ (g and h are marginal densities for $f(x, y; \theta)$) so

$$c_T(\theta) \leq 2 \int\limits_{-\infty}^{\infty} \int\limits_{-\infty}^{\infty} \ln \frac{f(x, y; \theta)}{g(x)\,h(y)}\, f(x, y; \theta)\, dx\, dy\,. \qquad (5.4.8)$$

Now we shall compare all considered statistics in the case of testing the independence of components for the bivariate Gaussian distribution.

Let $F(x, y; \theta)$ be the d.f. of the standard bivariate normal law with zero means, with variances of components equal to 1, and with the correlation coefficient θ. Computations of local exact slopes using formulas (5.4.2), (5.4.3), (5.4.4), and (5.4.7) (in the last case for $J_1(u) = J_2(u) = \Phi^{-1}(u)$, which corresponds to the normal scores test or van der Waerden correlation coefficient (see Hájek and Šidak (1967))) give the following results:

$$c_{B^1}(\theta) \sim \frac{9}{\pi^2} \cdot \theta^2 \approx 0.912 \cdot \theta^2,$$

$$c_{B^2}(\theta) \sim \frac{\pi^2}{12} \cdot \theta^2 \approx 0.822 \cdot \theta^2,$$

$$c_{\Gamma}(\theta) \sim \frac{4}{\pi^2} \cdot \theta^2 \approx 0.405 \cdot \theta^2,$$

$$c_M(\theta) \sim \frac{6}{\pi^2} \cdot \theta^2 \approx 0.608 \cdot \theta^2,$$

$$c_T(\theta) \sim \theta^2,$$

$$c_{\tau}(\theta) \sim \frac{9}{\pi^2} \cdot \theta^2 \approx 0.912 \cdot \theta^2.$$

Finally, calculating the right-hand side in inequality (5.4.8) we obtain, due to Kullback (1959, Chapter 1), that it equals

$$-\ln(1 - \theta^2) \sim \theta^2, \qquad \theta \to 0.$$

Thus the normal scores test as in the case of homogeneity testing turns out to be locally asymptotically optimal in the Bahadur sense. The same is true also for the Pearson sample correlation coefficient (see Woodworth (1970) and Groeneboom et al. (1976)). The statistics $B_{n,1,1}^1$ and $B_{n,1,1}^2$ are slightly less efficient. Using formulas (5.4.2) and (5.4.7), we can easily see that the Spearman rank correlation coefficient (when $J_1(u) = J_2(u) = \sqrt{12}\,(u - \frac{1}{2})$) is locally equivalent to the statistic $B_{n,1,1}^1$ and that the first component of the statistic $B_{n,1,1}^2$ in the Durbin–Knott sense (i.e., the statistic $B_{n,\,q_1,q_2}^1$ for $q_1(u) = q_2(u) = \sin \pi u$, introduced by Koziol and Nemec (1979)) is locally equivalent to the statistic $B_{n,1,1}^2$.

Accord on the question of describing the most natural model of alternative to the independence hypothesis is lacking in the statistical literature (see Hájek and Šidak (1967), Sections 6.2.6 and 7.2.4). Perhaps more frequently one encounters a model of the form

$$F(x, y; \theta) = G(x)\,H(y) + \theta \cdot \Omega\big(G(x), H(y)\big), \qquad (5.4.9)$$

where Ω is a real function on $[0, 1]^2$, satisfying some natural limitations. For example, Morgenstern (1956) and Farlie (1960) considered $\Omega_1(x, y) = x(1 - x)y(1 - y)$, whereas Konijn (1959) investigated $\Omega_2(x, y) = \min(x, y) - x\,y$. Other variants have been proposed by Plackett (1965), Ledwina (1984, 1986a,b), and Bajorski (1987). Surely, the value of θ must be sufficiently small in order for the function $F(x, y; \theta)$ in (5.4.9) to be a d.f.

For example, let us calculate local exact slopes of statistics under consideration in the case of Ω_1. Acting as in the Gaussian case (there we had $\Omega(x, y) = \varphi\big(\Phi^{-1}(x)\big)\varphi\big(\Phi^{-1}(y)\big)$) we obtain, as $\theta \to 0$, that

$$c_{B^1}(\theta) \sim \frac{1}{9} \cdot \theta^2 \approx 0.111 \cdot \theta^2,$$

$$c_{B^2}(\theta) \sim \frac{\pi^4}{900} \cdot \theta^2 \approx 0.108 \cdot \theta^2,$$

$$c_\Gamma(\theta) \sim \frac{1}{16} \cdot \theta^2 \approx 0.062 \cdot \theta^2,$$

$$c_M(\theta) \sim \frac{1}{12} \cdot \theta^2 \approx 0.083 \cdot \theta^2,$$

$$c_\tau(\theta) \sim \frac{1}{9} \cdot \theta^2 \approx 0.111 \cdot \theta^2,$$

$$c_T(\theta) \sim \frac{1}{9} \cdot \theta^2 \approx 0.111 \cdot \theta^2 \quad \text{for } J_1(u) = J_2(u) = \sqrt{12}\left(u - \tfrac{1}{2}\right).$$

At the same time the upper bound (5.4.8) for slopes is equal to $\frac{1}{9}\theta^2(1 + o(1))$, $\theta \to 0$, so the statistic $B^1_{n,1,1}$ and the rank correlation coefficients of Spearman and Kendall turn out to be locally asymptotically optimal.

The conditions of local asymptotic optimality will be considered more profoundly in the following chapter.

The approximate Bahadur efficiency of statistics we have studied may be calculated by the same methods as in previous chapters. One must take into account that the statistics τ_n and T_n are, as is well known, asymptotically normal (see, e.g., Kendall (1970) and Hettmansperger (1984)), and limiting distributions of all the others (see Blum et al. (1961)) are distributions of functionals of the Gaussian field $\xi(s, t)$ with zero mean and correlation function

$$E\xi(s, t)\,\xi(u, v) = \big(\min(s, u) - s\,u\big)\big(\min(t, v) - t\,v\big). \tag{5.4.10}$$

The rate of decreasing of the tails of these distributions, that is, the exponent a from (1.2.18), is found in the same way as in the case of hypotheses H_0, H_1, and H_2, namely, using Lemma 4.8.1 and the results by Borell (1975) and Kallianpur and Oodaira (1978). In all mentioned cases we find the coincidence of the local exact and local approximate Bahadur efficiencies for any pair of considered statistics. The same conclusion holds for the limiting Pitman ARE, which may be reduced to the local approximate Bahadur ARE on the basis of Theorems 1.4.1 and 1.4.2.

5.5 Large Deviations of Statistics for Testing Uniformity on a Square

We proceed to the investigation of the case when we are verifying the independence hypothesis H_3 for *known* continuous distributions of components X and Y. Then it is possible to make for these components of the initial sample a Smirnov transform giving as a result a new bivariate sample having under H_3 the uniform distribution on the unit square I^2. Therefore, testing the hypothesis H_3 is reduced to testing uniformity on a square, that is, to goodness-of-fit testing. The form of the alternative after such transform is easily found in every concrete case.

Let $F_n(s, t)$ be the empirical d.f. based on the transformed sample. For the solution of the problem under consideration it is quite natural to introduce the statistics

$$\Delta_n = \sup_{0 \le s, t \le 1} \left| F_n(s, t) - s\,t \right|, \tag{5.5.1}$$

$$\Omega^1_{n, q} = \int_0^1 \int_0^1 \left[F_n(s, t) - s\,t \right] q(s, t)\, ds\, dt, \tag{5.5.2}$$

$$\Omega^2_{n, q} = \int_0^1 \int_0^1 \left[F_n(s, t) - s\,t \right]^2 q(s, t)\, ds\, dt, \tag{5.5.3}$$

$$\mu_n = \sup_{0 \le s \le 1} \left| \int_0^1 F_n(s, t)\, dt - \tfrac{1}{2}\, s \right|. \tag{5.5.4}$$

The one-sided variants of these statistics are defined in a similar way. We suppose here that q is a nonnegative weight function that is subject to complementary conditions to be stated subsequently. The statistic Δ_n is the two-dimensional counterpart of the Kolmogorov statistic, and the statistics $\Omega^1_{n,q}$ and $\Omega^2_{n,q}$ are the two-dimensional counterparts of the integral statistics investigated in Chapter 2. Asymptotic properties of $\Omega_{n,q}^2$ for $q \equiv 1$ were studied by Krivyakova, Martynov, and Tyurin (1977) and then by Cotterill and Csörgö (1982). The statistic μ_n proposed by Durbin (1970) is specific for the two-dimensional case. In order to compute the local Bahadur efficiency of statistics (5.5.1)–(5.5.4) we must find for them rough large-deviation asymptotics. Our results are based on Nikitin (1984b) and Nikitin (1986b).

Theorem 5.5.1 *For any* $a \in (0, 1)$ *under the hypothesis* H_3 *one has*

$$\lim_{n\to\infty} n^{-1} \ln P\left(\Delta_n \geq a\right) = -g_1(a),$$

where $g_1(a)$ *has been defined in (2.2.2).*

The proof is omitted since it differs insignificantly from the proof of Theorem 2.2.1.

Denote for brevity for any $a > 0$ and $t \in (0, 1)$

$$\rho(a,t) := \inf_{s\geq 0} \left[\exp\left\{-s\left(a + \tfrac{1}{2}t\right)\right\} \left(\left(e^s - 1\right)t\,s^{-1} + 1 - t\right) \right]$$

$$w(a) := \sup_{0<t<1} \ln \rho(a,t).$$

Theorem 5.5.2 *For any* $a \in (0, 1)$ *under the hypothesis* H_3 *one has*

$$\lim_{n\to\infty} n^{-1} \ln P\left(\mu_n \geq a\right) = w(a),$$

where the function w *is continuous for small* $a > 0$ *and*

$$w(a) = -\tfrac{9}{2}a^2\left(1 + o(1)\right), \qquad a \to 0. \qquad (5.5.5)$$

Proof For the sake of simplicity we consider the one-sided statistic μ_n^+ which has the same large-deviation asymptotics. For a fixed $t \in (0, 1)$ put

$$\xi_j(t) := \mathbf{1}_{\{X_j<t\}} \cdot (1 - Y_j), \qquad j = 1, 2, \ldots.$$

We have

$$P\left(\mu_n^+ \geq a\right) \geq P\left(\sum_{j=1}^{n}\xi_j(t) \geq n\left(\tfrac{1}{2}t + a\right)\right)$$

and, moreover,

$$E\exp\left\{u\,\xi_j(s)\right\} = E_{Y_j}\left\{s\exp\left\{u\left(1-Y_j\right)\right\}+1-s\right\} = u^{-1}\left(e^u-1\right)s+1-s.$$

Applying Theorem 1.6.2 we obtain

$$\lim_{n\to\infty} n^{-1}\ln P\big(\mu_n^+ \geq a\big) \geq \sup_{0<t<1}\ \ln \rho(a,t) = w(a)\,. \qquad (5.5.6)$$

The reverse inequality will be obtained by the same manner as in Theorems 2.2.1 and 2.2.2:

$$P\big(\mu_n^+ \geq a\big) \leq P\Big(\max_{1\leq i\leq k}\Big(\int_0^1 F_n\Big(\tfrac{i}{k},\,u\Big)\,du - \tfrac{i-1}{k}\Big) \geq a\,\Big)$$

$$\leq \sum_{i=1}^{k} P\Big(\sum_{j=1}^{k}\xi_j\Big(\tfrac{i}{k}\Big) - \tfrac{i}{2k} \geq n\Big(a - \tfrac{1}{2k}\Big)\Big)$$

$$\leq \sum_{i=1}^{k} \rho^n\Big(a - \tfrac{1}{2k},\,\tfrac{i}{k}\Big)$$

$$\leq k \exp\Big\{n\,w\Big(a - \tfrac{1}{2k}\Big)\Big\}\,,$$

where k is a positive integer such that $0 < a - (2k)^{-1} < 1$. As the function w is continuous, we have

$$\overline{\lim_{n\to\infty}}\ n^{-1}\ln P\big(\mu_n^+ \geq a\big) \leq w(a)\,. \qquad (5.5.7)$$

From (5.5.6) and (5.5.7) follows the first conclusion of Theorem 5.5.2. To prove (5.5.5) we note that

$$\rho(a,s) = \exp\big\{-\tau\big(a + \tfrac{1}{2}s\big)\big\}\big(\tau^{-1}\,(e^\tau - 1)\,s + 1 - s\big),$$

where τ is determined by the following equation:

$$\frac{\big(\tau e^\tau - e^\tau + 1\big)\,s}{\tau^2\big(\tau^{-1}(e^\tau - 1)\,s + 1 - s\big)} - \frac{s}{2} = a\,.$$

For any $s \in (0, 1)$ and $a \to 0$ we obtain, using the Lagrange inversion formula (see, e.g., De Bruijn (1961), Section 2.2), that

$$\tau = \frac{12\,a}{4\,s - 3\,s^2} + o(a)\,,$$

$$\ln \rho(a,s) = -\frac{6\,a^2}{4\,s - 3\,s^2} + o(a^2)\,.$$

Since $\rho(a,s)$ is clearly jointly continuous in a and s, we have

$$\lim_{a\to 0}\Big[a^{-2}\sup_{0<s<1}\ln \rho(a,s)\Big] = \sup_{0<s<1}\Big[\lim_{a\to 0} a^{-2}\ln \rho(a,s)\Big] = -\tfrac{9}{2}\,,$$

which completes the proof of (5.5.5).   □

The statistics $\Omega^1_{n,q}$ given by formula (5.5.2) reduce to a sum of centered independent identically distributed random variables. Applying Theorem 1.6.2 we obtain, as in Section 2.4, the following result.

Theorem 5.5.3 *For sufficiently small* $a > 0$

$$\lim_{n\to\infty} n^{-1} \ln P\left(\Omega^1_{n,q} \geq a\right) = -\frac{a^2}{2\sigma_0^2} + \sum_{j\geq 3} l_j\, a^j,$$

where the series on the right-hand side is convergent and

$$\sigma_0^2 := \left(\int_0^1\!\int_0^1\!\int_0^1\!\int_0^1 K_2(s,t;\,u,v)\, q(s,t)\, q(u,v)\, ds\, dt\, du\, dv\right)^{-1},$$

$$K_2(s,t;\,u,v) := \min(s,u)\,\min(t,v) - s\,u\,t\,v, \qquad s,\,u,\,t,\,v \in [0,1].$$

$$(5.5.8)$$

We proceed to the analysis of the statistics $\Omega^2_{n,q}$.

Theorem 5.5.4 *Let the weight function* q *be continuous and positive on* $[0,1]^2$. *Then for sufficiently small* $a > 0$

$$\lim_{n\to\infty} n^{-1} \ln P\left(\Omega^2_{n,q} \geq a\right) = -\tfrac{1}{2}\lambda_0\, a + \sum_{j\geq 3} d_j\, a^{j/2},$$

where the series on the right-hand side converges and λ_0 *is the principal eigenvalue of the linear integral equation*

$$h(s,t) = \lambda \int_0^1\!\int_0^1 K_2(s,t;\,u,v)\, q(u,v)\, h(u,v)\, du\, dv. \qquad (5.5.9)$$

(The kernel K_2 *was given by formula (5.5.8)).*

Proof We carry out the proof briefly because of a strong similarity with the one-dimensional variant in Sections 2.3–2.4. Denote by $\mathcal{F}_2$ the set of all absolutely continuous d.f.s $f(s,t)$ on I^2 and by f_{st}, f_{sst}, f_{stt} corresponding densities and derivatives, if they exist.

Consider for any $a > 0$ the set

$$\mathbf{E}_a := \left\{ f\colon f \in \mathcal{F}_2,\ \int_0^1\!\int_0^1 \bigl(f(s,t) - s\,t\bigr)^2 q(s,t)\, ds\, dt \geq a \right\}$$

and put for any $A \in \mathcal{F}_2$

$$K(A) := \inf\left\{ \int_0^1\!\int_0^1 f_{st}\, \ln f_{st}\, ds\, dt\colon\ f \in A \right\}.$$

We show in the sequel that $K(\mathbf{E}_a)$ is a continuous function of a in a neighborhood of zero. Also, the statistics $\Omega_{n,q}^2$ are uniformly continuous and, a fortiriori, are τ-continuous functionals of F_n. By Theorem 1.6.7 we obtain

$$\lim_{n \to \infty} n^{-1} \ln P\left(\Omega_{n,q}^2 \geq a\right) = -K(\mathbf{E}_a) \qquad (5.5.10)$$

for points of continuity of the function $a \mapsto K(\mathbf{E}_a)$. Arguments similar to those of Lemma 2 show that we commit an error of order $O(\delta)$ if we calculate not $K(\mathbf{E}_a)$ but $K\left(\mathbf{E}_a \cap V_\delta\right)$, where V_δ has been given by (2.3.10).

Putting $f(s,t) - st = h(s,t)\,\varepsilon$ and $\varepsilon = \sqrt{a}$ we minimize the functional

$$\int_0^1\!\!\int_0^1 \left(1 + h_{st}\,\varepsilon\right) \ln\left(1 + h_{st}\,\varepsilon\right) ds\,dt \qquad (5.5.11)$$

on the set of absolutely continuous d.f.s h on I^2 that possess the following properties:

$$1 + h_{st}\,\varepsilon \geq \delta\,, \qquad \int_0^1\!\!\int_0^1 h^2(s,t)\,q(s,t)\,ds\,dt \geq 1\,, \qquad (5.5.12)$$

$$h(0,t) = h(s,0) = h(1,1) = 0\,, \qquad s,\,t \in [0,1]\,. \qquad (5.5.13)$$

Applying (after suitable formalization of this extremal problem) the Lagrange principle (Theorem 1.8.1) and calculating the variation, we omit some transformations and cite the final Euler–Lagrange equation

$$h_{sstt} - \mu\,h\,q\left(1 + h_{st}\,\varepsilon\right)^2 + \left(h_{sstt}\,h_{st} - h_{sst}\,h_{stt}\right)\varepsilon^2 = 0\,. \qquad (5.5.14)$$

This equation should be considered together with the normalization condition

$$\int_0^1\!\!\int_0^1 h^2(s,t)\,q(s,t)\,ds\,dt = 1\,, \qquad (5.5.15)$$

and boundary conditions (5.5.13) and also with the so-called natural boundary conditions (see, e.g., Buslaev (1980), Chapter 2, Section 3):

$$h_{sst}(1,t) = h_{sst}(s,1) = 0\,, \qquad s,\,t \in [0,1]\,. \qquad (5.5.16)$$

Consider now the linearization of equation (5.5.14) corresponding to the case $\varepsilon = 0$:

$$h_{sstt} = \mu\,h\,q\,. \qquad (5.5.17)$$

Under boundary conditions (5.5.13) and (5.5.16) equation (5.5.17) is tantamount to integral equation (5.5.9), as is easily proved with the aid of the formula of Gauss and Ostrogradskii. Integral equations of the form (5.5.9) have been investigated much more intensively than boundary-value problems for the partial differential equation (5.5.17).

Let λ_0 be the principal eigenvalue of equation (5.5.9) and $x_0(s,t)$ be the corresponding eigenfunction satisfying normalization condition (5.5.15). We emphasize that λ_0 has again multiplicity one, as under the conditions of Theorem 5.5.4 the linear integral operator in (5.5.9) is u_0-positive (see Krasnoselskii (1964), Theorem 2.10). Therefore, we rule out the multidimensional branching case.

Now we make the substitutions

$$x = h - x_0, \qquad \lambda = \mu - \lambda_0$$

and rewrite (5.5.14) in the form

$$x_{sstt} - \lambda_0\, x\, q = R_4(x,\varepsilon,\lambda), \qquad (5.5.18)$$

where $R_4(x,\varepsilon,\lambda)$ contains the terms with x, ε, and λ in powers greater than 1. As the operator on the left-hand side of (5.5.18) is noninvertible, we have the branching case when it is necessary to set up the Lyapunov–Schmidt branching equation for analytic operators. This may be done similarly to the one-dimensional case considered in Section 2.4. Considering the branching equation simultaneously with the normalization condition and using the theory of implicit functions, we have

$$h(s,t) = x_0(s,t) + \sum_{j \geq 1} e_j(s,t)\, \varepsilon^j, \qquad (5.5.19)$$

where the series on the right-hand side converges absolutely with respect to the natural norm in the space of all twice differentiable functions on I^2. Differentiating (5.5.19) term by term and substituting the result in the functional (5.5.11) under minimization, we obtain after elementary transforms the conclusion of Theorem 5.5.4. $\qquad\square$

The generalization of Theorem 5.5.4 for the case of the statistics $\Omega_{n,q}^k$, $k > 2$, meets with substantial difficulties because of the lack of information about the spectrum of the nonlinear boundary-value problem

$$h_{sstt} = \mu\, h^{k-1}\, q$$

under boundary conditions (5.5.13) and (5.5.16) and normalization condition (5.5.12), which replaces linear problem (5.5.17).

In conclusion we note that somewhat weaker results than Theorems

5.5.1 and 5.5.4 may be deduced from Singh (1981) where an entirely different approach has been used.

5.6 Local Exact Slopes

We have already seen in Section 5.5 that the hypothesis H_3 in the case of known marginal distributions of the initial sample can be replaced by the hypothesis of uniformity on a unit square with uniform marginals. We assume that the transformed alternative d.f. $\widetilde{F}(s, t; \theta)$ is sufficiently regular and

$$\widetilde{F}(s, t; 0) = s\,t\,, \qquad \widetilde{F}(s, 1; \theta) = s\,, \qquad \widetilde{F}(1, t; \theta) = t \quad \text{for all } \theta\,.$$

Using Theorems 5.5.1–5.5.4 and the Glivenko–Cantelli theorem we are now able to compute the principal part of the Bahadur exact slopes of statistics (5.5.1)–(5.5.4) as $\theta \to 0$. We obtain

$$c_\Delta(\theta) \sim 4 \sup_{0 \le s, t \le 1} \left| \widetilde{F}(s, t; \theta) - s\,t \right|^2,$$

$$c_\mu(\theta) \sim 9 \sup_{0 \le s \le 1} \left| \int_0^1 \widetilde{F}(s, t; \theta)\, dt - \tfrac{1}{2}\, s \right|^2,$$

$$c_{\Omega^1}(\theta) \sim \sigma_0^{-2} \left(\int_0^1 \!\! \int_0^1 \left(\widetilde{F}(s, t; \theta) - s\,t \right) q(s, t)\, ds\, dt \right)^2,$$

$$c_{\Omega^2}(\theta) \sim \lambda_0 \int_0^1 \!\! \int_0^1 \left(\widetilde{F}(s, t; \theta) - s\,t \right)^2 q(s, t)\, ds\, dt,$$

where the constants σ_0^2 and λ_0 have been determined in Theorems 5.5.3 and 5.5.4.

As an example let us perform computations in the simplest case of the standard Gaussian distribution in $\mathbf{R}^2$ with the correlation coefficient between components equal to θ. In this case

$$\widetilde{F}(s, t; \theta) = s\,t + \varphi\big(\Phi^{-1}(s)\big)\, \varphi\big(\Phi^{-1}(t)\big) \cdot \theta + o(\theta)\,, \qquad \theta \to 0\,.$$

We consider the statistics $\Omega^1_{n,q}$ and $\Omega^2_{n,q}$ for $q \equiv 1$. Durbin (1970) and Krivyakova et al. (1977) proved that $\sigma_0^2 = 7/144$ and $\lambda_0 = 15.814\dots$.

After simple calculations we obtain, as $\theta \to 0$, that

$$c_\Delta(\theta) \sim \pi^{-2} \cdot \theta^2 \approx 0.101\,\theta^2,$$

$$c_\mu(\theta) \sim \frac{9}{8}\,\pi^{-2} \cdot \theta^2 \approx 0.114\,\theta^2,$$

$$c_{\Omega^1}(\theta) \sim \frac{9}{7}\,\pi^{-2} \cdot \theta^2 \approx 0.130\,\theta^2,$$

$$c_{\Omega^2}(\theta) \sim \frac{\lambda_0}{12}\,\pi^{-2} \cdot \theta^2 \approx 0.134\,\theta^2.$$

Thus the local exact slopes of statistics under consideration are very close in their values, but still the sequence $\{\Omega_{n,1}^2\}$ is better and the sequence $\{\Delta_n\}$ is worse than others. The upper bound for exact slopes given by Theorem 1.2.3 in our case equals $\theta^2\big(1 + o(1)\big)$, $\theta \to 0$. It is clear that the test statistics under consideration are in no way inferior to their analogs studied in Sections 5.1–5.4. The situation is preserved for other common alternatives. Apparently it means that the tests based on the difference $F_n - G_n\,H_n$ are more sensitive and flexible and therefore deserve preference.

The expressions we have obtained for local exact slopes may be used also for the comparison of two-dimensional goodness-of-fit tests based on statistics (5.5.1)–(5.5.4). Let the observations (X_1, Y_1), (X_2, Y_2), ... , (X_n, Y_n) have continuous d.f. $F(s, t; \theta)$ depending on real parameter θ. It is necessary to test the hypothesis

$$H_0\colon\ \theta = 0$$

against the alternative hypothesis

$$A_0\colon\ \theta > 0\,.$$

The form of F is assumed to be completely known.

It is well known that the two-dimensional analogs of the Kolmogorov–Smirnov and ω^2 statistics for testing goodness-of-fit hypothesis H_0 have distributions depending on the distribution of the initial sample. In order to overcome this obstacle we transform this sample into a sample from the uniform distribution on the unit square using the so-called Rosenblatt transform (see Rosenblatt (1952a) and Martynov (1978)). Then d.f. $F(s, t; 0)$ will be transformed into d.f. $s\,t$, $0 \le s \le 1$, $0 \le t \le 1$, and d.f. $F(s, t; \theta)$ for $\theta \ne 0$ into some d.f. $\widetilde{F}(s, t; \theta)$ on I^2. Such a transform is widely used for multidimensional samples (see, e.g., Durbin (1970), Martynov (1978), and Tyurin (1978)).

Now the problem of testing H_0 against A_0 is reduced to testing uniformity on I^2. We can use with that end statistics (5.5.1)–(5.5.4) and for their comparison apply their local exact slopes obtained earlier.

We shall perform calculations again in the simplest case of the Gaussian standard distribution in $\mathbf{R}^2$ with the location alternative. The role of the function $\widetilde{F}(s,t;\theta)$ is now played by the function $\Phi\big(\theta+\Phi^{-1}(s)\big)\,\Phi\big(\theta+\Phi^{-1}(t)\big)$. Simple arguments show that

$$\widetilde{F}(s,t;\theta) = s\,t + \Big(s\,\varphi\big(\Phi^{-1}(t)\big) + t\,\varphi\big(\Phi^{-1}(s)\big)\Big)\cdot\theta + o(\theta), \qquad \theta \to 0.$$

Therefore

$$c_\Delta(\theta) \sim 4 \sup_{s,t}\ \big|\varphi(s)\,\Phi(t) + \varphi(t)\,\Phi(s)\big|^2\cdot\theta^2$$

$$= 16\,\big[\varphi(z_0)\,\Phi(z_0)\big]^2\cdot\theta^2 \approx 0.948\,\theta^2,$$

where $z_0 \approx 0.505$ is a root of the equation $\varphi(z) = z\,\Phi(z)$. Further,

$$c_\mu(\theta) \sim 9 \sup_{0\le s\le 1} \Big(s \int_{-\infty}^{\infty} \varphi^2(t)\,dt + \frac{1}{2}\,\varphi\big(\Phi^{-1}(s)\big)\Big)^2\cdot\theta^2$$

$$= \frac{9}{4}\Big(\frac{1}{\sqrt{\pi}}\,\Phi\Big(\frac{1}{\sqrt{\pi}}\Big) + \varphi\Big(\frac{1}{\sqrt{\pi}}\Big)\Big)^2\cdot\theta^2 \approx 1.242\,\theta^2,$$

$$c_{\Omega^1}(\theta) \sim \frac{144}{7}\Big(\int_{-\infty}^{\infty} \varphi^2(t)\,dt\Big)^2\cdot\theta^2 \approx 1.637\,\theta^2,$$

$$c_{\Omega^2}(\theta) \sim \lambda_0\Big(\frac{2}{3}\int_{-\infty}^{\infty} \varphi^3(t)\,dt + 2\Big(\int_{-\infty}^{\infty}\varphi^2(t)\,\Phi(t)\,dt\Big)^2\Big)\cdot\theta^2$$

$$= \lambda_0\Big(\frac{1}{3\pi\sqrt{3}} + \frac{1}{8\pi}\Big)\cdot\theta^2 \approx 1.598\,\theta^2.$$

It is interesting to compare these results with the maximum possible exact slope which equals $2\,K(\theta)$, due to Theorem 1.2.3, and is attained for the likelihood ratio test. After elementary calculations one obtains

$$2\,K(\theta) = 2\int_0^1\!\!\int_0^1 \ln \widetilde{F}_{st}(s,t;\theta)\ \widetilde{F}_{st}(s,t;\theta)\ ds\,dt \sim 2\,\theta^2, \qquad \theta \to 0.$$

Therefore in our example the best test is based on the statistic $\Omega^1_{n,1}$, having a local Bahadur efficiency of approximately 0.82. The statistics $\Omega^2_{n,1}$, μ_n, and Δ_n follow in that order and the ordering of statistics under consideration on the whole coincides with the one-dimensional case (cf. Section 2.6).

Finally we note that the local approximate Bahadur efficiency of considered statistics may be calculated in the same way as in Section 5.4

for statistics (5.1.1)–(5.1.4) and their variants. The main difference lies
in the fact that these statistics are functionals of the empirical field
$\eta_n(s,t) = \sqrt{n}\,(F_n(s,t) - s\,t)$ which weakly converges in the corresponding Skorokhod space $D(I^2)$ to the Gaussian random field $\eta(s,t)$ with zero
mean and correlation function (5.5.8). The limiting Pitman efficiency
also may be calculated on the basis of Theorems 1.4.1 and 1.4.3. For the
statistics studied in this section, all three types of ARE coincide locally.

5.7 Hodges–Lehmann Efficiency of Independence Tests

Let again, as at the beginning of Section 5.1, $(X_1, Y_1), (X_2, Y_2), \ldots,$
(X_n, Y_n) be a sample with absolutely continuous d.f. $F(x,y)$ and marginal d.f.s $G(x)$ and $H(y)$. We are testing the independence hypothesis
H_3 against the alternative A_3. Denote by Θ_0 the set of all absolutely
continuous d.f.s in the real plane satisfying the condition of independence
of components and by Θ_1 the set of all d.f.s in the real plane with positive
densities for which the aforementioned condition fails. We may assume
that we are testing the hypothesis

$$H_3\colon F \in \Theta_0$$

against the simple alternative

$$A_3^0\colon F = R \in \Theta_1\,.$$

Let $\{T_n\}$ be a sequence of nonparametric statistics and $1 - \beta_{3n}(\alpha; R)$
be the corresponding probability of error of the second kind at a level
$\alpha \in (0, 1)$. Theorem 1.3.1 immediately implies an estimate of the rate
of decreasing of $1 - \beta_{3n}(\alpha; R)$.

Denote

$$\mathfrak{J}(R) := \inf_{l,\,m}\; \int\limits_{-\infty}^{\infty} \int\limits_{-\infty}^{\infty} \ln\frac{l(x)\,m(y)}{r(x,y)}\; l(x)\,m(y)\,dx\,dy,$$

where r is the density of the alternative d.f. R and the infimum is taken
over all possible densities l and m on the real line. We are unable to
present an explicit expression for $\mathfrak{J}(R)$ only in terms of r.

Theorem 5.7.1 (Nikitin 1986c) *The inequality*

$$\lim_{n\to\infty}\; n^{-1}\ln\left[\,1 - \beta_{3n}(\alpha; R)\,\right] \geq -\mathfrak{J}(R) \tag{5.7.1}$$

holds.

It follows that the Hodges–Lehmann index $d_T(R)$ of the sequence $\{T_n\}$ should satisfy the condition

$$d_T(R) \leq 2\,\mathfrak{J}(R),$$

and this sequence is said to be *asymptotically optimal (AO)* in the *Hodges–Lehmann sense* if the equality holds.

The statistics Γ_n, M_n, and $B^2_{n,\,q_1,\,q_2}$ introduced in Sections 5.1 and 5.4 are the best known nonparametric statistics used for testing H_3 against A_3^0. We consider the following modification of statistic $B^k_{n,\,q_1,\,q_2}$:

$$\bar{B}^\lambda_{n,\,q_1,\,q_2} = \int\limits_{-\infty}^{\infty} \int\limits_{-\infty}^{\infty} \big|\, F_n(s,t) - G_n(s)\,H_n(t)\,\big|^\lambda$$

$$\times\, q_1\big(G_n(s)\big)\, q_2\big(H_n(t)\big)\, dG_n(t)\, dH_n(t)$$

where $\lambda \geq 1$ and weight functions q_1 and q_2 are positive and satisfy condition (5.3.1).

Theorem 5.7.2 (Nikitin 1986c) *The sequences of statistics* $\{\Gamma_n\}$, $\{M_n\}$, *and* $\bar{B}^\lambda_{n,\,q_1,\,q_2}$ *are Hodges–Lehmann asymptotically optimal.*

Proof As all statistics under consideration are ρ-continuous functionals of empirical distributions, we may apply here Theorem 1.6.8. If $\{T_n\}$ is any of these statistics, $T_n := T(F_n)$, we have

$$\varlimsup_{n\to\infty} n^{-1} \ln\big[\, 1 - \beta_{3n}(\alpha; R)\,\big]$$

$$\leq -\inf\left\{ \int\limits_{-\infty}^{\infty} \int\limits_{-\infty}^{\infty} \ln \frac{f(x,y)}{r(x,y)}\, f(x,y)\, dx\, dy\colon T(F) \leq 0 \right\}, \quad (5.7.2)$$

where f is the density of F.

But the condition $T(F) \leq 0$ is equivalent to the independence condition $F(x,y) = G(x)\,H(y)$ for almost all x and y with respect to the measure corresponding to the density $g(x)\,h(y)$. Therefore the right-hand side of inequality (5.7.2) coincides with the quantity $-\mathfrak{J}(R)$. Comparing the obtained inequality with (5.7.1) we establish the conclusion of Theorem 5.7.2. □

An analogous result is true for the two-sided statistics introduced in Section 5.5.

Clearly Theorem 5.7.2 admits a slight generalization along the lines of the paper by Kallenberg and Kourouklis (1992).

We are unaware of any results concerning the Hodges–Lehmann and Chernoff efficiencies of independence rank tests. There are strong reasons to believe that the situation is very close here to the situation with the rank tests of homogeneity and symmetry analyzed in previous chapters.

6

Local Asymptotic Optimality of Nonparametric Tests and the Characterization of Distributions

6.1 Introduction and Statement of the Problem

Let $X_1, X_2, \ldots, X_n$ be a sequence of independent observations with common d.f. $G(x; \theta)$, where parameter θ takes on values in an interval Θ of the form $[0, \theta_0)$, $0 < \theta_0 \leq \infty$. It is supposed that the family of d.f.s $\{G(x; \theta), \theta \in \Theta\}$ belongs to some functional class $\mathfrak{G}$ specified by some general regularity conditions that will be formulated in the sequel. Specifically, we always assume that the function $G(x; \theta)$ is differentiable with respect to both arguments and denote by $g(x; \theta)$ and $G'_\theta(x; \theta)$ the corresponding derivatives.

Consider the problem of testing the hypothesis

$$H_0: \theta = 0$$

against the alternative

$$A_0: \theta > 0$$

with the aid of a sequence of statistics $\{T_n\}$, $T_n := T_n(X_1, \ldots, X_n)$, large values of T_n being significant. Let $c_T(\theta)$ be the Bahadur exact slope of $\{T_n\}$ introduced and studied in the previous chapters. It follows from Theorem 1.2.3 that the following inequality is valid:

$$c_T(\theta) \leq 2\,K(\theta) = 2 \int\limits_{-\infty}^{\infty} \ln \frac{g(x; \theta)}{g(x; 0)}\, g(x; \theta)\, dx\,. \qquad (6.1.1)$$

We recall that the sequence $\{T_n\}$ is said to be Bahadur asymptotically optimal (AO) if for all $\theta \neq 0$ the inequality in (6.1.1) can be replaced by equality. As we have already seen in Chapter 1, the class of such statistics is rather narrow, though it contains under certain conditions the likelihood ratio statistics. A significantly less restrictive condition is the

208

condition of Bahadur *local asymptotic optimality* (LAO) (see Bahadur (1967)):

$$c_T(\theta) \sim 2\,K(\theta)\,, \qquad \theta \to 0\,. \qquad (6.1.2)$$

The class of those families $\{G(x;\theta),\ \theta \in \Theta\}$ from $\mathfrak{G}$ for which condition (6.1.2) holds is called the *Bahadur LAO domain* for a given sequence of statistics $\{T_n\}$ in $\mathfrak{G}$.

Another measure of the asymptotic efficiency of $\{T_n\}$ is the Hodges–Lehmann index $d_T(\theta)$, satisfying, as we know from Theorem 1.3.1, the inequality

$$d_T(\theta) \le 2\,K'(\theta) = 2 \int\limits_{-\infty}^{\infty} \ln \frac{g(x;0)}{g(x;\theta)}\, g(x;0)\,dx\,. \qquad (6.1.3)$$

The notions of AO and LAO in the Hodges–Lehmann sense are introduced as in the case of the Bahadur ARE; the only difference is the replacement of $K(\theta)$ by $K'(\theta)$.

Using inequalities (1.4.6) and (1.5.4), it is possible to introduce in certain cases the corresponding notions for the Pitman and Chernoff efficiencies.

A typical problem of the asymptotic theory of testing statistical hypotheses is to construct, for a given family of d.f.s $\{G(x;\theta),\ \theta \in \Theta\}$, a locally asymptotically optimal (in some sense) sequence of statistics $\{T_n\}$. We will be interested in the *opposite* problem: What is the structure of the LAO domains in the class $\mathfrak{G}$ for well-known nonparametric statistics such as the Kolmogorov–Smirnov ones, ω^2 statistics, rank statistics, and their variants. This problem is also of great interest due to the fact that the overwhelming majority of nonparametric statistics have been proposed for empirical reasons: In the opinion of their authors these statistics "must work" in one or another problem of testing statistical hypotheses. Therefore, the elucidation of the situation when they really work and give the locally asymptotically optimal result seems to be an important and worthwhile problem.

The statement of the problem just described dates back to Nikitin (1981, 1984a). Similar results, but under another definition of optimality and only for rank tests, were given by Behnen (1971, 1972), and Assatryan and Safaryan (1979). See also Gregory (1980).

Let $\mathbf{V}$ be a certain set of real functions on the interval $[0, 1]$. We call $\mathbf{V}$ a *leading set* for a given sequence $\{T_n\}$ in $\mathfrak{G}$ if the LAO condition

$$c_T(\theta) \sim 2\,K(\theta)\,, \qquad \theta \to 0\,,$$

is fulfilled if and only if

$$G'_\theta\big(G^{-1}(x;0);0\big) \in \mathbf{V}.\tag{6.1.4}$$

Elements of the leading set $\mathbf{V}$ are called *leading functions*.

As d.f. $G(x;\theta)$ is supposed to be differentiable with respect to the parameter, one has the expansion:

$$G(x;\theta) = G(x;0) + G'_\theta(x;0)\cdot\theta + o(\theta), \qquad \theta \to 0.$$

For any d.f. $G(x;\theta)$ in the LAO domain the following relationship holds due to (6.1.4):

$$G'_\theta(x;0) = v\big(G(x;0)\big)$$

for some leading function $v \in \mathbf{V}$. Hence the LAO domain for $\{T_n\}$ in $\mathfrak{G}$ contains precisely the families $\{G(x;\theta)\}$ for which, as $\theta \to 0$ and for $v \in \mathbf{V}$, we have

$$G(x;\theta) = G(x;0) + v\big(G(x;0)\big)\cdot\theta + o(\theta).\tag{6.1.5}$$

Thus, to describe the LAO domain it suffices to establish the equivalence between conditions (6.1.2) and (6.1.4) and to calculate the leading functions v or to find the leading set $\mathbf{V}$. One may say this set determines the "direction" in the class $\mathfrak{G}$ where the considered sequence of statistics has the maximum sensitivity.

If the nature of the dependence of the family $\{G(x;\theta)\}$ on θ is known (families with location or scale parameter, Lehmann families, etc.), then by solving differential equations of the form

$$G'_\theta(x;0) = v\big(G(x;0)\big), \qquad v \in \mathbf{V},\tag{6.1.6}$$

one can obtain a *characterization* of those d.f.s $G(x;\theta)$ for which a given sequence of statistics is LAO. Thus, for location families, Kolmogorov and ω^2 statistics for goodness-of-fit turn out to be LAO only in the case of the Laplace and "hyperbolic cosine" distributions, correspondingly. Similar characterizations arise also when analyzing the LAO conditions for homogeneity, symmetry, and independence tests.

Somewhat less substantial results arise from the description of the LAO domains in the Hodges–Lehmann sense. The main reason is that, as we have seen in previous chapters, the two-sided nonparametric tests in most cases are Hodges–Lehmann AO for a very broad class of distributions of the initial sample. For other tests (e.g., for linear rank tests for testing homogeneity or symmetry) the LAO domains are similar in structure to corresponding domains in the Bahadur sense, but usually are more limited because of restrictive regularity conditions imposed on the distributions of observations. The same concerns apply for the LAO domains in the Pitman and Chernoff senses when it is possible to describe them. However, the results for the Bahadur efficiency turn out to be the most interesting and mathematically attractive ones and we focus our attention on them.

Characterization theorems of mathematical statistics have a long history initiated by the works of Polya, Bernstein, and Linnik. The present state of investigations in this area is described in Kagan, Linnik, and Rao (1973), Galambos and Kotz (1978), and Yanushkevichius (1991). Our results given subsequently may be considered as characterizations of a new type not encountered previously in the statistical literature.

In the well-known review by Savage (1969) devoted specifically to applications of the Bahadur theory regarding nonparametric statistics we find the following statement : "Work has not begun on finding sequences of nonparametric procedures with exact slopes approaching the supremum of the available exact slopes. Selecting procedures because of their large exact slopes would parallel the current practice of selecting nonparametric procedures with large Pitman efficiency" (p. 119). The present chapter will be concerned with just the realization of the program outlined in this citation.

6.2 Domains of the Bahadur Local Asymptotic Optimality (LAO) for Goodness-of-Fit Tests

Consider again the problem of verifying H_0 against A_0 described at the beginning of Section 6.1. To elucidate the structure of the LAO domains for nonparametric goodness-of-fit tests we must formulate the regularity conditions imposed on families of d.f.s $\{G(x; \theta),\ \theta \in \Theta\}$ belonging to the class $\mathfrak{G}$. We shall need also in the sequel various subclasses of $\mathfrak{G}$ which will be specified by some additional conditions.

We assume that the distribution function $G(x; \theta)$ is defined on the set $\Delta \times \Theta$ where $\Delta = (\alpha, \beta)$ is a finite or infinite nondegenerate interval in $\mathbf{R}^1$ and Θ is an interval of the form $[0, \theta_0)$ with $0 < \theta_0 \leq \infty$.

Let $G(x; \theta)$ satisfy the following regularity conditions:

(R1) *Throughout the domain of definition d.f. $G(x; \theta)$ is differentiable in x and θ.*

(R2) *For almost all x the density $g(x; \theta)$ is absolutely continuous in θ and $g(x; 0)$ is strictly positive.*

(R3) The derivative $G'_\theta(x;\theta)$ for almost all $\theta \in \Theta$ is absolutely continuous in x, is not identically equal to 0 for $\theta = 0$, and, moreover,

$$\sup_x \sup_\theta |G'_\theta(x;\theta)| \le M,$$

$$\lim_{x\downarrow\alpha} G'_\theta(x;0) = \lim_{x\uparrow\beta} G'_\theta(x;0) = 0.$$

(R4) For almost all $x \in \Delta$

$$G''_{x\theta}(x;0) = G''_{\theta x}(x;0).$$

(R5) The Kullback–Leibler information number $K(\theta)$ behaves locally as follows:

$$K(\theta) = \int_\alpha^\beta \ln \frac{g(x;\theta)}{g(x;0)}\, g(x;\theta)\, dx \sim \tfrac{1}{2} I(g)\cdot\theta^2, \qquad \theta \to 0,$$

$$(6.2.1)$$

where

$$I(g) := \int_\alpha^\beta \frac{|g'_\theta(x;0)|^2}{g(x;0)}\, dx$$

is the Fisher information and $0 < I(g) < \infty$.

General condition (6.2.1) is very often used in various problems of asymptotic statistics. Sufficient conditions for (6.2.1) were given, for example, by Kullback (1959) and by Ibragimov and Khasminskii (1981, Chapter 1, Theorem 9.3.1). However, these conditions may turn out to be excessively rigid. For example, they are not fulfilled if $g(x;\theta) = \tfrac{1}{2}\exp(-|x-\theta|)$. At the same time, for this density, which plays an important role later, condition (6.2.1) is valid with $I(g) = 1$. Hence in order to restrict the class of distributions under consideration we prefer to use just (6.2.1).

It is necessary to note that the condition $G'_\theta(x;0) \not\equiv 0$ may fail in a number of interesting cases (the Rayleigh–Rice family (3.6.1)–(3.6.2) gives an example of such a situation). In these cases one should require the fulfillment of this condition for one of the subsequent derivatives (cf. analogous arguments about the calculation of the Pitman efficiency in Kendall and Stuart (1967, Chapter 25)).

The class of distribution families satisfying conditions (R1)–(R5) is rather broad. In Section 6.3 we will demonstrate it for particular cases of location, scale, and Lehmann families.

We turn now to the description of the LAO domains and leading sets **V** for the nonparametric goodness-of-fit statistics considered in Chapter 2. The exposition is based on the papers of Nikitin (1981, 1984a, 1987b) and Podkorytova (1990).

We start out with the Kolmogorov–Smirnov statistics and their variants. Consider the classical statistic D_n introduced in (2.1.1).

Theorem 6.2.1 *The leading set for the sequence* $\{D_n\}$ *in* $\mathfrak{G}$ *has the form*

$$\mathbf{V}(D) = \left\{ C \min\left(t, 1-t\right) \right\}, \quad 0 \le t \le 1, \quad C \in \mathbf{R}^1, \quad C \ne 0.$$

$$(6.2.2)$$

Proof It follows from Section 2.5 and the regularity conditions that the local exact slope of the sequence $\{D_n\}$ has the form

$$c_D(\theta) \sim 4 \left[\sup_x \left| G'_\theta(x;0) \right| \right]^2 \cdot \theta^2, \qquad \theta \to 0.$$

As the density $g(x;0)$ is supposed to be positive there exists the inverse d.f. $G^{-1}(x;0)$. Let us introduce the absolutely continuous function

$$v(x) = G'_\theta \big(G^{-1}(x;0); 0 \big).$$

$$(6.2.3)$$

Then we have

$$c_D(\theta) \sim 4 \left[\sup_{0 \le x \le 1} \left| v(x) \right| \right]^2 \cdot \theta^2, \qquad \theta \to 0,$$

$$(6.2.4)$$

and, in view of (6.2.1),

$$2\,K(\theta) \sim I(g) \cdot \theta^2 = \int_0^1 v'^2(x)\,dx \cdot \theta^2.$$

$$(6.2.5)$$

Therefore the problem is to establish when equality can hold in the inequality

$$4 \sup_{0 \le x \le 1} v^2(x) \le \int_0^1 v'^2(x)\,dx$$

$$(6.2.6)$$

resulting from (6.1.1). One should also add the boundary conditions

$$v(0) = v(1) = 0,$$

$$(6.2.7)$$

which follow from the regularity conditions imposed on elements of $\mathfrak{G}$. Put

$$v_1(x) := v(x) \left(\int_0^1 v'^2(x)\,dx \right)^{-1/2}.$$

The problem under consideration is equivalent to the problem of finding an absolutely continuous function v_1 satisfying (6.2.7) and the normalization condition

$$\int_0^1 v_1'^2(x)\,dx = 1,$$

for which the equality

$$\sup_{0 \le x \le 1} v_1^2(x) = \tfrac{1}{4}$$

is valid. Toward that end we search for such a function v_1 from this class for which the extremal value of $v_1(s_0)$ is attained (s_0 is an arbitrary point in $(0, 1)$). Using the representation

$$v_1(s_0) = \int_0^1 v_1'(t)\,\mathbf{1}_{[0,\,s_0]}(t)\,dt$$

and applying Theorem 1.8.1 with $\mathcal{X} = \mathring{\mathbf{W}}_{2,1}[0,1]$, we obtain for the extremal the Euler–Lagrange equation

$$v_1'(t) = \mu_1\,\mathbf{1}_{[0,\,s_0]}(t) + \mu_2,$$

where μ_1 and μ_2 are undetermined numerical multipliers. Integrating this equation and taking into account the boundary conditions and the normalization condition, we finally come to

$$v_1(t) = \pm \begin{cases} t\,\sqrt{(1-s_0)\,/\,s_0}\,, & 0 \le t \le s_0\,, \\[2mm] (1-t)\,\sqrt{s_0\,/(1-s_0)}\,, & s_0 \le t \le 1\,. \end{cases}$$

Therefore,

$$v_1(s_0) = \pm \sqrt{s_0\,(1-s_0)}$$

and the extremum is attained at the point $s_0 = \tfrac{1}{2}$. Hence

$$\sup_{0 \le s_0 \le 1} v_1^2(s_0) = \tfrac{1}{4}$$

and the upper bound is attained for the function $v_1(t) = \pm\min(t,\,1-t)$. This implies the conclusion of Theorem 6.2.1.

A shorter proof of (6.2.6) follows from the equality

$$v(x) = \int_0^1 \left(\mathbf{1}_{[0,\,x]}(t) - \tfrac{1}{2} \right) v'(t)\,dt$$

by applying to it the Cauchy–Bunyakovskii–Schwarz inequality, but this proof is not fit for generalizations. $\square$

Similar results can be established for the one-sided Smirnov statistics D_n^+ and D_n^- as well as for the Kuiper statistic V_n. Thus, in the case of the sequence $\{D_n^+\}$ the local index has the form

$$4 \left[\sup_{0 \le x \le 1} v(x) \right]^2.$$

The difference here from arguments used in the proof of Theorem 6.2.1 is that the condition

$$\sup_{0 \le x \le 1} v(x) > 0$$

should be true. Therefore, it is necessary to restrict the class $\mathfrak{G}$ down to

$$\mathfrak{G}_D^+ := \left\{ G \in \mathfrak{G}: \ \sup_x G_\theta'(x; 0) > 0 \right\}$$

and the leading set in this class again has the form (6.2.2). The same holds true for the sequence $\{D_n^-\}$ in the corresponding class $\mathfrak{G}_D^-$.

It is curious that for weighted Kolmogorov statistics (2.2.17) the leading set changes unessentially and in typical cases (when the function $g_q(a)$ in (2.2.18) is regular enough) has the form

$$\mathbf{V}(D_q) = \left\{ C \left(\min(t, \, x_0) - t \, x_0 \right) \right\}, \qquad C \ne 0, \quad x_0 \in \mathbf{X}_q,$$

where $\mathbf{X}_q$ is the set of those points at which the function $x(1-x)q(x)$ attains its maximum. The proof of this fact is very close to the proof of Theorem 6.2.1 (see Benson and Nikitin (1992)). Particularly, for weight (2.2.19), we see that the set $\mathbf{X}_q$ consists of two points

$$x_0' = \tfrac{1}{2} - \sqrt{\tfrac{1}{4} - e^{-2}} \quad \text{and} \quad x_0'' = \tfrac{1}{2} + \sqrt{\tfrac{1}{4} - e^{-2}}.$$

For the analysis of the Kuiper statistic V_n one should take into consideration that the Bahadur local index has now the form

$$4 \left[\sup_{0 \le x \le 1} v(x) - \inf_{0 \le x \le 1} v(x) \right]^2.$$

Arguing as in the proof of Theorem 6.2.1, we look for a function v_1 for which the maximum value of $\left[v_1(t_0) - v_1(s_0) \right]$ is attained. Let for the sake of definiteness $t_0 \ge s_0$; then the equation for extremals becomes

$$v_1'(t) = \mu_1 \, \mathbf{1}_{[s_0, \, t_0]}(t) + \mu_2,$$

from which it follows that $s_0 = 0$, $t_0 = \tfrac{1}{2}$, and the extremals themselves have again the form

$$v_1(t) = \pm \min(t, \, 1 - t).$$

Hence the leading set for $\{V_n\}$ is again given by (6.2.2).

To describe the leading sets of the Watson–Darling statistics (2.1.5)–(2.1.7) we introduce the family of functions

$$v^*(t,s) := \begin{cases} t^2 + t - 2st, & 0 \le t \le s, \\ t^2 - t + 2s - 2st, & s \le t \le 1, \end{cases} \qquad (6.2.8)$$

defined on a unit square and depending on parameter $s \in [0,1]$.

Together with the class $\mathfrak{G}$ we consider its subclasses

$$\mathfrak{G}_G^+ := \left\{ G \in \mathfrak{G}\colon \sup_x \left[G_\theta'(x;0) - \int_\alpha^\beta G_\theta'(x;0)\, dG(x;0) \right] > 0 \right\},$$

$$\mathfrak{G}_G^- := \left\{ G \in \mathfrak{G}\colon \sup_x \left[\int_\alpha^\beta G_\theta'(x;0)\, dG(x;0) - G_\theta'(x;0) \right] > 0 \right\}.$$

Theorem 6.2.2 *The leading sets of the sequences of statistics (2.1.5)–(2.1.7) in the classes $\mathfrak{G}$, $\mathfrak{G}_G^+$, and $\mathfrak{G}_G^-$, correspondingly, have the form*

$$\mathbf{V}(G) = \left\{ C\, v^*(t,s) \right\}, \qquad C \in \mathbf{R}^1,$$

$$\mathbf{V}(G^+) = \left\{ C\, v^*(t,s) \right\}, \qquad C > 0,$$

$$\mathbf{V}(G^-) = \left\{ C\, v^*(t,s) \right\}, \qquad C < 0.$$

Proof The LAO condition for the sequence $\{G_n^+\}$ in the class $\mathfrak{G}_G^+$ takes the form

$$12 \left\{ \sup_x \left[G_\theta'(x;0) - \int_\alpha^\beta G_\theta'(x;0)\, dG(x;0) \right] \right\}^2 = I(g).$$

Making the substitution $v(t) = G_\theta'\big(G^{-1}(t;0);0\big)$ we obtain

$$12 \left\{ \sup_{0 \le t \le 1} \left[v(t) - \int_0^1 v(s)\, ds \right] \right\}^2 = \int_0^1 v'^{\,2}(t)\, dt \qquad (6.2.9)$$

under the boundary conditions $v(0) = v(1) = 0$.

To understand when (6.2.9) is valid, let us fix an $s \in [0,1]$ and look for absolutely continuous functions v that maximize $v(s) - \int_0^1 v(t)\, dt$ for a constant value of $\int_0^1 v'^{\,2}(t)\, dt$. Applying the Lagrange principle we come to the equation for the extremals:

$$v'(t) = \mu_1 \left(\mathbf{1}_{[0,\,s]}(t) + t \right) + \mu_2,$$

from which it follows that the extremals have the form $C\,v^*(t,s)$, where v^* has been defined in (6.2.8). One should note additionally that

$$\sup_{0\leq x\leq 1}\left[v(x)-\int_0^1 v(t)\,dt\right]>0\,,$$

and therefore $C>0$. Simple inspection shows that for such v (6.2.9) is true. The sets $\mathbf{V}(G^-)$ and $\mathbf{V}(G)$ are described analogously. $\qquad\square$

Now we consider the statistic K_n defined in (2.1.12) and its one-sided variants $K_n^{\pm}$. Suppose the family $\{G(x;\theta)\}$ satisfies conditions A and B from Section 2.6.

It has been shown in Aki (1986) that under the mentioned conditions (and if the alternative hypothesis is true) one has

$$K_n\overset{P}{\longrightarrow}\sup_t\left|G(t;\theta)-\int_\alpha^t\frac{1-G(s;\theta)}{1-G(s;0)}\,dG(s;0)\right|:=b_K(\theta)\,.$$

Assume, additionally, that, as $\theta\to 0$,

$$b_K(\theta)\sim\sup_t\left|G_\theta'(t;0)+\int_\alpha^t\frac{G_\theta'(s;0)}{1-G(s;0)}\,dG(s;0)\right|\cdot\theta\,.\qquad(6.2.10)$$

It is possible to write out explicitly the regularity conditions ensuring the correctness of (6.2.10) (see Podkorytova (1990)). Now consider the class of families of d.f.s $\mathfrak{G}_K$ consisting of those elements of $\mathfrak{G}$ that satisfy conditions A and B and (6.2.10). Corresponding subclasses $\mathfrak{G}_K^+$ and $\mathfrak{G}_K^-$ may be determined on the basis of the corresponding quantities $b_{K^+}(\theta)$ and $b_{K^-}(\theta)$.

Theorem 6.2.3 *The leading set of the sequence $\{K_n\}$ in $\mathfrak{G}_K$ has the form*

$$\mathbf{V}(K)=\left\{\,C\,(1-t)\ln\,(1-t)\,\right\},\qquad C\neq 0\,.$$

The leading sets for $\{K_n^+\}$ and $\{K_n^-\}$ in $\mathfrak{G}_K^+$ and $\mathfrak{G}_K^-$ differ from $\mathbf{V}(K)$ in that the constant C should be negative or positive, correspondingly.

Proof The local index of the sequence $\{K_n\}$ in $\mathfrak{G}_K$ has the form

$$\sup_{0\leq y\leq 1}\left|v(y)+\int_0^y\frac{v(x)}{1-x}\,dx\right|^2,$$

where $v(y) = G'_\theta\big(G^{-1}(y;0);0\big)$. We are interested in the maximization problem

$$\sup_{0 \leq y \leq 1} \left| v(y) + \int_0^y \frac{v(x)}{1-x}\, dx \right|$$

on the set of $v \in \overset{\circ}{\mathbf{W}}_{2,1}[0,1]$ such that $\|v\| = 1$. The functional

$$h_y(v) := \left| v(y) + \int_0^y \frac{v(x)}{1-x}\, dx \right| \qquad (6.2.11)$$

is a linear continuous functional on $\overset{\circ}{\mathbf{W}}_{2,1}[0,1]$ with the norm

$$\| h_y \| := \sup \{ h_y(v) \colon \| v \| = 1 \}.$$

The next step is to prove that $\| h_y \| \leq \| h_1 \|$ for each $y \in [0,1]$. Indeed, let $\|h_y\| = r$. Then for any $\delta > 0$ there exists a function $v_\delta \in \overset{\circ}{\mathbf{W}}_{2,1}[0,1]$ such that $\|v_\delta\| = 1$ and

$$\left| v_\delta(y) + \int_0^y \frac{v_\delta(x)}{1-x}\, dx \right| > r - \delta.$$

Put

$$\widetilde{v}(x) := \begin{cases} v_\delta(x), & 0 \leq x \leq y, \\[2mm] v_\delta(y)\,(1-x)\,(1-y)^{-1}, & y \leq x \leq 1. \end{cases}$$

It is clear that $\widetilde{v} \in \overset{\circ}{\mathbf{W}}_{2,1}[0,1]$ and, moreover,

$$\| \widetilde{v} \| = \int_0^y v_\delta'^2(x)\, dx + \frac{v_\delta^2(y)}{1-y} \leq \int_0^1 v_\delta'^2(x)\, dx = \| v_\delta \| = 1.$$

In addition,

$$h_1(\widetilde{v}) = \left| \int_0^y \frac{v_\delta(x)}{1-x}\, dx + v_\delta(y) \right| = h_y(v_\delta) > r - \delta,$$

so $\|h_1\| \geq r - \delta$. In view of the arbitrariness of δ, one has

$$\| h_1 \| \geq r = \| h_y \|.$$

Thus we can maximize, instead of (6.2.11), a much simpler functional

$$h_1(v) = \left| \int\limits_0^1 \frac{v(y)}{1-y} \, dy \right|$$

subject to $v \in \overset{\circ}{\mathbf{W}}_{2,1}[0,1]$, $\|v\| = 1$. This classical variational problem has two solutions $v(x) = \pm(1-x)\ln(1-x)$ in $\overset{\circ}{\mathbf{W}}_{2,1}[0,1]$ and consequently in a narrower subclass $\mathfrak{G}_K$. It follows that the equality

$$\sup_{0 \le y \le 1} \left| v(y) + \int\limits_0^y \frac{v(x)}{1-x} \, dx \right|^2 = \int\limits_0^1 v'^2(x) \, dx$$

is attained for $v(x) = C(1-x)\ln(1-x)$, $C \in \mathbf{R}^1$. In the case of the one-sided statistics we obtain additionally limitations on the sign of the constant C. $\qquad\square$

Consider now the integral goodness-of-fit statistics and start out with the statistics $\omega_{n,q}^k$. We recall that for the analysis of large deviations of these statistics an important role was played by the problem

$$y'' = \lambda y^{k-1} q, \qquad y(0) = y(1) = 0, \qquad \int\limits_0^1 y^k q \, dt = 1. \qquad (6.2.12)$$

Let again $\lambda_0 = \lambda_0(q;k)$ be the principal eigenvalue of this problem and y_0 be a corresponding eigenfunction (possibly not unique.)

Let us introduce the subclasses $\mathfrak{G}_\omega(q;k)$ of the initial class $\mathfrak{G}$ according to the following definition: For even k they coincide with the class $\mathfrak{G}$ and for odd k

$$\mathfrak{G}_\omega(q;k) := \left\{ G \in \mathfrak{G}: \int\limits_\alpha^\beta \big(G_\theta'(x;0) \big)^k q\big(G(x;0) \big) \, dG(x;0) > 0 \right\}.$$

Theorem 6.2.4 *The leading set of the sequence $\{\omega_{n,q}^k\}$ in the class $\mathfrak{G}_\omega(q;k)$ has the form*

$$\mathbf{V}(\omega_q^k) = \big\{ C y_0(t) \big\}, \qquad C \ne 0.$$

Proof It follows from the results of Section 2.6 that the local exact slope of the sequence $\{\omega_{n,q}^k\}$ is equal to

$$c_{\omega^k}(\theta) \sim -\lambda_0 \left[\int\limits_\alpha^\beta \big(G(x;\theta) - G(x;0) \big)^k q\big(G(x;0) \big) \, dG(x;0) \right]^{2/k}.$$

Conditions (R1)–(R3) imply that

$$\theta^{-1}\left|G(x;\theta) - G(x;0)\right| \le \theta^{-1}\int_0^\theta \left|G_t'(x;t)\right|dt \le M \qquad \text{for } \theta > 0.$$

Using the Lebesgue theorem on dominated convergence, we obtain

$$\lim_{\theta\to 0} c_{\omega^k}(\theta)\,\theta^{-2} = -\lambda_0\left[\int_0^1 v^k(t)\,q(t)\,dt\right]^{2/k},$$

where the function v is again given by (6.2.3).

The LAO condition takes now the form

$$-\lambda_0\left[\int_0^1 v^k(t)\,q(t)\,dt\right]^{2/k} = \int_0^1 v'^2(t)\,dt \qquad (6.2.13)$$

under boundary conditions (6.2.7). (We enlarge again the set of admissible elements up to $\overset{\circ}{\mathbf{W}}_{2,1}[0,1]$.)

To see when (6.2.13) holds we look for the infimum of the functional $\int_0^1 v'^2(t)\,dt$ on the subset of the space $\overset{\circ}{\mathbf{W}}_{2,1}[0,1]$ defined by the condition

$$\int_0^1 v^k(t)\,q(t)\,dt = 1.$$

The existence of the solution follows from the lower semicontinuity of the considered functional in the topology of uniform convergence (see, e.g., Freidlin and Wentzell (1979), Chapter 3, Section 2). The necessary condition for an extremum is that the extremal $\hat{v}$ satisfies the equation

$$\hat{v}'' = \lambda\,\hat{v}^{k-1}\,q$$

under the additional conditions

$$\hat{v}(0) = \hat{v}(1) = 0, \qquad \int_0^1 \hat{v}^k(t)\,q(t)\,dt = 1.$$

It follows from the definition of y_0 that the infimum is actually attained for this function, and then (6.2.13) holds if and only if

$$v(t) = C\,y_0(t), \qquad C \neq 0. \qquad \square$$

Now we present some corollaries to Theorem 6.2.4. In the case of the classical Cramér–von Mises–Smirnov ω^2 statistic (where $q \equiv 1$, $k = 2$) the leading set consists of the functions $C \sin \pi t$, $C \neq 0$. It is interesting that the same set of functions, but for $C > 0$, turns out to be the leading set for the sequence $\{\kappa_n\}$ of the so-called first components in the Durbin–Knott sense of the statistics $\omega_{n,1}^2$ (see Durbin and Knott (1972)), where

$$\kappa_n = \int\limits_{\alpha}^{\beta} \left[\, F_n(x) - G(x;0) \,\right]\, \sin\, \pi G(x;0)\, dG(x;0)$$

(here we have $q(t) = \sin \pi t$, $k = 1$).

The case $q \equiv 1$, $k = 1$ corresponds to the so-called Chapman–Moses statistic (see Chapman (1958)). The leading set consists now of the functions $C\,t\,(1-t)$, $C > 0$.

We note that the leading set for the sequence $\{\omega_{n,q}^k\}$ has been found so far only for $q \in L^1$, and for nonsummable weights the local behavior of exact slopes is usually unknown. One rare exception is the case $q(t) = \left[t\,(1-t)\right]^{-1}$, $k = 2$, corresponding to the Anderson–Darling statistic (2.1.10).

It has been stated in Section 2.5 that

$$c_{A^2}(\theta) \sim 2\int\limits_{\alpha}^{\beta} \frac{\left(G(x;\theta) - G(x;0)\right)^2}{G(x;0)\left(1 - G(x;0)\right)}\, dG(x;0)\,, \qquad \theta \to 0\,.$$

Suppose, additionally, that $\{G(x;\theta)\}$ satisfies the condition

$$\int\limits_{\alpha}^{\beta} \frac{\left(G(x;\theta) - G(x;0)\right)^2 dG(x;0)}{G(x;0)\left(1 - G(x;0)\right)} \sim \int\limits_{\alpha}^{\beta} \frac{\left(G'_\theta(x;0)\right)^2 dG(x;0)}{G(x;0)\left(1 - G(x;0)\right)}\cdot \theta^2\,, \tag{6.2.14}$$

as $\theta \to 0$, and consider the class $\mathfrak{G}_{A^2}$ consisting of those elements from $\mathfrak{G}$ for which (6.2.14) holds. The LAO condition for the sequence $\{A_n^2\}$ in the class $\mathfrak{G}_{A^2}$ is equivalent to the equality

$$2\int\limits_{0}^{1} \frac{v^2(t)}{t\,(1-t)}\, dt = \int\limits_{0}^{1} v'^{\,2}(t)\, dt \tag{6.2.15}$$

under boundary conditions (6.2.7). But equality (6.2.15), due to Hardy's inequality (see Hardy, Littlewood, and Polya (1934), Theorem 262), can hold only in the case $v(t) = C\,t\,(1-t)$, $C \neq 0$. Hence we have proved the following statement.

Theorem 6.2.5 *The leading set of the sequence $\{A_n^2\}$ in the class $\mathfrak{G}_{A^2}$ is the set*

$$\mathbf{V}(A^2) = \left\{\, C\,t\,(1-t)\,\right\}, \qquad C \neq 0\,.$$

As the last integral statistic we consider the Watson statistic (2.1.11).

Theorem 6.2.6 *The leading set of the sequence $\{U_n^2\}$ in the class $\mathfrak{G}$ is the set*

$$\mathbf{V}(U^2) = \{\, C_1 \sin 2\pi t + C_2 \left(1 - \cos 2\pi t\right)\,\}, \qquad C_1^2 + C_2^2 > 0\,.$$

Proof Arguing just as for $\{\omega_{n,q}^k\}$ and using the results of Sections 2.5–2.6, we conclude that the LAO condition is equivalent to the equality

$$4\pi^2 \left\{ \int_0^1 v^2(t)\, dt - \left(\int_0^1 v(t)\, dt \right)^2 \right\} = \int_0^1 v'^2(t)\, dt \qquad (6.2.16)$$

under boundary conditions (6.2.7). Just as in the proof of Theorem 6.2.4 we must find the infimum of the right-hand side in (6.2.16) under the indicated boundary conditions and the normalization condition

$$\int_0^1 v^2(t)\, dt - \left(\int_0^1 v(t)\, dt \right)^2 = 1\,.$$

The existence of the solution is substantiated as in Theorem 6.2.4. The Euler–Lagrange equation takes now the form

$$\hat{v}'' = \lambda \left(\hat{v} - \int_0^1 \hat{v}\, dt \right), \qquad \hat{v}(0) = \hat{v}(1) = 0\,,$$

and its solutions are the pairs of functions $C_1 \sin 2k\pi t$ and $C_2\left(1 - \cos 2k\pi t\right)$, $k = 1, 2, \ldots$, corresponding to the eigenvalues $\lambda_k = -4k^2\pi^2$ of multiplicity 2. The principal eigenvalue $-4\pi^2$ corresponds to the eigenfunction

$$C_1 \sin 2\pi t + C_2\left(1 - \cos 2\pi t\right),$$

for which the equality in (6.2.16) holds. $\qquad\qquad\qquad\qquad\qquad\square$

As to the integral statistic L_n^2 given by (2.1.13), we have not succeeded in finding a simple description of the leading set.

6.3 Characterization of Distributions by the LAO Property of Goodness-of-Fit Tests under Concrete Alternatives

Let a sequence of statistics be given. As we have already noted in Section 6.1 the knowledge of its leading set enables one to describe explicitly the corresponding LAO domain under concrete alternatives. In typical cases these domains consist only of a single distribution. This possibility provides a way to determine the distribution uniquely by the LAO property for a given sequence of statistics.

We begin with location families. Let $G(x; \theta) = G(x + \theta)$, $x \in \mathbf{R}^1$, so $\Delta = (-\infty, +\infty)$. The functional class $\mathfrak{G}$ admits in this case a simpler description. Namely, we require that its elements possess positive absolutely continuous densities g such that condition (6.2.1) holds:

$$K(\theta) = \int\limits_{-\infty}^{\infty} \ln \frac{g(x + \theta)}{g(x)} \, g(x + \theta) \, dx \sim \tfrac{1}{2} I(g) \cdot \theta^2 \,, \qquad \theta \to 0 \,,$$

where

$$I(g) = \int\limits_{-\infty}^{\infty} \frac{g'^2(x)}{g(x)} \, dx \,, \qquad 0 < I(g) < \infty \,.$$

The other regularity conditions in the definition of the class $\mathfrak{G}$ follow from those listed previously (specifically, from the finiteness of the Fisher information $I(g)$). Let us denote by $\mathfrak{G}^l$ this class of d.f.s.

If the leading set $\mathbf{V}$ of some sequence of statistics is known, one can find the d.f.s G in the LAO domain, in conformity with (6.1.6), from the differential equation

$$G' = v(G) \,, \qquad v \in \mathbf{V} \,, \qquad (6.3.1)$$

under the additional conditions

$$G' \geq 0 \,, \qquad G(-\infty) = 0 \,, \qquad G(+\infty) = 1 \,.$$

Relying on Theorems 6.2.1–6.2.6 we obtain now the descriptions of the LAO domains in $\mathfrak{G}^l$ for various goodness-of-fit statistics. These domains consist of d.f.s having the form $G\big((x - \gamma)/\sigma\big)$, $x, \gamma \in \mathbf{R}^1$, $\sigma > 0$, but for convenience we shall point out only the corresponding densities $g(x)$.

Theorem 6.3.1 *The sequences $\{D_n\}$ and $\{D_n^+\}$ are LAO in $\mathfrak{G}^l$ only for the Laplace distribution with the density $g(x) = \tfrac{1}{2} e^{-|x|}$ and the sequence $\{D_n^-\}$ is never LAO in the class $\mathfrak{G}^l$.*

Proof From Theorem 6.2.1 it follows that we must integrate the equation

$$G' = C \min\big(G, 1 - G\big) \,,$$

which has no solutions for $C < 0$ and has as its unique solution for $C > 0$ the Laplace d.f.

$$G(x) = \begin{cases} \dfrac{1}{2}\exp\left\{\dfrac{x-\gamma}{\sigma}\right\} & \text{if } x \leq \gamma, \\[4mm] 1 - \dfrac{1}{2}\exp\left\{-\left(\dfrac{x-\gamma}{\sigma}\right)\right\} & \text{if } x > \gamma. \end{cases} \qquad \square$$

In the case of the weighted Kolmogorov statistic (2.2.17) we obtain asymmetric analogs of the Laplace distribution. For example, for the weight given by (2.2.19) one has the pair of densities $\sigma^{-1} f_i\big((x-\gamma)/\sigma\big)$, $i = 1, 2$, where

$$f_1(x) = \begin{cases} \exp\left\{-x_0' \, x - 2\right\}, & x \geq 0, \\[2mm] \exp\left\{(1 - x_0')\, x - 2\right\}, & x < 0, \end{cases}$$

$$f_2(x) = \begin{cases} \exp\left\{-x_0'' \, x - 2\right\}, & x \geq 0, \\[2mm] \exp\left\{(1 - x_0'')\, x - 2\right\}, & x < 0, \end{cases}$$

and the numbers x_0' and x_0'' were defined after the proof of Theorem 6.2.1.

Theorem 6.3.2 *The sequence $\{G_n^+\}$ is LAO in $\mathfrak{G}^l$ only for the distribution with the density $\frac{1}{2}\left(1 + |x|\right)^{-2}$, the sequence $\{G_n^-\}$ is LAO in $\mathfrak{G}^l$ only for the logistic distribution, and the sequence $\{G_n\}$ is LAO in $\mathfrak{G}^l$ only for both mentioned distributions.*

Proof In conformity with Theorem 6.2.2 one should look for the solutions of the equation

$$G' = C\, v^*(G, s), \qquad 0 \leq s \leq 1. \tag{6.3.2}$$

But the function $v^*(t, s)$ given by (6.2.8) preserves the sign only for three values: $s = 0$, $s = \frac{1}{2}$, and $s = 1$. Also, $v^*(t, \frac{1}{2}) \geq 0$, $v^*(t, 0) \leq 0$, and $v^*(t, 1) \leq 0$ for all t. Therefore in the case of G_n^+ one obtains the equation

$$G' = C\left(\min\left(G, \, 1 - G\right)\right)^2, \qquad C > 0,$$

having as the solution d.f. G with the density

$$g(x) = \frac{1}{2\sigma}\left(1 + \frac{|x-\gamma|^2}{\sigma^2}\right)^{-1},$$

and in the case of G_n^-

$$G' = C\,G\,(1 - G)\,, \qquad C > 0\,,$$

the solution of which is the logistic distribution function. In the case of the two-sided statistic G_n both solutions are suitable. $\qquad\Box$

Denote by $\mathfrak{G}_K^l$ the class $\mathfrak{G}^l \cap \mathfrak{G}_K$, where $\mathfrak{G}_K$ was introduced before the formulation of Theorem 6.2.3.

Theorem 6.3.3 *The sequences $\{K_n\}$ and $\{K_n^+\}$ are LAO in $\mathfrak{G}_K^l$ only for the distribution of extremal value with the density*

$$g(x) = \exp\left\{x - e^x\right\};$$

the sequence $\{K_n^-\}$ is never LAO in this class.

Proof The equation (6.3.1) takes now the form

$$G' = C\,(1 - G)\,\ln\,(1 - G)\,, \qquad (6.3.3)$$

so it has no solutions in the case of statistic K_n^- when $C > 0$. On the other hand, in the case $C < 0$, suitable for the other two statistics, this equation has the solution

$$G(x) = 1 - \exp\left\{-\exp\left(\frac{x - \gamma}{\sigma}\right)\right\}, \qquad \gamma \in \mathbf{R}^1,\ \sigma > 0\,. \qquad \Box$$

We proceed now to the investigation of integral statistics.

Theorem 6.3.4 *The sequence of statistics $\{\omega_{n,1}^k\}$ is LAO in $\mathfrak{G}^l$ only if*

$$g(x) = 2^{2/k-1}\left(B\left(\tfrac{1}{k},\,\tfrac{1}{k}\right)\cosh^{2/k}x\right)^{-1},$$

where $B\,(x,y) = \int_0^1 t^{x-1}\,(1 - t)^{y-1}\,dt$ is the Euler beta function.

Before giving the proof, we state some corollaries to this theorem.

Corollary 1 *The LAO property in $\mathfrak{G}^l$ of the sequence $\{\omega_{n,1}^2\}$ characterizes the "hyperbolic cosine" distribution with the density $g(x) = (\pi \cosh x)^{-1}$*

Corollary 2 *The LAO property in $\mathfrak{G}^l$ of the sequence $\{\kappa_n\}$ of the Durbin–Knott statistics characterizes the "hyperbolic cosine" distribution too.*

Corollary 3 *The LAO property in $\mathfrak{G}^l$ of the sequence $\{\omega_{n,1}^1\}$ characterizes the logistic distribution.*

Proof of Theorem 6.3.4 It follows from the proof of Theorem 6.2.4 that the LAO condition may be fulfilled only if the function $v(t) = g\big(G^{-1}(t)\big)$ for some constant $C \neq 0$ satisfies the equation

$$v'' = C\,v^{k-1} \tag{6.3.4}$$

under boundary conditions (6.2.7), where we are interested only in non-negative solutions.

Multiplying both parts of (6.3.4) by v and integrating by parts, we get

$$\int_0^1 v\,v''\,dt = -\int_0^1 v'^{\,2}(t)\,dt = C\int_0^1 v^k(t)\,dt\,,$$

hence $C < 0$. Further let us multiply both parts of (6.3.4) by v' and integrate this equation. One obtains

$$v'^{\,2}(t) = -C_1^2\,v^k(t) + C_2^2\,. \tag{6.3.5}$$

Note now that v is a concave function on $[0, 1]$ as $v'' \leq 0$. The equation (6.3.4) under boundary conditions (6.2.7) has together with $v(t)$ the solution $v(1-t)$ too. But as has been noted in Section 2.4, the solution of the studied problem is unique, so

$$v(t) = v(1-t)\,, \qquad 0 \leq t \leq 1\,,$$

that is, the function $v(t)$ is symmetric with respect to the line $t = \frac{1}{2}$. Therefore it follows that

$$v'(t) = \begin{cases} C_3^2\,\sqrt{1 - C_4^2\,v^k(t)}\,, & 0 \leq t \leq \frac{1}{2}\,, \\[2mm] -C_3^2\,\sqrt{1 - C_4^2\,v^k(t)}\,, & \frac{1}{2} \leq t \leq 1\,. \end{cases} \tag{6.3.6}$$

Now return from the function v to the density g and consider on the interval $(-\infty,\ G^{-1}(\frac{1}{2}))$ the equation into which turns (6.3.6), namely,

$$g' = C_3^2\,g\,\sqrt{1 - C_4^2\,g^k}\,.$$

Integrating it with the aid of the substitution $h = C_4^2\,g^k$ we have

$$\frac{dh}{h\,\sqrt{1-h}} = C_5^2\,dt\,,$$

and consequently

$$g(t) = \zeta_k\,\sigma^{-1}\,\cosh^{-2/k}\left(\frac{t-\gamma}{\sigma}\right),$$

where

$$\zeta_k = 2^{2/k-1} \left(B\left(\tfrac{1}{k}, \tfrac{1}{k} \right) \right)^{-1}, \qquad \gamma \in \mathbf{R}^1, \ \sigma > 0.$$

The interval $(G^{-1}(\tfrac{1}{2}), +\infty)$ can be considered analogously, and the sought-for density has there the same form. □

Setting k equal to 2 and 1 we obtain from this theorem Corollaries 1 and 3. Corollary 2 follows from the structure of the leading sets for the sequence $\{\omega_{n,1}^2\}$ and $\{\kappa_n\}$ described in Section 6.2.

Closely connected with the investigation of the LAO properties of the statistics $\{\omega_{n,q}^k\}$ is the question of an "asymptotically optimal" choice of the weight q for a given family $\{G(x;\theta)\}$, that is, the choice for which the sequence $\{\omega_{n,q}^k\}$ would be LAO in $\mathfrak{G}$. The arguments in the proof of Theorem 6.3.4 imply that this weight q^* can be obtained by the formula

$$q^*(t) = C \, v''(t) \, v^{1-k}(t) \tag{6.3.7}$$

where again $v(t) = G'_\theta \big(G^{-1}(t;0);0 \big)$, provided $q^* \in L^1[0,1]$; otherwise a special study is necessary of the principal part of the exact slope for $\{\omega_{n,q^*}^k\}$ as $\theta \to 0$. Formula (6.3.7) also shows that the class of leading functions is sufficiently broad. In any case, any twice differentiable concave function v satisfying conditions (6.2.7) will be a leading function for the sequence $\{\omega_{n,q}^1\}$ in $\mathfrak{G}$ for $q(t) = C \, v''(t)$.

Consider now the class of d.f.s $G(x;\theta) = G(x+\theta)$ in $\mathfrak{G}^l$ satisfying also condition (6.2.14), and denote it by $\mathfrak{G}^l_{A^2}$.

Theorem 6.3.5 *The sequence of statistics $\{A_n^2\}$ is LAO in $\mathfrak{G}^l_{A^2}$ only in the case of the logistic distribution.*

Proof It follows from Theorem 6.2.5 that d.f. G from the LAO domain must satisfy the equation

$$G' = C \, G \, (1 - G), \qquad C > 0.$$

Integrating it we obtain

$$G(t) = \left(1 + \exp\left(-\frac{t - \gamma}{\sigma} \right) \right)^{-1},$$

that is, the logistic distribution. □

Theorem 6.3.6 *The sequence of statistics $\{U_n^2\}$ is LAO in $\mathfrak{G}^l$ only in the case of the Cauchy distribution.*

Proof The differential equation we must solve has now, due to Theorem 6.2.6, the form

$$G' = C_1 \left(1 - \cos 2\pi G\right) + C_2 \sin 2\pi G, \qquad (6.3.8)$$

where clearly $C_1 > 0$ (otherwise (6.3.8) has no solutions in the space of d.f.s). If $C_2 = 0$, the equation may be easily solved and its solution is the Cauchy distribution with d.f.

$$G(t) = \frac{1}{2} + \frac{2}{\pi} \; \text{arctg} \left(\frac{t - \gamma}{\sigma} \right).$$

It remains to consider the case $C_2 \neq 0$ and to show that then there are no probabilistic solutions. Suppose, for definiteness, that $C_2 > 0$. Equation (6.3.8) may be rewritten in the form

$$G' = C_1 + \rho \sin \left(2\pi G + \varphi\right),$$

where

$$\rho = \sqrt{C_1^2 + C_2^2}\,, \qquad \sin \varphi = -\frac{C_1}{\rho}\,, \qquad \cos \varphi = \frac{C_2}{\rho}\,.$$

Integrating it with the aid of formula 2.551.3 in Gradshteyn and Ryzhik (1971), we have

$$\ln \frac{C_1 \tan \left(\pi G + \frac{1}{2}\,\varphi\right) + \rho - C_2}{C_1 \tan \left(\pi G + \frac{1}{2}\,\varphi\right) + \rho + C_2} = 2\,\pi\, C_2\, x + C_3\,.$$

As $\tan \left(\frac{1}{2}\,\varphi\right) = (C_2 - \rho)/C_1$, we obtain from here that

$$\frac{C_1^2 \tan \pi G + (C_2 - \rho)^2 \tan \pi G}{C_1^2 \tan \pi G + 2\,C_1\,C_2 - (C_2^2 - \rho^2)\tan \pi G} = \exp\left\{2\,\pi\, C_2\, x + C_3\right\}.$$

If $x \to \infty$ the right-hand side tends to infinity whereas the left-hand side tends to zero. The obtained contradiction excludes the case $C_2 > 0$. The case $C_2 < 0$ is treated analogously. □

Now we turn to the families with a scale parameter. Let $G(x; \theta) = G\left(x e^\theta\right)$, $x \geq 0$, $\theta \in \Theta$, that is, $\Delta = (0, +\infty)$. In this case the regularity conditions determining the class $\mathfrak{G}$ reduce, just as for the location case, to a lesser number of conditions. Namely, it suffices to assume that the density $g = G'$ is absolutely continuous and positive and satisfies condition (6.2.1) and also the condition

$$x\, g(x) \to 0 \qquad \text{as } x \to 0 \text{ and as } x \to \infty\,.$$

We denote by $\mathfrak{G}^s$ the class of corresponding d.f.s. Just as for the case of a location parameter we introduce a subclasses of $\mathfrak{G}^s$, for example, $\mathfrak{G}^s_{A^2}$.

The LAO condition for a sequence of statistics with the leading function v reduces in the class $\mathfrak{G}^s$ (or a suitable subclass of it) to the problem

$$x\,G'(x) = v\big(G(x)\big)\,, \qquad G(0+) = 0\,, \qquad G(\infty) = 1\,, \qquad G' \geq 0\,.$$
$$(6.3.9)$$

Comparing (6.3.9) with (6.3.1) we see that the solutions of (6.3.1) can be transformed into those of (6.3.9) by changing the argument x to $\ln x$. Hence all the characterizations indicated in Theorems 6.3.1–6.3.6 for a location parameter have obvious analogs for a scale parameter.

For example, the sequence $\{\omega^2_{n,1}\}$ is LAO in the class $\mathfrak{G}^s$ only for the densities having the form $\sigma^{-1}g_\lambda(x/\sigma)$, where $\sigma > 0$, $\lambda > 0$, and

$$g_\lambda(x) = \begin{cases} \dfrac{2\,\lambda}{\pi}\ \dfrac{x^{\lambda-1}}{x^{2\lambda}+1} & \text{if } x \geq 0, \\[2mm] 0 & \text{if } x < 0. \end{cases}$$

(the so-called log-logistic density). Likewise, the sequence $\{K_n\}$ is LAO in $\mathfrak{G}^s_K$ only for the Weibull densities ($\sigma > 0$, $\nu > 0$):

$$g_\nu(x) = \begin{cases} \nu\ x^{\nu-1}\sigma^{-\nu}\ \exp\big\{-(x/\sigma)^\nu\big\} & \text{if } x \geq 0, \\[2mm] 0 & \text{if } x < 0. \end{cases}$$

A similar situation arises for the family of d.f.s $G(x;\theta) = G\big(x^{1+\theta}\big)$, $x \geq 1$, $\theta \geq 0$. Here the required transition is from d.f. $G(x)$ corresponding to the location families to d.f. $G(\ln\ln x)$, $x \geq 1$.

Behnen and Neuhaus (1989) (see also Neuhaus (1987)) investigated the class of alternatives of *generalized shift and scale*. Instead of the model $G(x;\theta) = G(x-\theta)$ they proposed

$$G_1(x;\theta) = G\big(x - \theta\,\mathcal{D}(x)\big)\,, \qquad x \in \mathbf{R}^1\,,$$

where the shift function $\mathcal{D}\colon \mathbf{R}^1 \mapsto \mathbf{R}^1$ is assumed to be bounded with $\mathcal{D} \geq 0$, $\mathcal{D} \not\equiv 0$, and a bounded derivative $d\colon \mathbf{R}^1 \mapsto \mathbf{R}^1$. Obviously the side condition $\theta\,d(x) \leq 1$ for all $x \in \mathbf{R}^1$ has to be fulfilled in order to make $G_1(x)$ a proper d.f. The alternative of generalized scale is introduced analogously. In this case

$$G_2(x;\theta) = G\big(x\,\exp\{-\theta\,\mathcal{D}(x)\}\big)\,, \qquad x \in \mathbf{R}^1\,,$$

where $G_2(x;\theta)$ is a proper d.f. if $|\theta|\cdot\sup\big\{\,|x\,d(x)|\colon x \in \mathbf{R}^1\big\} \leq 1$. The "ideal" location and scale models arise for $\mathcal{D}(x) \equiv 1$. The reason for introducing such alternatives is the justified remark by Behnen and

Neuhaus (1989) that the models of *strict* shift and scale often are not realistic in practice and it seems more plausible that, for example, the extreme parts of a population react in quite another way to a treatment than the central part of that population. When considering models of such a kind it seems to be an interesting but for the present unexplored problem.

Now we consider the Lehmann alternatives. These are purely nonparametric alternatives proposed first by Lehmann (1953). Consider for $\theta \geq 0$ the following two families of d.f.s:

$$G(x;\theta) = \{\, G(x)\,\}^{1+\theta}, \qquad G(x;\theta) = 1 - \{\, 1 - G(x)\,\}^{1+\theta}. \qquad (6.3.10)$$

Assume G has absolutely continuous and positive density and denote by $\mathfrak{G}^L$ the class of such d.f.s. It is not hard to verify that

$$G'_\theta\big(\, G^{-1}(x;0);0\,\big) \equiv \begin{cases} x \ln x & \text{in the first case,} \\[2mm] (1-x)\ln(1-x) & \text{in the second case.} \end{cases}$$

$$(6.3.11)$$

Therefore, the LAO condition cannot hold for all the sequences of statistics studied earlier (except for the sequences $\{K_n\}$ and $\{K_n^-\}$) since their leading sets do not contain functions (6.3.11) and consequently their LAO domains in $\mathfrak{G}$ are empty. On the contrary, the sequences $\{K_n\}$ and $\{K_n^-\}$ are LAO in $\mathfrak{G}_K^L$ for the second family of d.f.s in (6.3.10) as the function $(1-x)\ln(1-x)$ belongs to their leading sets found in Theorem 6.2.3. (It is easy to see that the class $\mathfrak{G}_K$ includes the second family of d.f.s from (6.3.10) for $G \in \mathfrak{G}^L$.)

This situation is quite similar for another type of alternative also introduced by Lehmann (1953):

$$G(x;\theta) = (1 - \theta) \cdot G(x) + \theta \cdot G^{r+1}(x), \qquad r > 0, \qquad (6.3.12)$$

where d.f. G belongs to $\mathfrak{G}^L$ again. Clearly

$$G'_\theta\big(\, G^{-1}(x;0);0\,\big) = -x\,(1-x)^r, \qquad (6.3.13)$$

and therefore the LAO domain for the majority of statistics considered here is again empty for $r \neq 1$. The exception is the statistic $-\omega_{n,q}^1$ if we choose the weight function q according to formula (6.3.7), that is, take $q^*(t) = t^{r-1}$. Then the sequence $\{-\omega_{n,q}^1\}$ is LAO for all elements of $\mathfrak{G}^L$ at once.

But if $r = 1$, four sequences of statistics $\{-\omega_{n,q}^1\}$, $\{A_n^2\}$, $\{G_n\}$ and $\{G_n^-\}$ possess the LAO property for all elements of corresponding sub-classes of $\mathfrak{G}^L$. The situation is quite similar also for the so-called proportional odds model (see, e.g., Pettitt (1984) and Dabrowska, Doksum, and Miura (1989)) when

$$G(x;\theta) = \frac{G(x)}{1+\theta\left(1-G(x)\right)}$$

and, consequently,

$$G'_\theta\left(G^{-1}(x;0);0\right) = x\left(1-x\right).$$

6.4 The LAO Domains for Homogeneity Tests

Let $X_1, X_2, \ldots, X_m$ and $Y_1, Y_2, \ldots, Y_n$ be two independent samples with continuous d.f.s G_1 and G_2. The homogeneity hypothesis H_1 is that G_1 and G_2 are identical, whereas under the alternative $G_1(x) = G(x;0)$ and $G_2(x) = G(x;\theta)$, $\theta > 0$. Here $\{G(x;\theta)\}$ is a given family of absolutely continuous d.f.s such that $G(x;\theta) \neq G(x;0)$ for $\theta \neq 0$. For testing H_1 against A_1 we may use homogeneity tests based on the two-sample variants of statistics considered in Sections 6.2–6.3 (we have studied their large deviations and the Bahadur efficiency in Chapter 3). The main feature of the two-sample case is that the hypothesis H_1 is composite and hence inequality (6.1.1) has a more complex form. Let m and n tend to infinity in such a way that

$$\lim_{m,n\to\infty} \frac{m}{n+m} = \rho_1, \qquad 0 < \rho_1 < 1.$$

As has been already shown in Section 3.3, the exact slope $c_T(\theta)$ of any sequence of two-sample statistics $\{T_{m,n}\}$ used for testing H_1 against A_1 must satisfy the inequality

$$\tfrac{1}{2}c_T(\theta) \leq K^*(\theta) = \rho_1 \int\limits_\alpha^\beta \ln \frac{g(x;0)}{h(x;\theta)}\, g(x;0)\, dx$$

$$+ \rho_2 \int\limits_\alpha^\beta \ln \frac{g(x;\theta)}{h(x;\theta)}\, g(x;\theta)\, dx,$$

where $g(x;\theta)$ is the density of $G(x;\theta)$, $\rho_1 + \rho_2 = 1$, and

$$h(x;\theta) = \rho_1\, g(x;0) + \rho_2\, g(x;\theta).$$

We impose on the class of families of d.f.s $\{G(x; \theta)\}$ the same restrictions as those on $\mathfrak{G}$ with one change: An analog of (6.2.1) should be valid, namely,

$$K^*(\theta) \sim \tfrac{1}{2} \rho_1 \rho_2 I(g) \cdot \theta^2, \qquad \theta \to 0. \qquad (6.4.1)$$

We denote the class of such families by $\mathfrak{G}_1$.

We say that the sequence $\{T_{m,n}\}$ satisfies the LAO condition if

$$c_T(\theta) \sim 2 K^*(\theta), \qquad \theta \to 0,$$

and the set of families $\{G(x; \theta)\}$ in $\mathfrak{G}_1$ or some subclass of it, for which this is so, we call the LAO domain in this class.

Now we consider the two-sample variant of the statistic $\omega_{n,q}^k$, introduced in (3.1.7):

$$W_{m,n,q}^k = \left(\frac{m\,n}{(m+n)^2} \right)^{k/2} \int\limits_{\alpha}^{\beta} \left[F_m(x) - G_n(x) \right]^k q\big(H_{m+n}(x)\big)\, dH_{m+n}(x),$$

where F_m and G_n are the empirical d.f.s constructed with the help of the initial samples, k is a positive integer, and H_{m+n} is the empirical d.f. of the combined sample. It is also supposed that the nonnegative weight function q has a bounded derivative. Under these conditions, from the results of Sections 3.2–3.3 and the regularity conditions determining the class $\mathfrak{G}_1 \cap \mathfrak{G}_\omega(q; k)$, one deduces

$$c_{W^k}(\theta) \sim -\rho_1 \rho_2 \lambda_0 \left[\int\limits_0^1 v^k(t)\, q(t)\, dt \right]^{2/k} \cdot \theta^2, \qquad \theta \to 0,$$

where again

$$v(t) = G_\theta'\big(G^{-1}(t; 0); 0 \big).$$

With the aid of (6.4.1) we now establish that the LAO condition of the sequence $\{W_{m,n,q}^k\}$ is equivalent to equality (6.2.5) under boundary conditions (6.2.7). Hence the leading functions of the sequence $\{W_{m,n,q}^k\}$ in $\mathfrak{G}_1 \cap \mathfrak{G}_\omega(q; k)$ and the ones of $\{\omega_{n,q}^k\}$ in $\mathfrak{G}_\omega(q; k)$ coincide. Similar statements hold for other two-sample statistics, for example, the Smirnov statistic

$$D_{m,n} = \frac{\sqrt{m\,n}}{m+n} \sup_{-\infty < x < \infty} |F_m(x) - G_n(x)|.$$

The corresponding characterization results for location families differ from those of Theorems 6.3.1–6.3.6 only in the description of the subclasses of $\mathfrak{G}_1$. The changes arising in the cases of the scale, Lehmann, and proportional odds models are also insignificant.

An extensive class of statistics often used for testing H_1 against A_1 is the class of simple linear rank statistics (3.1.9), that is,

$$S_N = N^{-1} \sum_{i=1}^{m} a_N \big(R_i / (N+1) \big), \qquad (6.4.2)$$

where $N = m+n$, R_i is the rank of the ith observation of the first sample in the general variational series, and a_N is a real function. As usual, it is assumed that $a_N(u)$ converges in the mean square, as $N \to \infty$, to a score function J such that $0 < \int_0^1 J^2(u)\,du < \infty$. Without loss of generality, one may assume that

$$\int\limits_0^1 J(u)\,du = 0\,, \qquad \int\limits_0^1 J^2(u)\,du = 1\,. \qquad (6.4.3)$$

In addition, we suppose that condition (3.1.14) is fulfilled.

To describe the LAO domain of statistics (6.4.2) we need to impose additional restrictions on the score function, and to determine some subclass of $\mathfrak{G}_1$ with the aid of conditions on the properties of the family $\{G(x;\theta)\}$ given by properties of J. The reason for this is that the expression for the exact slope of the sequence of statistics (6.4.2) generally has a complicated structure (see Woodworth (1970) and Kremer (1979a,b)) and may be simplified only under additional restrictions.

First we assume that the score function J is differentiable on $(0, 1)$. Further, consider the quantity

$$t(\theta, \boldsymbol{\rho}) := \rho_1 \int\limits_{\alpha}^{\beta} J\big(\rho_1\, G(x;0) + \rho_2\, G(x;\theta) \big)\, dG(x;0)$$

and require the following conditions to be valid:

$$t(\theta, \boldsymbol{\rho}) > 0 \qquad \text{for } \theta > 0\,, \qquad\qquad (6.4.4)$$

$$t(\theta, \boldsymbol{\rho}) \sim \theta\, \rho_1\, \rho_2 \int\limits_{\alpha}^{\beta} J'\big(G(x;0)\big)\, G'_{\theta}(x;0)\, dG(x;0)\,, \quad \theta \to 0\,. \quad (6.4.5)$$

The right-hand side of (6.4.5) is a natural approximation of the expression for $t(\theta, \boldsymbol{\rho})$; however, in general, condition (6.4.5) restricts the class of families of d.f.s $\{G(x;\theta)\}$ subject to properties of J. Condition (6.4.4) is connected with the requirement of consistency of a test based on $\{S_N\}$. Under certain additional restrictions Kremer (1979a,b) proved

234   6 *Local Asymptotic Optimality of Nonparametric Tests*

the following asymptotic formula for the exact slope $c_S(\theta)$ of the sequence $\{S_N\}$:

$$c_S(\theta) \sim (\rho_1\rho_2)^{-1}t^2(\theta,\rho)\,, \qquad \theta \to 0\,. \qquad (6.4.6)$$

Together with (6.4.5) this implies that

$$c_S(\theta) \sim \rho_1\rho_2 \left(\int\limits_0^1 J'(u)\,v(u)\,du \right)^2 \cdot \theta^2\,, \qquad \theta \to 0\,, \qquad (6.4.7)$$

where v is again given by (6.2.3). Supposing additionally the validity of the conditions

$$\lim_{u\to 0+} J(u)\,v(u) = \lim_{u\to 1-} J(u)\,v(u) = 0\,, \qquad (6.4.8)$$

we may perform integration by parts in (6.4.7) and obtain the relation

$$c_S(\theta) \sim \rho_1\rho_2 \left(\int\limits_0^1 J(u)\,v'(u)\,du \right)^2 \cdot \theta^2\,, \qquad \theta \to 0\,. \qquad (6.4.9)$$

Condition (6.4.8) presupposes that v tends to 0 at the endpoints of $(0, 1)$ at a rate determined by the score function, and, in the general case, additionally narrows down the class of admissible families of d.f.s.

We prefer to operate directly with representation (6.4.9) instead of formulating explicitly the conditions making it possible to simplify the expression for the exact slope. Denote by $\mathfrak{G}_1^J$ the subclass of $\mathfrak{G}_1$ for which relations (6.4.4) and (6.4.9) hold.

Theorem 6.4.1 *The leading set of the sequence of statistics $\{S_N\}$ in $\mathfrak{G}_1^J$ has the form*

$$\mathbf{V}(S) = \left\{ C \int\limits_0^t J(u)\,du \right\}, \qquad C \neq 0\,.$$

Proof It follows from (6.4.9) and (6.4.1) that the LAO condition is equivalent to the equality

$$\left(\int\limits_0^1 J(u)\,v'(u)\,du \right)^2 = \int\limits_0^1 v'^2(u)\,du\,.$$

The Cauchy–Bunyakovskii–Schwarz inequality and conditions (6.4.3) imply that this is possible only if $v'(u) = C\,J(u)$. Integration of the obtained equation completes the proof.   □

Let us consider some examples and corollaries of this theorem.

1. Let

$$J(u) = \frac{r+1}{r} \sqrt{2r+1} \left(u^r - \frac{1}{r+1} \right), \qquad 0 \le u \le 1, \quad r > 0.$$

The linear rank statistics with such score function have been intensively investigated. Thus, the case $r = 1$ corresponds to the classical Wilcoxon test, $r = 2$ to the Taha statistic, and the general case of positive integer r to the Tamura one (see Hájek and Šidak (1967)). The function J satisfies all the conditions imposed on the score function. It follows that the leading set consists of the functions having the form $Ct(1 - t^r)$, $C \ne 0$.

2. Let $J(u) = \sqrt{2} \cos \pi u$. The restrictions on J hold again and the leading functions, as in the cases of $\{\omega_{n,1}^2\}$ and $\{\kappa_n\}$, have the form $C \sin \pi t$, $C \ne 0$.

3. Consider $J(u) = -1 - \ln(1-u)$ and $J(u) = 1 + \ln u$. Such a choice corresponds to the Savage test (see again Hájek and Šidak (1967), Chapter 3). In this case the class $\mathfrak{G}_1^J$ is somewhat narrower than the class $\mathfrak{G}_1$ and the leading functions have the forms $C(1 - t) \ln(1 - t)$ and $Ct \ln t$, correspondingly.

4. Let $J(u) = \Phi^{-1}(u)$; this corresponds to the normal scores and van der Waerden tests. In this case the class $\mathfrak{G}_1^J$ is again restricted in comparison with $\mathfrak{G}_1$ and the LAO domain in this class is determined by the leading function $C \int_0^t \Phi^{-1}(u)\, du$, $C \ne 0$.

We proceed to the problem of the characterization of distributions by the LAO property of linear rank statistics under concrete alternatives. We begin with the case of location parameters and introduce the class $\mathfrak{G}_1^l$ of $G(x + \theta)$. The difference with $\mathfrak{G}^l$ is that relation (6.4.1) should be fulfilled instead of (6.2.1). Using the leading functions of the most well-known linear rank tests found in Examples 1–4, it is easy to integrate differential equations of type (6.3.1). Here we give some conclusions.

1. If

$$J(u) = \frac{r+1}{r} \sqrt{2r+1} \left(u^r - \frac{1}{r+1} \right), \qquad 0 < r < \infty,$$

the conditions determining the class $\mathfrak{G}_1^J$ are clearly fulfilled for elements from $\mathfrak{G}_1^l$. Therefore, the LAO property of linear rank statistics with this score function characterizes the distribution with the density $\sigma^{-1} g((x - \gamma)/\sigma)$, where

$$g(x) = r^{-1} e^{-x} \left(1 + e^{-x} \right)^{-(r+1)/r}. \tag{6.4.10}$$

In the special case $r = 1$ (the Wilcoxon statistic) we obtain the logistic distribution.

2. In the case $J(u) = \sqrt{2} \cos \pi u$ the arguments are analogous and we get the characterization of the "hyperbolic cosine" distribution.

3. Let $J(u) = 1 + \ln u$. Denote by $\mathfrak{G}_1^{J,\,l}$ the subclass of $\mathfrak{G}_1^J$ corresponding to a location parameter. Then the LAO property of linear rank tests with this score function characterizes in $\mathfrak{G}_1^{J,\,l}$ the density

$$g(x) = \exp\left\{ -x - e^{-x} \right\}.$$

Analogously, in the case $J(u) = -1 - \ln(1 - u)$ we obtain the characterization of the density $\exp\{x - e^x\}$.

4. For the score function $J(u) = \Phi^{-1}(u)$ we again consider the corresponding class $\mathfrak{G}_1^{J,\,l}$. Then the normal scores and van der Waerden tests are LAO in that class only for the normal distribution. To see this, note that equation (6.3.1) takes the form

$$G'(x) = C \int\limits_0^{G(x)} \Phi^{-1}(u)\, du\,,$$

from which it follows that for some $C_1 \neq 0$

$$\int\limits_0^x \left(\int\limits_0^{G(t)} \Phi^{-1}(u)\, du \right)^{-1} dG(t) = -C_1^2\, x\,.$$

As $\Phi^{-1}(u) = -\left[\varphi\left(\Phi^{-1}(u) \right) \right]'$, we obtain

$$\int\limits_0^x \left(\int\limits_0^{G(t)} \Phi^{-1}(u)\, du \right)^{-1} dG(t) = - \int\limits_0^x \frac{dG(t)}{\varphi\left(\Phi^{-1}\left(G(t) \right) \right)}$$

$$= -\Phi^{-1}\left(G(x) \right) + C_2\,.$$

Consequently $\Phi^{-1}\left(G(x) \right) = C_1^2\, x - C_2$, so $G(x) = \Phi\left((x-\gamma)/\sigma \right)$, which proves the desired statement.

As has been stated already in Section 6.3, the case of a scale parameter is very similar to the case of a location parameter; one should only make in the solution of the corresponding equation (see (6.3.9)) the substitution of $\ln x$ for x. In the same way as before let us introduce the classes of d.f.s $\mathfrak{G}_1^s$ and $\mathfrak{G}_1^{J,\,s}$. Then all the characterization results for a location parameter have obvious analogs for a scale one, but in suitable classes. Thus, the linear rank test with the score function

$$J(u) = \frac{r+1}{r} \sqrt{2r+1} \left(u^r - \frac{1}{r+1} \right)$$

is LAO in the class $\mathfrak{G}_1^s$ only for the densities of the form $\sigma^{-1} g_\delta(x/\sigma)$, where for $\delta > 0$

$$g_\delta(x) = \delta x^{\delta/r-1} \left[r (1 + x^\delta)^{(r+1)/r} \right]^{-1}, \qquad x \geq 0.$$

For $J(u) = -1 - \ln(1 - u)$ we obtain the characterization of the Weibull distribution and for $J(u) = \Phi^{-1}(u)$ the characterization of the lognormal one (in the corresponding class $\mathfrak{G}_1^{J,s}$). The situation is quite similar for the family $G(x; \theta) = G(x^{1+\theta})$ as well.

In the case of the Lehmann alternatives (6.3.10) and (6.3.12) the results for the two-sample homogeneity tests are very close to the results for goodness-of-fit ones. The LAO domain for the majority of tests considered in this section is empty because their leading functions are different from (6.3.11). By contrast, the linear rank tests with the score functions $J(u) = 1 + \ln u$ and $J(u) = -1 - \ln(1 - u)$ have leading sets which include functions (6.3.11) so their LAO domains are the corresponding classes $\mathfrak{G}^L \cap \mathfrak{G}_1^J$.

For alternative (6.3.12) the LAO domain is again empty for all tests considered earlier except the linear rank tests with the score functions

$$J(u) = C_r \left(u^r - \frac{1}{r+1} \right),$$

for which this domain consists of the whole class $\mathfrak{G}^L \cap \mathfrak{G}^J$.

The main results of this section have been published by Nikitin (1984a).

In conclusion we briefly discuss the problem of the description of the LAO domains for other types of ARE. As has already been noted in Section 6.1 and Chapter 3, many homogeneity tests (e.g., of ω^2 type or Smirnov-type) are Hodges–Lehmann AO under general conditions regarding the structure of the alternative. Linear rank tests do not possess this property, but their Hodges–Lehmann indices (under somewhat more restrictive conditions; see Section 3.5) are locally equivalent to the Bahadur exact slopes. As the upper bound for the Hodges–Lehmann indices (see inequality (3.4.5)) usually admits asymptotics (6.4.1), the structure of the LAO domains in the Bahadur and Hodges–Lehmann senses is determined by the common leading function, but in different given a priori classes $\mathfrak{G}$. The corresponding classes are narrower in the case of the Hodges–Lehmann efficiency as they conform to the relatively restrictive conditions formulated in Theorem 3.5.1.

When calculating the Pitman efficiency of nonparametric procedures one usually compares their efficiency with that of locally most powerful parametric tests constructed under concrete alternatives. Sometimes this approach enables one to make an asymptotically optimal choice of a nonparametric test. The results of such nature obtained by Chernoff and Savage (1958), Mielke (1972), and Goria (1980) for linear rank tests agree well with the results of the present section.

Mielke (1972) has found certain distributions for which the linear rank tests with the score functions

$$J(u) = C_r \left(u^r - \frac{1}{r+1} \right), \qquad r > 0,$$

are Pitman AO under the location and scale alternatives. It follows from our results that the class of such distributions is in fact much wider and includes more general families of distributions (6.4.10)–(6.4.11), which can be proved easily using the methods of Chernoff and Savage (1958) and Mielke (1972).

It may be concluded that the Pitman AO domains in the sense specified earlier are very similar for the statistics under consideration to the Bahadur LAO domains. The differences are connected mainly with the regularity conditions determining the class $\mathfrak{G}$; also, the Pitman ARE for the statistics with non-Gaussian limiting distributions should be interpreted in the spirit of the paper by Wieand (1976) (see Section 1.4). Some results in this direction for the ω^2-type statistics were given by Gregory (1980).

6.5 The LAO Domains for Symmetry Tests

The description of the LAO domains for symmetry tests, introduced in Section 4.1, is very similar to the description for goodness-of-fit and homogeneity tests. Let $X_1, X_2, \ldots, X_n$ be a sample with continuous d.f. F that under null hypothesis H_2 is symmetric with respect to zero, that is, satisfies (4.1.1). Under the alternative A_2 this sample has a d.f. $G(x;\theta)$, $\theta \geq 0$, that is symmetric only for $\theta = 0$. We assume $G(x;\theta)$ satisfies all regularity conditions introduced in Section 6.2, but we suppose that $\Delta = (\alpha, \beta) = \mathbf{R}^1$ and, instead of condition (6.2.1), we require asymptotics (4.4.5) to be fulfilled, where $0 < I_1(g) < \infty$. The class of such families of d.f.s is denoted by $\mathfrak{G}_2$. The LAO condition is introduced on the basis of inequality (4.4.4).

Using the expressions for the local indices of nonparametric symmetry tests found in Section 4.4, we see that the leading sets of the sequences of statistics (4.1.3)–(4.1.9) in the class $\mathfrak{G}_2$ or some suitable subclasses coincide with the leading sets of corresponding goodness-of-fit statistics described in Section 6.2. As to sign statistic (4.1.2), it is easy to see that the LAO condition in the subclass of $\mathfrak{G}_2$ satisfying the additional condition $G'_\theta(0,0) > 0$ is equivalent to the equality

$$4\,v^2(\tfrac{1}{2}) = \int\limits_0^1 v'^2(x)\,dx\,.$$

Arguing as in the proof of Theorem 6.2.1, we see that this equality may hold only if $v(t) = C \min(t, 1-t)$, $C \neq 0$. Other details are contained in Nikitin (1982), Abbakumov (1987), and Abbakumov and Nikitin (1993).

By the analysis of signed rank statistic (4.1.10) we will assume that the conditions of Theorem 4.2.3 are fulfilled and

$$\int\limits_0^\infty J\big(\,G(x;\theta) - G(-x;\theta)\big)\,dG(x;\theta) - \frac{1}{2}\int\limits_0^1 J(u)\,du > 0 \qquad \text{for } \theta > 0\,.$$

Then the exact slope $c_Z(\theta)$ admits, as has been observed in Section 4.4, the local representation

$$c_Z(\theta) \sim \left[\int\limits_0^\infty J\big(\,G(x;\theta) - G(-x;\theta)\big)\,dG(x;\theta) - \tfrac{1}{2}\int\limits_0^1 J(u)\,du\right]^2,$$

$$\theta \to 0\,.$$

To simplify further this expression we suppose that

$$\int\limits_0^\infty J\big(\,G(x;\theta) - G(-x;\theta)\big)\,dG(x;\theta) - \frac{1}{2}\int\limits_0^1 J(u)\,du$$

$$\sim \frac{1}{2}\int\limits_0^\infty J\big(2G(x;0) - 1\big)\,d\big[\,G'_\theta(x;0) + G'_\theta(-x;0)\big]\cdot\theta\,, \qquad (6.5.1)$$

as $\theta \to 0$. We denote by $\mathfrak{G}_2^J$ the subclass of families $\{G(x;\theta)\}$ from $\mathfrak{G}_2$ satisfying all mentioned conditions.

The approximation (6.5.1) is quite natural but imposes additional restrictions on admissible families $\{G(x;\theta)\}$, connecting them with properties of the score function J.

The LAO condition of the sequence $\{Z_n\}$ in $\mathfrak{G}_2^J$ takes the form

$$\left[\int_0^\infty J\left(2G(x;0) - 1 \right) d\left(G_\theta'(x;0) + G_\theta'(-x;0) \right) \right]^2 = 2\,I_1(g)\,. \quad (6.5.2)$$

Applying the Cauchy–Bunyakovskii–Schwarz inequality to the left-hand side of (6.5.2) we obtain, due to (4.2.9), that

$$\left[\int_0^\infty J\left(2G(x;0) - 1 \right) d\left(G_\theta'(x;0) + G_\theta'(-x;0) \right) \right]^2$$

$$\leq \int_0^\infty J^2\left(2G(x;0) - 1 \right) dG(x;0) \cdot I_1(g) = 2\,I_1(g)\,.$$

Therefore, equality (6.5.2) is valid if and only if

$$J\left(2G(x;0) - 1 \right) = C\,\frac{g_\theta'(x;0) - g_\theta'(-x;0)}{g(x;0)}\,, \qquad x \geq 0\,, \quad (6.5.3)$$

for some constant $C \neq 0$. This relation may be rewritten in terms of the function $v(t) = G_\theta'\left(G^{-1}(t;0);0 \right)$ as

$$J(2t - 1) = C\left(v'(t) - v'(1 - t) \right)\,, \qquad \tfrac{1}{2} \leq t \leq 1\,, \quad (6.5.4)$$

or, equivalently,

$$v(t) + v(1 - t) = C\int_{1/2}^t J(2u - 1)\,du\,, \qquad \tfrac{1}{2} \leq t \leq 1\,. \quad (6.5.5)$$

Thus we have that the leading set in $\mathfrak{G}_2^J$ for the sequence of statistics (4.1.10), with the score function satisfying the conditions of Theorem 4.2.3, is determined by relation (6.5.4) or (6.5.5).

As an example we consider the location alternative, that is, $G(x;\theta) = G(x + \theta)$. Denote by $\mathfrak{G}_2^l$ the class of d.f.s on $\mathbf{R}^1$ with even, differentiable, and positive densities satisfying condition (4.4.5), which in our case, as $\theta \to 0$, takes the form

$$\int_{-\infty}^\infty \ln\frac{2\,g(x + \theta)}{g(x + \theta) + g(-x + \theta)}\,g(x + \theta)\,dx \sim \frac{1}{2}\int_{-\infty}^\infty \frac{g'^2(x)}{g(x)}\,dx \cdot \theta^2\,.$$

Within the scope of suitable subclasses of $\mathfrak{G}_2^l$ we may obtain the description of those densities for which the Bahadur LAO condition is fulfilled. For example, this condition for sign test (4.1.2) characterizes the Laplace distribution and for the Hill–Rao test (4.1.9) it characterizes the Cauchy distribution.

For the Wilcoxon signed rank test, which corresponds to $J(t) = \sqrt{12}\, t$, equation (6.5.5) has the form

$$g\big(G^{-1}(x;0) \big) = C \int\limits_{1/2}^{x} (2u - 1)\, du\,, \qquad \tfrac{1}{2} \le x \le 1\,, \qquad C > 0\,,$$

or equivalently, by the symmetry of G,

$$G'(x) = C\, G(x) \big(1 - G(x) \big)\,, \qquad 0 \le x \le \infty\,, \qquad C > 0\,.$$

As we have seen in Section 6.3, the solution of this equation is the logistic d.f. (the solution is extended to the negative semi-axis by symmetry).

In the case of the normal scores or Fraser's tests we have $J(t) = \Phi^{-1}\big((1 + t)/2\big)$ and the equation

$$g(x) = C_1 \int\limits_{0}^{G(x)} \Phi^{-1}(u)\, du\,, \qquad x \ge 0\,, \qquad C_1 > 0\,,$$

which may be solved in the same manner as used in Section 6.3. The unique solution of this equation is the normal law.

As far as the Hodges–Lehmann, Pitman, and Chernoff efficiencies are concerned, the remarks and statements made for goodness-of-fit and homogeneity tests are still true.

6.6 The LAO Domains for Independence Tests

Let (X_1, Y_1), (X_2, Y_2), $\dots$, (X_n, Y_n) be a sample of size n with continuous d.f. $F(x, y)$ and marginal d.f.s $G(x)$ and $H(y)$. We are testing, as in Chapter 5, the independence hypothesis

$$H_3\colon\ F(x, y) = G(x)\, H(y) \qquad \text{for all } x \text{ and } y$$

against the alternative A_3' under which the d.f. of the inital sample has the form $F(x, y; \theta)$, $x, y \in \mathbf{R}^1$, $\theta \in \Theta$, where Θ is some interval $[0, \theta_0]$, $\theta_0 > 0$. Assume $F(x, y; \theta) = G(x)\, H(y)$ only for $\theta = 0$.

We introduce now a class of regular families of d.f.s $\{F(x, y; \theta)\}$ within the scope of which we describe the LAO conditions of independence tests.

Assume $F(x, y; \theta)$ are absolutely continuous with corresponding densities $f(x, y; \theta)$, which are differentiable in θ for all $\theta \in \Theta$. Denote for brevity

$$w(x, y) := F_\theta'\big(G^{-1}(x), H^{-1}(y); 0 \big). \tag{6.6.1}$$

The following regularity conditions enable one to select a special class $\mathfrak{F}$ of families of d.f.s $\{F(x, y; \theta)\}$:

For almost all x, y, and θ the function $f(x, y; \theta) > 0$;
for almost all x, y the function $w(x, y)$ from (6.6.1) satisfies the boundary condition $w\big|_{\partial I^2} = 0$ and possesses generalized mixed derivative w_{xy} for almost all x and y;

$$\mathcal{K}(\theta) \sim \tfrac{1}{2} I_2(w) \cdot \theta^2, \qquad \theta \to 0, \tag{6.6.2}$$

where

$$\mathcal{K}(\theta) := \int\limits_{-\infty}^{\infty} \int\limits_{-\infty}^{\infty} \ln \frac{f(x, y; \theta)}{f(x, y; 0)} \, f(x, y; \theta) \, dx \, dy,$$

$$I_2(w) := \int\limits_{0}^{1}\int\limits_{0}^{1} w_{xy}^2(x, y) \, dx \, dy, \qquad 0 < I_2(w) < \infty.$$

Condition (6.6.2), which seems to be restrictive, is the direct analog of conditions (6.2.1), (6.4.1), and (4.4.5) and is quite natural for this circle of problems.

In accordance with (5.4.8) we define the LAO condition for any sequence of statistics $\{T_n\}$ by way of the relation

$$c_T(\theta) \sim 2\mathcal{K}(\theta), \qquad \theta \to 0.$$

The leading set is introduced as in Section 6.1.

Theorem 6.6.1 (Nikitin and Pankrashova 1988) *The leading set in the class $\mathfrak{F}$ of the sequence of statistics $\{\Gamma_n\}$, given by (5.1.1), has the form*

$$\mathbf{V}(\Gamma) = \Big\{ C \min(x, 1 - x) \, \min(y, 1 - y) \Big\}, \qquad C \neq 0. \tag{6.6.3}$$

The difference between the leading sets of the sequences $\{\Gamma_n^+\}$ and $\{\Gamma_n^-\}$ and (6.6.3) is that the constant C must be positive or negative, respectively.

Proof It follows from the definition of the class $\mathfrak{F}$ and formula (5.4.3) that the LAO condition in $\mathfrak{F}$ of the sequence $\{\Gamma_n\}$ is equivalent to the equality

$$16 \sup_{0 \leq x, y \leq 1} w^2(x, y) = \int_0^1 \int_0^1 w_{xy}^2(x, y)\, dx\, dy \qquad (6.6.4)$$

under the boundary condition

$$w\big|_{\partial I^2} = 0. \qquad (6.6.5)$$

Consider the Hilbert space $\mathfrak{W}$ of real functions on the unit square with quadratically summable generalized mixed derivatives and satisfying boundary conditions (6.6.5). This space is supplied with the norm

$$\| w \|_{\mathfrak{W}} := \left\{ \int_0^1 \left(|w|^2 + |w_x|^2 + |w_y|^2 + |w_{xy}|^2 \right) dx\, dy \right\}^{1/2}.$$

Fix $x_0 \in [0, 1]$ and $y_0 \in [0, 1]$ and look for an upper bound (in w) of the value $w(x_0, y_0)$ in the space $\mathfrak{W}$, fixing the value of the integral $\int_0^1 \int_0^1 w_{xy}^2\, dx\, dy$. It is convenient to write $w(x_0, y_0)$ as follows

$$w(x_0, y_0) = \int_0^1 \int_0^1 w_{xy}(x, y)\, \mathbf{1}_{[0, x_0]}(x)\, \mathbf{1}_{[0, y_0]}(y)\, dx\, dy.$$

Using the Lagrange principle (Theorem 1.8.1) we conclude that the extremal $\widehat{w}$ must maximize the functional

$$\int_0^1 \int_0^1 \left[\widehat{w}(x, y)\, \mathbf{1}_{[0, x_0]}(x)\, \mathbf{1}_{[0, y_0]}(y) + \lambda\, \widehat{w}_{xy}^2(x, y) \right] dx\, dy, \qquad (6.6.6)$$

where λ is the indeterminate Lagrange multiplier.

Let δ be an absolutely continuous function on I^2 satisfying condition (6.6.5). The variation of functional (6.6.6) along δ has the form

$$\int_0^1 \int_0^1 \left[\mathbf{1}_{[0, x_0]}(x)\, \mathbf{1}_{[0, y_0]}(y) + \lambda\, \widehat{w}_{xy}(x, y) \right] \delta_{xy}(x, y)\, dx\, dy. \qquad (6.6.7)$$

Now choose as $\delta(x, y)$ the elements of the two-dimensional trigonometric system

$$\left\{ (1 - \cos 2\pi m x)\, \sin 2\pi n y \right\}_{m, n \geq 1}.$$

Elementary properties of multiple Fourier series (see, e.g., Stein and

Weiss (1971), Chapter 7) yield that if variation (6.6.7) is equal to 0, then

$$\widehat{w}_{xy}(x, y) = \mu \, \mathbf{1}_{[0,x_0]}(x) \, \mathbf{1}_{[0,y_0]}(y) + u(x) + v(y),$$

where u and v are some absolutely continuous functions.

Integrating this relation and denoting

$$U(x) := \int\limits_0^x u(t) \, dt, \qquad V(y) := \int\limits_0^y v(s) \, ds,$$

we get the equality

$$\widehat{w}(x, y) = \mu \min(x, x_0) \cdot \min(y, y_0) + U(x) \, y + V(y) \, x.$$

Taking into account the boundary conditions, we have

$$\widehat{w}(x, y) = \mu \left(\min(x, x_0) - x \, x_0 \right) \left(\min(y, y_0) - y \, y_0 \right). \qquad (6.6.8)$$

It is clear now that $\sup \left\{ \widehat{w}(x_0, y_0): 0 \le x_0, y_0 \le 1 \right\}$ is attained at $x_0 = y_0 = \frac{1}{2}$. Consequently equality (6.6.4) may be valid if and only if (6.6.8) holds with $x_0 = y_0 = \frac{1}{2}$. Clearly this is true not only for all of the space $\mathfrak{W}$, but also for that part of it where the function w from (6.6.1) is generated by the elements of class $\mathfrak{F}$. This implies the assertion of Theorem 6.6.1 for $\{\Gamma_n\}$. The changes when considering $\{\Gamma_n^+\}$ and $\{\Gamma_n^-\}$ are obvious. $\square$

Now we proceed to the investigation of the statistics M_n and $M_n^\pm$ introduced in Section 5.4.

Theorem 6.6.2 (Nikitin and Pankrashova 1988) *The leading set for the sequence $\{M_n\}$ in $\mathfrak{F}$ has the form*

$$\mathbf{V}(M) = \left\{ C \min(x, 1 - x) \, y \, (1 - y) \right\}, \qquad C \ne 0.$$

In the case of the sequences $\{M_n^+\}$ and $\{M_n^-\}$ only the cases $C > 0$ and $C < 0$ are admissible in corresponding classes of d.f.s families.

Proof The definition of the class $\mathfrak{F}$ and (5.4.5) imply that the LAO condition is equivalent to the equality

$$48 \left[\sup_{0 \le x \le 1} \left| \int\limits_0^1 w(x, y) \, dy \right| \right]^2 = \int\limits_0^1 \int\limits_0^1 w_{xy}^2(x, y) \, dx \, dy \qquad (6.6.9)$$

for w generated by the elements of $\mathfrak{F}$. We consider this equality on the entire space $\mathfrak{W}$. Arguing as in the proof of the previous theorem, let us look for an upper bound of the functional

$$\int\limits_0^1 w(x_0, y)\, dy = -\int\limits_0^1 \int\limits_0^1 y\, w_{xy}(x, y)\, \mathbf{1}_{[0, x_0]}(x)\, dx\, dy$$

for an arbitary $x_0 \in [0, 1]$ and a fixed value of the integral

$$\int\limits_0^1 \int\limits_0^1 w_{xy}^2(x, y)\, dx\, dy.$$

The Euler–Lagrange equation for the extremal $\widehat{w}$ has the form

$$\widehat{w}_{xy}(x, y) = \mu\, y\, \mathbf{1}_{[0, x_0]}(x) + u(x) + v(y), \qquad (6.6.10)$$

where u and v are some absolutely continuous functions. Integrating (6.6.10) and using the boundary conditions we finally obtain

$$\widehat{w}(x, y) = \mu\, y\, (1 - y)\, \big(\min(x, x_0) - x\, x_0\big), \qquad \mu \neq 0. \qquad (6.6.11)$$

Now it is clear that

$$\sup_{0 \leq x_0 \leq 1}\ \Big|\ \int\limits_0^1 \widehat{w}(x_0, y)\, dy\ \Big|$$

is attained at $x_0 = \frac{1}{2}$. Therefore equality (6.6.9) may hold only if w has the form (6.6.11) with $x_0 = \frac{1}{2}$.

Considering $\{M_n^+\}$ one should take into account that the condition

$$\sup_{0 \leq x \leq 1}\ \int\limits_0^1 w(x, y)\, dy > 0$$

has been used for calculating the local exact slope. $\square$

We proceed to the analysis of the LAO conditions for the integral statistics B_{n, q_1, q_2}^k. First of all we introduce the functional class

$$\mathfrak{F}_B(q_1, q_2; k) :=$$

$$\mathfrak{F} \cap \Big\{ F\colon \int\limits_{-\infty}^{\infty} \int\limits_{-\infty}^{\infty} F_\theta'^k(x, y; 0)\, q_1\big(G(x)\big)\, q_2\big(H(y)\big)\, dG(x)\, dH(y) > 0 \Big\}.$$

Assuming Theorems 5.3.2 and 5.3.1 to be true, using (5.4.1) and (6.6.1), we reduce the LAO condition to the equality

$$\lambda_{0k} \left[\int_0^1 \int_0^1 w^k(x,y)\, q_1(x)\, q_2(y)\, dx\, dy \right]^{2/k} = \int_0^1 \int_0^1 w_{xy}^2(x,y)\, dx\, dy$$

(6.6.12)

under the additional condition (6.6.5).

The same arguments as in the proof of Theorem 6.2.4 enable one to conclude that relation (6.6.12) in the space $\mathfrak{W}$ may hold only if $w = C\, x_{0k}$, where x_{0k} is the principal solution of (5.3.30) under the additional conditions (5.3.20)–(5.3.21). Moreover it is true in a narrower set of functions corresponding to class $\mathfrak{F}$. Thus we have proved the following result.

Theorem 6.6.3 (Nikitin 1986a) *The leading set of the sequence of statistics $\{B_{n,q_1,q_2}^k\}$ in the class $\mathfrak{F}_B(q_1,q_2;k)$ under the conditions of Theorem 6.2.4 has the form*

$$\mathbf{V}\big(B_{q_1,q_2}^k\big) = \big\{\, C\, x_{0k} \,\big\}, \qquad C \neq 0.$$

Corollary 1 *The leading set in the class $\mathfrak{F}$ of a sequence of Blum–Kiefer–Rosenblatt statistics consists of the functions*

$$\big\{\, C\, \sin \pi x \, \sin \pi y \,\big\}, \qquad C \neq 0. \tag{6.6.13}$$

Now consider the sequence of the first components of the statistic $B_{n,1,1}^2$ in the Durbin–Knott sense considered by Koziol and Nemec (1979). This sequence coincides with the sequence $\{B_{n,q_1,q_2}^1\}$ for a special choice of the weights $q_1(x) = q_2(x) = \sin \pi x$.

Corollary 2 *The leading set in the corresponding subclass of $\mathfrak{F}$ of a sequence of the Koziol–Nemec statistics consists of functions (6.6.13) for $C > 0$.*

Thus, we observe again the effect of a partial coincidence of leading sets and, consequently, of the LAO domains for the ω^2-type statistic and its first component, already mentioned in Section 6.2.

If we return to the model of alternative introduced by (5.4.9), it is possible to describe a given a priori class of d.f.s $\mathfrak{F}(\Omega)$, corresponding to the class $\mathfrak{F}$, in terms of regularity conditions imposed directly on Ω. It is clear that in all cases considered earlier the LAO condition implies that the function Ω should belong to the corresponding leading set.

In particular, the sequence of statistics $\{B^1_{n,1,1}\}$ turns out to be LAO in model (5.4.9) for $\Omega(x,y) = C\,x\,(1-x)\,y\,(1-y)$, $C > 0$, and the sequence $\{B^2_{n,1,1}\}$ is LAO for $\Omega(x,y) = C\,\sin\pi x\,\sin\pi y$, $C \neq 0$. The sequence of Kolmogorov-type statistics $\{\Gamma_n\}$ possesses this property for $\Omega(x,y) = C\,\min(x, 1-x)\cdot\min(y, 1-y)$, $C \neq 0$, and so forth.

In conclusion we shall describe the conditions ensuring the LAO property for linear rank statistics (5.1.5), assuming that the conditions of Theorems 1.6.11–1.6.12 and, in particular, equality (5.4.6) are fulfilled. We require, additionally, that $F \in \mathfrak{F}$ is such that for $\theta > 0$

$$b_T(\theta) = \int_{-\infty}^{\infty} \int_{-\infty}^{\infty} J_1\big(F(x,\infty;\theta) \big)\, J_2\big(F(\infty,y;\theta) \big)\, dF(x,y;\theta) > 0$$

and, as $\theta \to 0$,

$$b_T(\theta) \sim \int_{-\infty}^{\infty} \int_{-\infty}^{\infty} J_1\big(G(x) \big)\, J_2\big(H(y) \big)\, dF'_\theta(x,y;0)\cdot\theta.$$

The last relation may be rewritten, due to (6.6.1), in the form

$$b_T(\theta) \sim \int_0^1 \int_0^1 J_1(u)\, J_2(v)\, w_{uv}(u,v)\, du\, dv \cdot \theta. \tag{6.6.14}$$

Let us denote by $\mathfrak{F}^{J_1,\,J_2}$ the class of d.f.s satisfying all mentioned conditions. The LAO condition in this class takes the form

$$\left(\int_0^1 \int_0^1 J_1(u)\, J_2(v)\, w_{uv}(u,v)\, du\, dv \right)^2 = \int_0^1 \int_0^1 w^2_{uv}(u,v)\, du\, dv.$$

The Cauchy–Bunyakovskii–Schwarz inequality implies the following equality

$$w(u,v) = C \int_0^u J_1(t)\, dt \cdot \int_0^v J_1(s)\, ds, \qquad C > 0, \tag{6.6.15}$$

which is necessary and sufficient for the LAO property. We may formulate the following result, which seems to be published for the first time here.

Theorem 6.6.4 *The sequence of linear rank statistics (5.1.5) for independence testing under the conditions of Theorems 1.6.11–1.6.12 has in the class $\mathfrak{F}^{J_1, J_2}$ the leading set that consists of functions (6.6.15).*

Specifically, for the Spearman rank correlation coefficient (5.1.6) we have $J_r(u) = \sqrt{12}\,(u - \frac{1}{2})$, $r = 1$, 2, so the leading set is the same as for the sequence $\{B^1_{n,1,1}\}$ and consists of the functions $C\,x\,(1-x)\,y(1-y)$, $C > 0$. In the class of families (5.5.9) the LAO domain is specified by the function

$$\Omega(x, y) = C\,x\,(1 - x)\,y\,(1 - y), \qquad C > 0.$$

Bibliography

Abbakumov, V. L. (1986). Large deviations and asymptotic efficiency of the Watson type test for testing symmetry (in Russian). Vestnik Leningrad. Un-ta: Matematika 4: 98–100.

Abbakumov, V. L. (1987). Asymptotic Efficiency of Nonparametric Symmetry Tests (in Russian). Ph.D. Thesis, Leningrad University, Leningrad.

Abbakumov, V. L., and Ya. Yu. Nikitin (1993). Bahadur efficiency and local optimality of a new nonparametric test of symmetry. In *Statistics and Probability, R. R. Bahadur Festschrift*, eds. J. K. Ghosh, S. K. Mitra, K. R. Parthasarathy, and B. S. L. Prakasa Rao, pp. 1–12. New Delhi: Wiley Eastern.

Abrahamson, I .G. (1967). Exact Bahadur efficiencies for the Kolmogorov–Smirnov and Kuiper one- and two-sample statistics. *Ann. Mathem. Statist.* 38: 1475–90.

Aki, S. (1986). Some test statistics based on the martingale term of the empirical distribution function. *Ann. Inst. Statist. Mathem.*, 38: 1–21.

Aki, S., and N. Kashiwagi (1989). Asymptotic properties of some goodness-of-fit tests based on the L^1-norm. *Ann. Inst. Statist. Mathem.* 41: 753–64.

Akimov, P. S., V. S. Efremov, and A. N. Kubasov (1978). On the stability of nonparametric tests under noncoherent treatment (in Russian). *Radiotekhnika i Elektronika* 23: 1164–73.

Albers, W. (1974). *Asymptotic Expansions and the Deficiency Concept in Statistics.* Mathem. Center Tracts, no. 58. Amsterdam: Mathematisch Centrum.

Alexeev, V. M., V. M. Tikhomirov, and S. V. Fomin (1979). *Optimal Control* (in Russian). Moscow: Nauka.

Anderson, T. W., and D. A. Darling (1952). Asymptotic theory of certain "goodness-of-fit" criteria based on stochastic processes. *Ann. Mathem. Statist.* 23: 193–212.

Anderson, T. W., and D. A. Darling (1954). A test of goodness-of-fit. *J. Amer. Statist. Assoc.* 49: 765–9.

Arbuthnott, J. (1710). An argument for divine providence taken from the constant regularity observed in the birth of both sexes. *Philos. Trans.* 27: 186–90.

249

Archambault, W. A. T., and P. W. Mikulski (1979). On bounds for the asymptotic power and on Pitman efficiencies of the Cramér–von Mises test. *Zastosov. Mathem.* 16: 429–44.

Artamonov, A. F., and I. F. Shishkin (1972). Contrast reception by nonlinear receiver (in Russian). *Radiotekhnika* 26: 94–7.

Assatryan, D. G., and I. A. Safaryan (1979). On distribution functions for which the rank test is locally most powerful (in Russian). *Doklady Armyan. Akadem. Nauk* 69: 193–7.

Atkinson, F. V. (1964). *Discrete and Continuous Boundary Problems.* New York: Academic Press.

Aubin, J. P., and I. Ekeland (1984). *Applied Nonlinear Analysis.* New York: Wiley.

Azencott, R. (1980). *Grandes déviations et applications.* Lect. Notes in Mathem. 774, pp. 1–176. New York: Springer-Verlag.

Bahadur, R. R. (1960a). On the asymptotic efficiency of tests and estimates. *Sankhyā* 22: 229–52.

Bahadur, R. R. (1960b). Stochastic comparison of tests. *Ann. Mathem. Statist.* 31: 276–95.

Bahadur, R. R. (1965). An optimal property of the likelihood ratio statistic. In *Proc. 5th Berkeley Sympos. on Probab. Theory and Mathem. Statist.*, vol. 1, eds. L. Le Cam and J. Neyman, pp. 13–26. Berkeley and Los Angeles: Univ. of California Press.

Bahadur, R. R. (1967). Rates of convergence of estimates and test statistics. *Ann. Mathem. Statist.* 38: 303–24.

Bahadur, R. R. (1971). *Some Limit Theorems in Statistics.* Philadelphia: SIAM.

Bahadur, R. R., and J. C. Gupta (1985). Distribution optimality and second-order efficiency of test procedures. In *Adaptive Statistical Procedures and Related Topics*, ed. J. van Ryzin, pp. 315–31. New York: Upton.

Bahadur, R. R., and M. Raghavachari (1972). Some asymptotic properties of likelihood ratios on general sample spaces. In *Proc. 6th Berkeley Sympos. on Probab. Theory and Mathem. Statist.*, vol. 1, eds. L. Le Cam, J. Neyman, and E. L. Scott, pp. 129–52. Berkeley and Los Angeles: Univ. of California Press.

Bahadur, R. R., and R. Ranga Rao (1960). On deviations of the sample mean. *Ann. Mathem. Statist.* 31: 1015–27.

Bahadur, R. R., and S. L. Zabell (1979). Large deviations of the sample mean in general vector spaces. *Ann. Probab.* 7: 587–621.

Bahadur, R. R., T. K. Chandra, and D. Lambert (1982). Some further properties of likelihood ratios on general sample spaces. In *Proc. Indian Statist. Instit., Golden Jubilee Intern. Confer. on Statistics: Applications and New Directions*, pp. 1–19. Calcutta: Indian Statist. Institute.

Bahadur, R. R., J. C. Gupta, and S. L. Zabell (1980). Large deviations, tests and estimates. In *Asymptotic Theory of Statistical Tests and Estimates*, ed. I. M. Chakravarti, pp. 33–65. New York: Academic Press.

Bajorski, P. (1987). Local Bahadur optimality of some rank tests of independence. *Statistics and Probab. Letters* 5: 255–62.

Baringhaus, L. (1987). Asymptotic optimality of multivariate linear hypothesis test. *J. Multivar. Anal.* 23: pp. 303–11.

Behnen, K. (1971). Asymptotic optimality and ARE of certain rank-order tests under contiguity. *Ann. Mathem. Statist.* 42: 325–9.

Behnen, K. (1972). A characterization of certain rank-order tests with bounds for the asymptotic relative efficiency. *Ann. Mathem. Statist.* 43: 1839–51.

Behnen, K., and G. Neuhaus (1989). *Rank Tests with Estimated Scores and Their Application.* Stuttgart: Teubner.

Benson, Y. E., and Ya. Yu. Nikitin (1992). On conditions of Bahadur local optimality of weighted Kolmogorov–Smirnov statistics (in Russian). *Zapiski Nauch. Semin. POMI* 194: 21–7.

Berk, R. H. (1976). Asymptotic efficiencies of sequential tests. *Ann. Statist.* 4: 891–911.

Berk, R. H. (1984). Stochastic bounds for attained levels. *J. Multivar. Anal.* 14: 376–89.

Berk, R. H., and L. D. Brown (1978). Sequential Bahadur efficiency. *Ann. Statist.* 6: 567–81.

Berk, R. H., and A. Cohen (1979). Asymptotically optimal methods of combining tests. *J. Amer. Statist. Assoc.* 74: 812–14.

Berk, R. H., and D. H. Jones (1978). Relatively optimal combinations of test statistics. *Scand. J. Statist.* 5: 158–62.

Billingsley, P. (1968). *Convergence of Probability Measures.* New York: Wiley.

Blum, J. R., J. Kiefer, and M. Rosenblatt (1961). Distribution-free tests of independence based on the sample distribution function. *Ann. Mathem. Statist.* 32: 485–98.

Bolthausen, E. (1984). On the probability of large deviations in Banach spaces. *Ann. Probab.* 12: 427–35.

Bönner, N., and H. P. Kirschner (1977). Notes on conditions of weak convergence of von Mises differentiable functions. *Ann. Statist.* 5: 405–7.

Borell, C. (1975). The Brunn–Minkowski inequality in Gauss space. *Invent. Mathem.* 30: 207–16.

Borovkov, A. A. (1967). Boundary-value problems for random walks and large deviations in functional spaces. *Probab. Theory Appl.* 12: 575–94.

Borovkov, A. A. (1986). *Probability Theory* (in Russian), 2d ed. Moscow: Nauka.

Borovkov, A. A. (1984). *Mathematical Statistics*, (in Russian). Moscow: Nauka.

Borovkov, A. A., and A. A. Mogulskii (1978). Probabilities of large deviations in topological spaces, Part I (in Russian). *Siberian Mathem. J.* 19: 988–1004.

Borovkov, A. A., and A. A. Mogulskii (1980). Probabilities of large deviations in topological spaces. Part II (in Russian). *Siberian Mathem. J.* 21: 12–26.

Borovkov, A. A., and A. A. Mogulskii (1992). *Large Deviations and Testing Statistical Hypotheses* (in Russian). Novosibirsk: Nauka.

Borovkov, A. A., and N. M. Sycheva (1970). On asymptotically optimal nonparametric criteria. In *Nonparametric Techniques in Statistical Inference*, ed. M. L. Puri, pp. 259–66. Cambridge: Cambridge Univer. Press.

Bretagnolle, J. (1979). Formule de Chernoff pour les lois empiriques de variables à valeurs dans les espaces généraux. *Astérisque* 68: 33–52.

Brown, L. D. (1971). Nonlocal asymptotic optimality of appropriate likelihood ratio tests. *Ann. Mathem. Statist.* 42: 1206–40.

Brown, L. D., F. H. Ruymgaart, and D. K. Truax (1984). Hodges–Lehmann efficacies for likelihood ratio tests in curved bivariate normal families. *Statist. Neerlandica* 38: 21–36.

Buslaev, V. S. (1980). *Calculus of Variations* (in Russian). Leningrad: Leningrad Univer. Press.

Caperaà, P. (1988). Tail ordering and asymptotic efficiency of rank tests. *Ann. Statist.* 16: 470–8.

Chandra, T. K. (1989). A stochastic representation of the logarithm of P-values and related results. *Sankhyā, Ser. A* 51: 205–11.

Chandra, T. K., and J. K. Ghosh (1978). Comparison of tests with same Bahadur efficiency. *Sankhyā, Ser. A* 40: 253–77.

Chapman, D. (1958). A comparative study of several one-sided goodness-of-fit tests. *Ann. Mathem. Statist.* 29: 655–74.

Chatterjee, S. K., and P. K. Sen (1973). On Kolmogorov–Smirnov's type tests for symmetry. *Ann. Inst. Statist. Mathem.* 25: 287–300.

Chentsov, N. N. (1958). Applying the Methods of the Theory of Random Processes to Statistical Criteria (in Russian). Ph.D. Thesis, Moscow.

Chernoff, H. (1952). A measure of asymptotic efficiency for tests of a hypothesis based on sums of observations. *Ann. Mathem. Statist.* 23: 493–507.

Chernoff, H., and I. R. Savage. (1958) Asymptotic normality and efficiency of certain nonparametric tests. *Ann. Mathem. Statist.* 29: 972–94.

Chibisov, D. M. (1961). On the tests of fit based on sample spacings. *Probab. Theory Appl.* 6: 325–9.

Chibisov, D. M. (1983). Asymptotic expansions and deficiencies of tests. In *Proc. Intern. Mathem. Congress, Warszawa*, eds. Z. Ciesielski and C. Olech, pp. 1063–79. Warszawa: PWN; Amsterdam: North-Holland.

Cochran, W. G. (1952). The χ^2 goodness-of-fit test. *Ann. Mathem. Statist.* 23: 315–45.

Cohen, A., J. I. Marden, and K. Singh (1982). Second order asymptotic and non-asymptotic optimality properties of combined tests. *J. Statist. Planning and Inference* 6: 253–76.

Cotterill, D., and M. Csörgö (1982). On the limiting distribution of and critical values for the multivariate Cramér–von Mises statistics. *Ann. Statist.* 10: 233–44.

Cotterill, D., and M. Csörgö (1985). On the limiting distribution of and critical values for the Hoeffding, Blum, Kiefer, Rosenblatt independence criterion. *Statistics & Decisions* 3: 1–48.

Cramér, H. (1938). Sur un nouveau théorème limite de la théorie des probabilités. N 736, Actual. Sci. et Industr., Paris: Hermann.

Csörgö, S. (1979a). Érdös–Rényi laws. *Ann. Statist.* 7: 772–87.

Csörgö, S. (1979b). Bahadur efficiency and Érdös–Rényi maxima. *Sankhyā, Ser. A* 41: 141–4.

Dabrowska, D., K. Doksum, and R. Miura (1989). Rank estimates in a class of semiparametric two-sample models. *Ann. Inst. Statist. Mathem.* 41: 63–79.

Dacunha-Castelle, D. (1979). Les fondements probabilistes de la théorie des grandes déviations. *Astérisque* 68: 3–18.

Darling, D. (1957). The Kolmogorov–Smirnov, Cramér–von Mises tests. *Ann. Mathem. Statist.* 28: 823–38.

Darling, D. (1983a). On the asymptotic distribution of Watson's statistic. *Ann. Statist.* 11: 1263–6.

Darling, D. (1983b). On the supremum of a certain Gaussian process. *Ann. Probab.* 11: 803–6.

Dasgupta, R. (1984). On large deviation probabilities of *U*-statistics in non-i.i.d. case. *Sankhyā, Ser. A* 46: 110–16.

Davenport, W. B., and W. L. Root (1958). *An Introduction to the Theory of Random Signals and Noises.* New York: McGraw Hill.

De Bruijn, N. G. (1961). *Asymptotic Methods in Analysis*, 2d ed. Amsterdam: North-Holland.

Deheuvels, P. (1980). *Nonparametric tests of independence.* Lect. Notes in Mathem., vol. 821, pp. 95–108. New York: Springer.

Deheuvels, P. (1981). A Kolmogorov–Smirnov type test for independence and multivariate samples. *Revue Roum. Mathem. Pures et Appl.* 26: 213–26.

Deheuvels, P. (1982). Some applications of the dependence functions to statistical inference: nonparametric estimates of extreme value distributions and a Kiefer type universal bound for the uniform test of independence. In *Nonparametric Statistical Inference*, eds. B. V. Gnedenko, M. L. Puri, and I. Vincze, Volume 1, pp. 183–201. Amsterdam: North-Holland.

Denker, M. (1985). *Asymptotic Distribution Theory in Nonparametric Statistics.* Braunschweig: Vieweg.

Deuschel, J. D., and D. W. Stroock (1989). *Large Deviations.* New York: Academic Press.

Dmitrieva (Pankrashova), A. G. (1988). Asymptotic behavior of Durbin's type statistic for testing independence (in Russian). *Vestnik Leningrad. Un-ta: Matematika* 1: 102–4.

Donsker, M. D., and S. R. S. Varadhan (1975a). Asymptotic evaluation of certain Markov process expectations for large time, Part I. *Commun. Pure and Appl. Mathem.* 28: 1–47.

Donsker, M. D., and S. R. S. Varadhan (1975b). Asymptotic evaluation of certain Markov process expectations for large time, Part II. *Commun. Pure and Appl. Mathem.* 28: 279–301.

Donsker, M. D., and S. R. S. Varadhan (1976). Asymptotic evaluation of certain Markov process expectations for large time, Part III. *Commun. Pure and Appl. Mathem.* 29: 389–461.

Doob, J. L. (1949). Heuristic approach to the Kolmogorov–Smirnov theorems. *Ann. Mathem. Statist.* 20: 393–403.

Durbin, J. (1970). Asymptotic distributions of some statistics based on the bivariate sample distribution functions. In *Nonparametric Techniques in Statistical Inference*, ed. M. L. Puri, pp. 435–51. Cambridge: Cambridge Univer. Press.

Durbin, J., and M. Knott (1972). Components of Cramér–von Mises statistics, Part I. *J. Roy. Statist. Soc., Ser. B* 34: 290–307.

Durbin, J., and M. Knott (1975). Components of Cramér–von Mises statistics, Part II. *J. Roy. Statist. Soc., Ser. B* 37: 216–37.

Ellis, R. S. (1984). *Entropy, Large Deviations and Statistical Mechanics.* Berlin: Springer.

Ermakov, M. S. (1990). Asymptotic minimaxity of usual goodness-of-fit tests.

In *Proc. 5th Vilnius Conf. on Probab. Theory and Mathem. Statist.*, vol. 1, eds. Yu. V. Prohorov et al., pp. 323–31. Utrecht: VNU Sci Press.

Eubank, R. L., V. N. La Riccia, and R. B. Rosenstein (1987). Test statistics derived as components of Pearson's phi-squared distance measure. *J. Amer. Statist. Soc.* 8:, pp. 816–25.

Farlie, D. (1960). The performance of some correlation coefficients for a general bivariate distribution. *Biometrika* 47: 307–23.

Feller, W. (1971). *An Introduction to Probability Theory and Its Applications*, 2d ed. Volume 2, New York: Wiley.

Fernique, X. (1971). Regularité de processus Gaussiens. *Invent. Mathem.* 12: 304–20.

Freidlin, M. I., and A. D. Wentzell (1979). *Fluctuations in Dynamic Systems under the Effect of Small Random Perturbations* (in Russian) Moscow: Nauka.

Fu, J. C. (1985). On probabilities of large deviations for empirical distributions. *Statistics & Decisions* 3: 115–42.

Galambos, J., and S. Kotz (1978). *Characterizations of Probability Distributions: A Unified Approach with Emphasis on Exponential and Related Models*. Lect. Notes in Mathem., vol. 675. New York: Springer.

Gibbons, J. D., and J. W. Pratt (1975). *P*-values: interpretation and methodology. *Amer. Statist.* 29: 20–4.

Gnedenko, B. V. (1954). Testing homogeneity of distributions in two independent samples (in Russian). *Mathem. Nachrichten* 12: 29–66.

Goria, M. N. (1980). Some locally most powerful generalized rank tests. *Biometrika* 67: 497–500.

Govindarajulu, Z., L. Le Cam, and M. Raghavachari (1967). Generalizations of theorems of Chernoff and Savage on asymptotic normality of nonparametric test statistics. In *Proc. 5th Berkeley Symp. on Probab. Theory and Mathem. Statist.*, vol. 1, eds. L. Le Cam and J. Neyman, pp. 609–38. Berkeley and Los Angeles: Univ. of California Press.

Gradshteyn, I. S., and I. M. Ryzhik (1971). *Tables of Integrals, Sums, Series and Products* (in Russian), 5th ed. Moscow: Nauka.

Gregory, G. (1977). Cramér–von Mises type tests for symmetry. *South Africa Statist. J.* 11: 49–61.

Gregory, G. (1980). On efficiency and optimality of quadratic tests. *Ann. Statist.* 8: 116–31.

Groeneboom, P. (1980). *Large Deviations and Asymptotic Efficiencies.* Math. Center Tracts, no. 118. Amsterdam: Mathematisch Centrum.

Groeneboom, P., and J. Oosterhoff (1977). Bahadur efficiency and probabilities of large deviations. *Statist. Neerlandica* 31: 1–24.

Groeneboom, P., and J. Oosterhoff (1981). Bahadur efficiency and small sample efficiency. *Intern. Statist. Review* 49: 127–41.

Groeneboom, P., and G. Shorack (1981). Large deviations of goodness-of-fit statistics and linear combinations of order statistics. *Ann. Probab.* 9: 971–87.

Groeneboom, P., Y. Lepage, and F. H. Ruymgaart (1976). Rank tests for independence with best strong exact Bahadur slope. *Z. Warsch. Verw. Geb.* 36: 119–27.

Groeneboom, P., J. Oosterhoff, and F. H. Ruymgaart (1979). Large deviation theorems for empirical probability measures. *Ann. Probab.* 7: 553–86.

Hájek, J. (1968). Asymptotic normality of simple linear rank statistics under alternatives. *Ann. Mathem. Statist.* 39: 325–46.

Hájek, J. (1974). Asymptotic sufficiency of the vector of ranks in the Bahadur sense. *Ann. Statist.* 2: 75–83.

Hájek, J., and Z. Šidak (1967). *Theory of Rank Tests.* Prague: Academia.

Hardy, G. H., J. E. Littlewood, and G. Polya (1934). *Inequalities.* Cambridge: Cambridge Univer. Press.

Hettmansperger, T. P. (1973). On the Hodges–Lehmann approximate efficiency. *Ann. Inst. Statist. Mathem.* 25: 279–86.

Hettmansperger, T. P. (1984). *Statistical Inference Based on Ranks.* New York: Wiley.

Hill, D. L., and P. V. Rao (1977). Tests of symmetry based on Cramér–von Mises statistics. *Biometrika* 64: 484–94.

Ho, N. V. (1974). Asymptotic efficiency in the Bahadur sense for the signed rank tests. In *Proc. Prague Symp. on Asympt. Statistics*, vol. 2, pp. 127–56. Prague: Charles Univer. Press.

Hoadley, A. B. (1967). On the probability of large deviations of functions of several empirical c.d.f.'s. *Ann. Mathem. Statist.* 38: 360–81.

Hodges, J., and E. L. Lehmann (1956). The efficiency of some nonparametric competitors of the *t*-test. *Ann. Mathem. Statist.* 26: 324–35.

Hodges, J., and E. L. Lehmann (1961). Comparison of the normal scores and Wilcoxon tests. In *Proc. 4th Berkeley Symp. on Probab. Theory and Mathem. Statist.*, vol. 1, ed. J. Neyman, pp. 307–17. Berkeley and Los Angeles: Univ. of California Press.

Hodges, J., and E. L. Lehmann (1970). Deficiency. *Ann. Mathem. Statist.* 41: 783–801.

Hoeffding, W. (1948). A nonparametric test of independence. *Ann. Mathem. Statist.* 19: 546–57.

Hoeffding, W. (1965). On probabilities of large deviations. In *Proc. 5th Berkeley Symp. on Probab. Theory and Mathem. Statist.*, vol. 1, ed. L. Le Cam and J. Neyman, pp. 203–19. Berkeley and Los Angeles: Univ. of California Press.

Hollander, M., and D. Wolfe (1973). *Nonparametric Statistical Methods.* New York: Wiley.

Hwang, T.-Y. (1976). On the probability of a large deviation for type A Chernoff–Savage statistics. *Tamkang J. Mathem.* 7: 233–46.

Hwang, T.-Y. (1978). The Chernoff efficiency of the Wilcoxon rank test. *Commun. Statist. – Theory and Meth., Ser. A* 7: 543–55.

Ibragimov, I. A. (1956). On the composition of unimodal distributions. *Probab. Theory Appl.* 1: 255–60.

Ibragimov, I. A., and R. Z. Khasminskii (1981). *Statistical Estimation: Asymptotic Theory.* New York: Springer.

Inglot, T., and T. Ledwina (1990). On probabilities of excessive deviations for Kolmogorov–Smirnov, Cramér–von Mises and chi-square statistics. *Ann. Statist.* 18: 1491–5.

Inglot, T., and T. Ledwina (1993). Moderately large deviations and expansions of large deviations for some functionals of weighted empirical process. *Ann. Probab.* 21: 1691–705.

Inglot, T. , W. C. M. Kallenberg, and T. Ledwina (1992). Strong Moderate Deviation Theorems. *Ann. Probab.* 20: 987–1003.

Ioffe, A. D., and V. M. Tikhomirov (1979). *Theory of Extremal Problems.* Amsterdam: North-Holland.

Jardine, N., and R. Sibson (1971). *Mathematical Taxonomy.* New York: Wiley.

Jeurnink, G. A. M., and W. C. M. Kallenberg (1990). Limiting values of large deviation probabilities of quadratic statistics. *J. Multivar. Anal.* 35: 168–85.

Jones, D. H. (1978). Limiting Pitman efficient combinations of test statistics. *Scand. J. Statist.* 5: 209–12.

Kagan, A. M., Yu. V. Linnik, and C. R. Rao (1973). *Characterization Problems in Mathematical Statistics.* New York: Wiley.

Kallenberg, W. C. M. (1981). Bahadur deficiency of likelihood ratio tests in exponential families. *J. Multivar. Anal.* 11: 506–31.

Kallenberg, W. C. M. (1982). Chernoff efficiency and deficiency. *Ann. Statist.* 10: 583–94.

Kallenberg, W. C. M. (1983a). Intermediate efficiency, theory and examples. *Ann. Statist.* 11: 170–82.

Kallenberg, W. C. M. (1983b). Asymptotic efficiency and deficiency of tests. In *Proc. 44th Session of the Intern. Statist. Instit., Madrid,* Vol. L, pp. 1173–89.

Kallenberg, W. C. M., and S. Kourouklis (1992). Hodges–Lehmann optimality of tests. *Statistics and Probab. Letters* 14: 31–8.

Kallenberg, W. C. M., and T. Ledwina (1987). On local and nonlocal measures of efficiency. *Ann. Statist.* 15: 1401–20.

Kallianpur, J., and H. Oodaira (1978). Freidlin–Wentzell type estimates for abstract Wiener spaces. *Sankhyā, Ser. A* 40: 116–37.

Kamke, E. (1959). *Differentialgleichungen: Lösungsmethoden und Lösungen,* 4th ed. Leipzig: Akadem. Verlag.

Kendall, M. G. (1970). *Rank Correlation Methods.* London: Griffin.

Kendall, M. G., and A. Stuart (1967). *The Advanced Theory of Statistics,* vol. 2, 2d ed. London: Griffin.

Khmaladze, E. V. (1981). Martingale approach in the theory of goodness-of-fit tests. *Probab. Theory Appl.* 26: 240–59.

Kiefer, J. (1959). *K*-sample analogues of the Kolmogorov–Smirnov and Cramér–von Mises tests. *Ann. Mathem. Statist.* 30: 420–47.

Klotz, J. (1965) Alternative efficiencies for signed rank tests. *Ann. Mathem. Statist.* 36: 1759–66.

Kolmogorov, A. N. (1933). Sulla determinazione empirica di una legge di distribuzione. *Giorn. dell'Instit. Ital. degli Att.* 4: 1–11.

Komlós, J., P. Major, and G. Tusnády (1975). An approximation of partial sums of independent r.v.'s and sample d.f., Part I. *Z. Warsch. Verv. Geb.* 32: 111–33.

Komlós, J., P. Major, and G. Tusnády (1976). An approximation of partial sums of independent r.v.'s and sample d.f., Part II. *Z. Warsch. Verv. Geb.* 34: 33–58.

Konijn, H. S. (1959). Positive and negative dependence of two random variables. *Sankhyā* 21: 269–80.

Kopylev, L., and Ya. Yu. Nikitin (1992). On Chernoff efficiency of some nonparametric tests. In *Statistique des processus en milieu médical. Séminaire 92 Bru Huber Prum, Paris V,* pp. 225–36. Paris: University of Paris.

Korolyuk, V. S., and Yu. V. Borovskikh (1984). *Asymptotic Analysis of Distributions of Statistics* (in Russian). Kiev: Naukova Dumka.

Korolyuk, V. S., and Yu. V. Borovskikh (1993). *Theory of U-Statistics*. Dordrecht: Kluwer.

Kourouklis, S. (1984). Bahadur optimality of sequential experiments for exponential families. *Ann. Statist.* 12: 1522–7.

Kourouklis, S. (1988). Hodges–Lehmann efficacies of certain tests in multivariate analysis and regression analysis. *Canad. J. Statist.* 16: 87–95.

Kourouklis, S. (1989). On the relation between Hodges–Lehmann efficiency and Pitman efficiency. *Canad. J. Statist.* 17: 311–8.

Kourouklis, S. (1990). A relation between the Chernoff index and the Pitman efficacy. *Statistics and Probab. Letters* 9: 391–395.

Koziol, J. A. (1980). On a Cramér–von Mises type statistic for testing symmetry. *J. Amer. Statist. Assoc.* 75: 161–7.

Koziol, J. A. (1989). A modification of Watson's statistic for goodness-of-fit. *Commun. Statist. – Theory and Meth.* 18: 3739–47.

Koziol, J. A., and A. F. Nemec (1979). On a Cramér–von Mises type statistic for testing bivariate independence. *Canad. J. Statist.* 7: 43–52.

Krasnoselskii, M. A. (1964). *Positive Solutions of Operator Equations.* Groningen: Nordhoff.

Krasnoselskii, M. A., and P. P. Zabreiko (1975). *Geometrical Methods of Nonlinear Analysis* (in Russian). Moscow: Nauka.

Kremer, E. (1979a). Approximate and local Bahadur efficiency of linear rank tests in the two-sample problem. *Ann. Statist.* 7: 1246–55.

Kremer, E. (1979b). Lokale Bahadur–Effizienz Linearer Rangtests (in German). Ph.D. Thesis, Hamburg.

Kremer, E. (1981). Local Bahadur efficiency of rank tests for the independence problem. *J. Multivar. Anal.* 11: 532–43.

Kremer, E. (1982). Local comparison of linear rank tests in the Bahadur sense. *Metrika* 29: 159–73.

Kremer, E. (1983). Bahadur-Efficiency of linear rank tests – a survey. *Acta Univer. Carol. Mathem. et Phys.* 24: 61–76.

Krishnaiah, P. R., and P. K. Sen, eds. (1984). *Handbook of Statistics.* Vol. 4, *Nonparametric Methods.* Amsterdam: North-Holland.

Krivyakova, E. N., G. N. Martynov, and Yu. N. Tyurin (1977). On the distribution of the ω^2-statistic in the multidimensional case. *Probab. Theory Appl.* 22: 406–11.

Kuiper, N. (1960). Tests concerning random points on a circle. *Proc. Konink. Ned. Acad. van Wettenschaften, Ser. A* 63: 38–47.

Kullback, S. (1959). *Information Theory and Statistics.* New York: Wiley.

Lai, T. L. (1975). On Chernoff–Savage statistics and sequential rank tests. *Ann. Statist.* 3: 825–45.

Lambert, D., and W. Hall (1982). Asymptotic lognormality of P-values. *Ann. Statist.* 10: 44–64.

Ledwina, T. (1984). O efektywnosci Bahadura pewnych testów niezaleznosci (in Polish). *Matemat. Stosowana* 25: 5–26.

Ledwina, T. (1986a). Large deviations and Bahadur slopes of some rank tests of independence. *Sankhyā, Ser. A* 48: 188–207.

Ledwina, T. (1986b). On the limiting Pitman efficiency of some rank tests of independence. *J. Multivar. Anal.* 20: 265–71.

Ledwina, T. (1987). An expansion of the index of linear rank tests. *Statistics and Probab. Letters* 5: 247–8.

Lehmann, E. L. (1951). Consistency and unbiasedness of certain nonparametric tests. *Ann. Mathem. Statist.* 22: 165–79.

Lehmann, E. L. (1953). The power of rank tests. *Ann. Mathem. Statist.* 24: 23–43.

Lehmann, E. L. (1959). *Testing Statistical Hypotheses*. New York: Wiley.

Lehmann, E. L. (1975). *Nonparametrics: Statistical Methods Based on Ranks*. San Francisco: Holden Day.

Leont'yev. R. S. (1987). Asymptotic behaviour of P-values for Kolmogorov–Smirnov tests of fit (in Russian). *Vestnik Leningrad. Un-ta: Matematika* 1: 30–2.

Leont'yev, R. S. (1988). Asymptotics of P-values for omega-square goodness-of-fit statistics (in Russian). *Zapiski Nauch. Semin. LOMI* 166: 67–72; transl.: *J. Soviet Mathem.* 52 (1990): 2908–10.

Leont'yev, R. S. (1990). Asymptotic Behavior of P-values of Nonparametric Goodness-of-Fit Statistics (in Russian). Ph.D. Thesis, Leningrad University, Leningrad.

Levin, B. R. (1976). *Theoretical Foundations of Statistical Radiotechnics* (in Russian), vol. 3. Moscow: Sovetskoye Radio.

Littell, R. C., and J. L. Folks (1971). Asymptotic optimality of Fisher's method of combining tests, Part I. *J. Amer. Statist. Assoc.* 66: 802–6.

Littell, R. C., and J. L. Folks (1973). Asymptotic optimality of Fisher's method of combining tests, Part II. *J. Amer. Statist. Assoc.* 68: 193–4.

Loh, W.-Y. (1984). Bounds on ARE's for restricted classes of distributions defined via tail-orderings. *Ann. Statist.* 12: 685–701.

Maag, U. R. (1966). A k-sample analogue of Watson's U^2-statistic. *Biometrika* 53: 570–83.

Maag, U. R., and M. A. Stephens (1968). The V_{NM} two-sample test. *Ann. Mathem. Statist.* 39: 923–35.

Marcus, M. B., and L. A. Shepp (1972). Sample behaviour of Gaussian processes. In *Proc. 6th Berkeley Symp. on Probab. Theory and Mathem. Statist.*, vol. 2, eds. L. Le Cam, J. Neyman, and E. L. Scott, pp. 423–41. Berkeley and Los Angeles: Univ. of California Press.

Martynov, G. N. (1978). *Omega-Squared Tests* (in Russian). Moscow: Nauka.

Mason, D. A. (1984). Bahadur efficiency comparison between one and two-sample rank statistics and their sequential rank statistics analogues. *J. Multivar. Anal.* 14: 181–200.

Mielke, P. W. (1972). Asymptotic behaviour of the two-sample tests based on powers of ranks for detecting scale and location alternatives. *J. Amer. Statist. Assoc.* 67: 850–4.

Mikulski, P. W. (1976). On comparison of different approaches to the efficiency of statistical procedures. *Sankhyā, Ser. A* 38: 131–42.

Mogulskii, A. A. (1977). Remarks on large deviations for the ω^2-statistic. *Probab. Theory Appl.* 22: 166–71.

Mogulskii, A. A. (1984). Probabilities of large deviations for trajectories of random walks. In *Limit Theorems for Sums of Random Variables* (in Russian), ed. A. A. Borovkov, pp. 93–124. Novosibirsk: Nauka.

Mood, A. M. (1954). On the asymptotic efficiency of certain nonparametric two-sample tests. *Ann. Mathem. Statist.* 25: 514–20.

Morgenstern, D. (1956). Einfache Beispiele zweidimensionaler Verteilungen. *Mitt. Mathem. Statist.* 8: 234–5.

Müller–Funk, U. (1983). A quantitative SLLN for linear rank statistics. *Statistics & Decisions* 1: 371–8.

Neuhaus, G. (1982). H_0-contiguity in nonparametric testing problems and sample Pitman efficiency. *Ann. Statist.* 10: 575–82.

Neuhaus, G. (1987). Local asymptotics for linear rank statistics with estimated scores functions. *Ann. Statist.* 15: 491–512.

Neyman, J. (1980). Some memorable incidents in probabilistic / statistical studies. In *Asymptotic Theory of Statistical Tests and Estimates*, ed. I. M. Chakravarti, pp. 1–32. New York: Academic Press.

Nikitin, Ya. Yu. (1976). Bahadur relative asymptotic efficiency of statistics based on the empirical distribution function (in Russian). *Dokl. Akad. Nauk SSSR* 231: 800–3; transl.: *Soviet Mathem. Reports* 17: 1645–9.

Nikitin, Ya. Yu. (1979). Large deviations and asymptotic efficiency of statistics of integral type (in Russian), Part I. *Zapiski Nauch. Semin. LOMI* 85: 175–87; transl.: *J. Soviet Mathem.* 20 (1980): 2224–31.

Nikitin, Ya. Yu. (1980). Large deviations and asymptotic efficiency of statistics of integral type (in Russian), Part II. *Zapiski Nauch. Semin. LOMI* 97: 151–75; transl.: *J. Soviet Mathem.* 24 (1984): 585–603.

Nikitin, Ya. Yu. (1981). Characterization of distributions by the local asymptotic optimality property of test statistics (in Russian). *Zapiski Nauch. Semin. LOMI* 108: 120–34; transl.: *J. Soviet Mathem.* 25 (1984): 1186–96.

Nikitin, Ya. Yu. (1982). Bahadur asymptotic efficiency of integral tests of symmetry (in Russian). *Zapiski Nauch. Semin. LOMI* 119: 181–94; transl.: *J. Soviet Mathem.* 27 (1984): pp. 3124–34.

Nikitin, Ya. Yu. (1984a). Local asymptotic Bahadur optimality and characterization problems. *Probab. Theory Appl.* 29: 79–92.

Nikitin, Ya. Yu. (1984b). Large deviations of the Durbin statistics for testing uniformity on a square (in Russian). *Zapiski Nauch. Semin. LOMI* 136: 165–7; transl.: *J. Soviet Mathem.* 33 (1986): 799–801.

Nikitin, Ya. Yu. (1984c). Integral test of independence: large deviations, asymptotic efficiency, local asymptotic optimality. *Probab. Theory Appl.* 29: 806–8.

Nikitin, Ya. Yu. (1985). Hodges–Lehmann asymptotic efficiency of the Kolmogorov and Smirnov goodness-of-fit tests (in Russian). *Zapiski Nauch. Semin. LOMI* 142: 119–123; transl.: *J. Soviet Mathem.* 36 (1987): 517–20.

Nikitin, Ya. Yu. (1986a). Large deviations and asymptotic efficiency of integral statistics for testing independence. In *Probability Distributions and Mathematical Statistics* (in Russian), eds. V. M. Zolotarev et al., pp. 388–406. Tashkent: FAN; transl.: *J. Soviet Mathem.* 38 (1987): 2382–91.

Nikitin, Ya. Yu. (1986b). Asymptotic comparison of bidimensional goodness-of-fit tests (in Russian). In *Rings and Modules. Limit Theorems of Probability Theory*, vol. 1, eds. Z. I. Borevich and V. V. Petrov, pp. 193–200. Leningrad: Leningrad Univer. Press.

Nikitin, Ya. Yu. (1986c). Hodges–Lehmann efficiency of nonparametric tests. In *Proc. 4th Vilnius Confer. on Probab. Theory and Mathem. Statist.*, vol. 2, eds. Yu. V. Prohorov and V. V. Sazonov, pp. 391–408. Utrecht: VNU Sci. Press.

Nikitin, Ya. Yu. (1987a). On the Hodges–Lehmann asymptotic efficiency of nonparametric tests of goodness-of-fit and homogeneity. *Probab. Theory Appl.* 32: 77–85.

Nikitin, Ya. Yu. (1987b). Bahadur efficiency of Watson–Darling goodness-of-fit tests (in Russian). *Zapiski Nauch. Semin. LOMI* 158: 138–45; transl.: *J. Soviet Mathem.* 43 (1988): 2833–8.

Nikitin, Ya. Yu. (1987c). On Hodges–Lehmann indices of nonparametric tests. In *Proc. 1st World Congress of Bernoulli Soc.*, vol. 2, eds. Yu. V. Prohorov and V. V. Sazonov, pp. 275–8. Utrecht: VNU Sci. Press.

Nikitin, Ya. Yu. (1988). On large deviations of the Watson goodness-of-fit statistic. In *Rings and Modules. Limit Theorems of Probability Theory* (in Russian), vol. 2, eds. Z. I. Borevich and V. V. Petrov, pp. 181–4. Leningrad: Leningrad Univer. Press.

Nikitin, Ya. Yu. (1989). Chernoff efficiency of the sign and Wilcoxon tests for testing the symmetry hypothesis (in Russian). *Zapiski Nauch. Semin. LOMI* 177: 101–7; transl.: *J. Soviet Mathem.* 61 (1992): pp. 1891–6.

Nikitin, Ya. Yu. (1990a). Local Chernoff efficiency of linear rank tests. In *Proc. 5th Vilnius Confer. on Probab. Theory and Mathem. Statist.*, vol. 2, eds. Yu. V. Prohorov et al., pp. 234–43. Utrecht: VNU Sci. Press.

Nikitin, Ya. Yu. (1990b). Chernoff and Hodges–Lehmann efficiency of linear rank symmetry tests (in Russian). *Zapiski Nauch. Semin. LOMI* 184: 215–26.

Nikitin, Ya. Yu. (1991). Hodges–Lehmann and Chernoff efficiency of linear rank tests. *J. Statist. Planning and Inference* 29: 309–23.

Nikitin, Ya. Yu., and R. P. Filimonov (1978). Relative asymptotic efficiency of certain two-sample rules of signal detection (in Russian). *Radiotekhnika i Elektronika* 23, 201–4.

Nikitin, Ya. Yu., and R. P. Filimonov (1979). Asymptotic efficiency of some nonparametric rules of detection in the two-channel model (in Russian). *Avtometriya* 4: 99–104.

Nikitin, Ya. Yu., and R. P. Filimonov (1981). Locally most powerful rank test for noncoherent detection of signals in noise of unknown level (in Russian). *Radiotekhnika i Elektronika* 26: 87–91.

Nikitin, Ya. Yu., and R. P. Filimonov (1984). Rules of detection of weak signals in non-Gaussian noise of unknown power (in Russian). *Radiotekhnika i Elektronika* 29: 914–19.

Nikitin, Ya. Yu., and R. P. Filimonov (1989). Hodges–Lehmann relative efficiency of certain rules of signal detection in noise of unknown level (in Russian). *Radiotekhnika i Elektronika* 34: 1397–402.

Nikitin, Ya. Yu., and A. G. Pankrashova (1988). Bahadur efficiency and local asymptotic optimality of certain nonparametric tests for independence (in Russian). *Zapiski Nauch. Semin. LOMI* 166: 112–27; transl.: *J. Soviet Mathem.* 52 (1990): 2942–55.

Noether, G. (1950). Asymptotic properties of the Wald–Wolfowitz test of randomness. *Ann. Mathem. Statist.* 21: 231–46.

Noether, G. (1955). On a theorem of Pitman. *Ann. Mathem. Statist.* 26: 64–8.

Oosterhoff, J. (1969). *Combination of One-Sided Statistical Tests.* Mathem. Centre Tracts, no. 28. Amsterdam: Mathematisch Centrum.

Orlov, A. I. (1972). On testing the symmetry of distributions. *Probab. Theory Appl.* 17: 357–62.

Pallini, A. (1990). Asymptotically optimal combinations of dependent test statistics. In *Small Sample Asymptotics and Related Problems*, pp. 149–61. Padova: Cleup.

Pesarin, F., and M. Zucchetto (1986). Ottimalitá asintotica nei test combinati. In *Scritti in onore di F. Brambilla* (in Italian), vol. 2, pp. 599–604. Milano: Universitá Commerciale "L. Bocconi."

Petrov, V. V. (1975). *Sums of Independent Random Variables*. Berlin: Springer.

Pettitt, A. N. (1976). A two-sample Anderson–Darling rank statistic. *Biometrika* 63: 161–8.

Pettitt, A. N. (1979). Two-sample Cramér–von Mises type rank statistics. *J. Roy. Statist. Soc., Ser. B* 41: 46–53.

Pettitt, A. N. (1984). Proportional odds model for survival data and estimates using ranks. *J. Roy. Statist. Soc., Ser. C* 33: 169–75.

Pitman, E. J. G. (1949). Lecture Notes on Nonparametric Statistical Inference. Columbia University, New York, Mimeographed.

Pitman, E. J. G. (1979). *Some Basic Theory for Statistical Inference*. New York: Wiley.

Plackett, R. L. (1965). A class of bivariate distributions. *J. Amer. Statist. Assoc.* 60: 516–22.

Podkorytova, O. A. (1990). Large deviations and Bahadur efficiency of Khmaladze–Aki statistic (in Russian). *Zapiski Nauch. Semin. LOMI* 184: 227–33.

Podkorytova, O. A. (1993). Asymptotic efficiency of integral Khmaladze–Aki statistic. In *Rings and Modules. Limit Theorems of Probability Theory*, vol. 3, eds. Z. I. Borevich and V. V. Petrov, pp. 221–9. St. Petersburg: St. Petersburg Univer. Press.

Pratt, J. W., and J. D. Gibbons (1981). *Concepts of Nonparametric Theory*. New York: Springer.

Prokof'yev, V. N. (1973). Invariant rule of noncoherent detection of signal in noise of unknown level (in Russian). *Radiotekhnika i Elektronika* 18: 547–53.

Raghavachari, M. (1970). On a theorem of Bahadur on the rate of convergence of test statistics. *Ann. Mathem. Statist.* 41: 1695–9.

Raghavachari, M. (1983). On the computation of Hodges–Lehmann efficiency of test statistics. In *Festschrift for E. L. Lehmann*, eds. P. J. Bickel, K. A. Doksum, and J. L. Hodges, Jr., pp. 367–78. Belmont: Wadsworth.

Rao, C. R. (1962). Efficient estimates and optimum inference procedures in large samples. *J. Roy. Statist. Soc., Ser. B* 24: 46–72.

Rao, C. R. (1963). Criteria of estimation in large samples. *Sankhyā, Ser. A* 25: 189–206.

Rao, C. R. (1965). *Linear Statistical Inference and Its Applications*. New York: Wiley.

Rao, C. R. (1977). Applications of cluster analysis to the mixing of races in human populations. In *Classification and Clustering*, ed. J. van Ryzin, pp. 148–67. New York: Academic Press.

Rao, J. S. (1972). Bahadur efficiencies of some tests for uniformity on the circle. *Ann. Mathem. Statist.* 43: 468–79.

Rényi, A. (1961). On measures of entropy and information. In *Proc. 4th Berkeley Symp. on Probab. Theory and Mathem. Statist.*, vol. 1, ed. J. Neyman, pp. 547–61. Berkeley: Univ. of California Press.

Ronzhin, A. F. (1985). Efficiency of Chernoff type for goodness-of-fit tests based on empirical distribution functions. *Probab. Theory Appl.* 30: 404–6.

Rosenblatt, M. (1952a). Remarks on a multivariate transformation. *Ann. Mathem. Statist.* 23: 470–2.

Rosenblatt, M. (1952b). Limit theorems associated with variants of the von Mises statistics. *Ann. Mathem. Statist.* 23: 617–23.

Rothe, G. (1981). Some properties of the asymptotic relative Pitman efficiency. *Ann. Statist.* 9: 663–9.

Rothmann, E., and M. Woodroofe (1972). A Cramér–von Mises type statistic for testing symmetry. *Ann. Mathem. Statist.* 43: 2035–8.

Rubin, H., and J. Sethuraman (1965a). Probabilities of moderate deviations. *Sankhyā, Ser. A* 27: 325–40.

Rubin, H., and J. Sethuraman (1965b). Bayes risk efficiency. *Sankhyā, Ser. A* 27: 347–56.

Rublik, F. (1989). On optimality of the LR tests in the sense of exact slopes, Part I, General case. *Kybernetika* 25: 13–25.

Sanov, I. N. (1957). On probabilities of large deviations of random variables (in Russian). *Matem. Sbornik* 42 (84): 11–44.

Savage, I. R. (1956). Contributions to the theory of rank order statistics. *Ann. Mathem. Statist.* 27: 590–615.

Savage, I. R. (1962). *Bibliography of Nonparametric Statistics.* Cambridge, MA: Harvard Univer. Press.

Savage, I. R. (1969). Nonparametric statistics: a personal review. *Sankhyā, Ser. A* 31: 107–44.

Saulis, L., and V. A. Statulevicius (1991). *Limit Theorems for Large Deviations.* Dordrecht: Kluwer.

Sethuraman, J. (1964). On the probability of large deviations of families of sample means. *Ann. Mathem. Statist.* 35: 1304–16.

Serfling, R. J. (1980). *Approximation Theorems of Mathematical Statistics.* New York: Wiley.

Shiryaev, A. N. (1984). *Probability.* New York: Springer.

Shorack, G. R., and J. A. Wellner (1986). *Empirical Processes with Applications to Statistics.* New York: Wiley.

Siegmund, D. (1982). Large deviations for boundary crossing probabilities. *Ann. Probab.* 10: 581–8.

Sievers, G. L. (1969). On the probability of large deviations and exact slopes. *Ann. Mathem. Statist.* 40: 1908–21.

Sievers, G. L. (1976). Probabilities of large deviations for empirical measures. *Ann. Statist.* 4: 766–70.

Sinclair, C., B. Spurr, and M. Ahmad (1990). Modified Anderson–Darling test. *Commun. in Statistics: Theory and Meth.* 19: 3677–86.

Singh, K. (1981). Large deviation probabilities for certain dependent processes. *J. Multivar. Anal.* 11: 354–67.

Singh, K. (1984). Asymptotic comparison of tests. A review. In *Handbook of Statistics*, eds. P. R. Krishnaiah and P. K. Sen, vol. 4: 173–84. Amsterdam: North-Holland.

Sinha, B. K., and H. S. Wieand (1977). Bounds on the efficiencies of four commonly used nonparametric tests of location. *Sankhyā, Ser. B* 39: 121–9.

Smirnov, N. V. (1939). The estimate of divergence of two empirical distribution functions in two samples (in Russian). *Moscow Univer. Mathem. Bullet.* 2: 3–14.

Smirnov, N. V. (1944). Approximation of distributions of random variables by empirical data (in Russian). *Uspekhi Mathem. Nauk* 19: 179–206.

Smirnov, N. V. (1947). On the test of symmetry for the distribution of random variable (in Russian). *Dokl. Akad. Nauk SSSR* 56: 13–16.

Srinivasan, R., and P. Godio (1974). A Cramér–von Mises type statistic for testing symmetry. *Biometrika* 61: 196–8.

Stein, E. M., and G. Weiss (1971). *Introduction to Fourier Analysis on Euclidean Spaces*. Princeton, NJ: Princeton Univer. Press.

Steinebach, J. (1980). Large deviation probabilities and some related topics. *Carleton Mathem. Lect. Notes* 28: 1–84.

Stephens, M. A. (1974). Components of goodness-of-fit statistics. *Ann. Instit. H. Poincaré, Ser. B* 10: 37–54.

Stone, M. (1967). Extreme tail probabilities of the two-sample Wilcoxon statistics. *Biometrika* 54: 629–40.

Stone, M. (1968). Extreme tail probabilities for sampling without replacement and exact Bahadur efficiency of two-sample normal scores test. *Biometrika* 55: 371–5.

Stone, M. (1974). Large deviations of empirical probability measures. *Ann. Statist.* 2: 362–6.

Strassen, V. (1964). An invariance principle for the law of the iterated logarithm. *Z. Warsch. Verw. Geb.* 3: 211–26.

Stroock, D. W. (1984). *An Introduction to the Theory of Large Deviations*. Berlin: Springer.

Tusnády, G. (1977). On asymptotically optimal tests. *Ann. Statist.* 5: 385–93.

Tyurin, Yu. N. (1978). *Nonparametric Statistical Methods* (in Russian). Moscow: Soc. Znanie Publ.

Vainberg, M. M., and V. A. Trenogin (1974). *Theory of Branching of Solutions of Nonlinear Equations*. Leiden: Noordhoof.

Varadhan, S. R. (1984). *Large Deviations and Applications*. Philadelphia: SIAM.

Voshtchenko, V. S., Ya. Yu. Nikitin, and R. P. Filimonov (1990). Detection of weak signals in non-Gaussian noises by the phase method (in Russian). *Radiotekhnika i Elektronika* 35: 2072–80.

Wald, A., and J. Wolfowitz (1940). On a test whether two samples are from the same population. *Ann. Mathem. Statist.* 11: 147–62.

Walsh, J. E. (1962,1965,1968). *Handbook of Nonparametric Statistics*, vols. 1, 2, 3. Princeton: Van Nostrand.

Watson, G. S. (1961). Goodness-of-fit tests on a circle, Part I. *Biometrika* 48: 109–14.

Watson, G. S. (1962). Goodness-of-fit tests on a circle, Part II. *Biometrika* 49: 57–63.

Watson, G. S. (1976). Optimal invariant tests for uniformity. In *Studies in Probability and Statistics, Papers in honour of E. J. G. Pitman*, ed. E. J. Williams, pp. 121–7. Amsterdam: North-Holland.

Weissfeld, L. A., and H. S. Wieand (1984). Bounds on efficiencies for some two-sample nonparametric statistics. *Commun. Statist. – Theory and Meth.* 13: 1741–57.

Wentzell, A. D. (1990). *Limit Theorems on Large Deviations for Markov Stochastic Processes*. Dordrecht: Kluwer.

de Wet, T. (1980). Cramér–von Mises tests for independence. *J. Multivar. Anal.* 10: 38–50.

de Wet, T., and J. H. Venter (1973). Asymptotic distributions for quadratic forms with applications to tests of fit. *Ann. Statist.* 1: 380–7.

Wieand, H. S. (1976). A condition under which the Pitman and Bahadur approaches to efficiency coincide. *Ann. Statist.* 4: 1003–11.

Woodworth, G. (1970). Large deviations and Bahadur efficiency of linear rank statistics. *Ann. Mathem. Statist.* 41: 251–83.

Yanushkevichius, R. (1991). *Stability for characterizations of distributions.* Vilnius: Mokslas Publishers.

Young, L. C. (1969). *Lectures on the Calculus of Variations and Optimal Control Theory.* Philadelphia: Saunders.

Yu, C. S. (1971). Pitman efficiencies of Kolmogorov–Smirnov tests. *Ann. Mathem. Statist.* 42: 1595–605.

Zabreiko, P. P., A. I. Koshelev, M. A. Krasnoselskii, S. G. Mikhlin, L. A. Rakovshchik, and V. Ya. Stetsenko (1968). *Integral Equations* (in Russian). Moscow: Nauka.

Zolotarev, V. M. (1961). Concerning a certain probability problem. *Probab. Theory Appl.* 6: 201–4.

Index

265

274   *Index*

U-statistic
degree of, 24
kernel of, 24–5, 153
large deviations of, 25, 132
nondegenerate, 24
U-statistics, 23, 25, 153

van der Waerden correlation coefficient,
194
van der Waerden statistic, 96, 104,
235–6
Vainberg, M. M., xiv, 38–9, 51, 61, 71,
141–2
Varadhan, S. R., 21
variance, 11
asymptotic, 12, 73
variation, 52, 99, 107, 158, 200, 243–4
variational calculus, 50, 52, 56, 77, 107
variational method, 73, 110, 136
variational problems, xiii, 167, 219
variational series, 31, 96, 233
Venter, J. H., 80
von Mises functionals, 24–5, 132
Voshtshenko, V. S., 126

Wald, A., xi
Walsh, J. E., x
Watson, G. S., 41, 72, 77, 84, 95, 108
Watson statistic, 42, 77, 95, 129, 222
large deviation of, 66, 70
for symmetry testing, 71
Watson–Darling statistics, 41, 45, 84,
95, 128, 216
Weibull density, 229–30
Weierstrass–Bernstein theorem, 175
weight, 4, 9, 70, 101, 137, 142, 170
asymptotically optimal choice of, 227

infinitely differentiable, 141
nonsummable, 66, 80, 221
weight function, 48–9, 66, 84, 95, 128,
149, 182, 199, 206, 230
nonnegative, 41, 170, 232
summable, xiii, 51, 84, 86
weighted statistics, xiii
Weiss, G., 186, 244
Weissfeld, L. A., 16
Wellner, J. A., 42
Wentzell, A. D., 21, 28, 220
de Wet, T., 80, 170
Wieand, H. S., xi, 16, 18, 81, 238
Wieand's condition, 105, 167–8
Wieand's sequence, 16–17
Wilcoxon rank test, ix, xv, 15, 104, 125,
235, 241
Wilcoxon statistic, 32, 96, 110, 125, 129,
146, 151, 236
Wilcoxon signed rank statistic, 152
Wilcoxon–Mann–Whitney statistic, 121
Wolfe, D., x, xi, 127
Wolfowitz, J., xi
Woodroofe, M., 128–9
Woodworth, G., 31–3, 96, 103, 135,
192–4, 233

Yanushkevichius, R. V., 211
Young, L. C., 56
Young–Fenchel conjugate, 91
Yu, C. S., 16

Zabell, S. L., 21–2, 28
Zabreiko, P. P., 58, 72
Zolotarev, V. M., 77
Zucchetto, M., 81